Chromatin Structure and Gen

Frontiers in Molecular Biology

SERIES EDITORS

B. D. Hames

*Department of Biochemistry
and Molecular Biology
University of Leeds, Leeds LS2 9JT, UK*

D. M. Glover

*Department of Genetics,
University of Cambridge, UK*

TITLES IN THE SERIES

Chromatin Structure and Gene Expression

Second Edition

EDITED BY

Sarah C. R. Elgin

Department of Biology
Washington University
St Louis, Missouri, USA

and

Jerry L. Workman

Howard Hughes Medical Institute
Department of Biochemistry and Molecular Biology
The Pennsylvania State University
University Park, Pennsylvania, USA

OXFORD

UNIVERSITY PRESS

This book has been printed digitally and produced in a standard specification
in order to ensure its continuing availability

OXFORD
UNIVERSITY PRESS

Great Clarendon Street, Oxford OX2 6DP

Oxford University Press is a department of the University of Oxford.
It furthers the University's objective of excellence in research, scholarship,
and education by publishing worldwide in

Oxford New York

Auckland Cape Town Dar es Salaam Hong Kong Karachi
Kuala Lumpur Madrid Melbourne Mexico City Nairobi
New Delhi Shanghai Taipei Toronto
With offices in
Argentina Austria Brazil Chile Czech Republic France Greece
Guatemala Hungary Italy Japan South Korea Poland Portugal
Singapore Switzerland Thailand Turkey Ukraine Vietnam

Oxford is a registered trade mark of Oxford University Press
in the UK and in certain other countries

Published in the United States
by Oxford University Press Inc., New York

ISBN 978-0-19-963890-1

Printed and bound by CPI Antony Rowe, Eastbourne

Preface

It has been 5 years since publication of the previous edition of *Chromatin structure and gene expression*. Since then dramatic advances have occurred, significantly improving our understanding of the relationship between chromatin structure and gene regulation. These include a high-resolution structure of the nucleosome core, discovery of the enzymes and complexes that mediate histone acetylation and deacetylation, discovery of novel ATP-dependent chromatin remodelling complexes, and more. Many of these advances have resulted from a synergistic coupling of genetic and biochemical approaches that is somewhat unique within eukaryotic molecular biology. For example, many, if not most, of the identified subunits in novel chromatin-modifying complexes represent the products of genes previously characterized genetically in yeast, *Drosophila*, or even mammals. During the past 5 years biochemical studies have confirmed several predictions of genetic studies or provided molecular insights into genetic observations. In turn, continued genetic analyses point to the *in vivo* functions of proteins acting on chromatin and provide direction for further biochemical analysis. This combination of biochemical and genetic approaches has clearly demonstrated the central role of chromatin in the regulation of gene expression. Indeed, many of the co-activators and co-repressors required for regulation of gene expression have been found to be chromatin-modifying complexes. In addition, continuing improvements in light microscopy, coupled with genetic manipulation, have added to our appreciation of the spatial organization in the nucleus, and the role of this parameter in gene regulation. In light of these advances it was clearly time for a second edition of *Chromatin structure and gene expression*, both to update previous chapters and to include additional chapters that introduce new concepts in the process of gene regulation in chromatin.

As in the previous edition, authors from at least two of the leading laboratories studying each particular topic prepared each chapter. This format has allowed the chapters to come directly from the leaders in the field while providing broader insights than might come from a single author. Moreover, the authors of at least two other chapters provided critiques and queries for the 'Final comments' section of each chapter. These discussion sections provide the reader with insights regarding the controversial and unanswered questions in each topic. They serve as a look forward into where the field may be headed.

The book begins with chapters describing nucleosome and chromatin structure and assembly (Chapters 1 and 2). These provide insights into the structure and dynamics of the nucleosome, the process of nucleosome assembly, and its linkage with DNA replication or repair. Chapter 3 describes the connection between chromatin and the regulation of genes in eukaryotic cells and introduces several important concepts that will be further examined in later chapters. Chapter 4 is devoted to genetic studies in order to illustrate how genetic analysis has advanced our understanding of histone function and identified many of the important protein components of complexes involved in chromatin modification or remodelling. The next four chapters focus on four groups of chromatin-modifying complexes that have been analysed extensively in the past 5 years. These include two groups of ATP-dependent chromatin remodelling complexes (Chapters 5 and 6), histone acetyltransferase complexes (Chapter 7), and histone deacetylase complexes (Chapter 8). The next several chapters explore the contributions of chromatin to the developmental regulation of gene expression (Chapter 9) and to epigenetic regulation of transcription in yeast (Chapter 10), *Drosophila* (Chapter 11), and mammals (Chapter 12). The final two chapters focus on the higher-order organization of chromatin and chromosomes. Chapter 13 focuses on chromatin boundaries, which may organize functional chromatin domains, while Chapter 14 describes the organization of chromosomes in the nucleus and the relationship to chromatin function.

In writing our chapters we have tried to make the material accessible to scientists unfamiliar with the chromatin field and to students beginning graduate studies. We assumed that the readers will have a basic knowledge of chromatin structure and transcription similar to that provided by a general text, such as *Molecular biology of the cell* by B. Alberts, D. Bray, J. Lewis, M. Raff, K. Roberts, and J.D. Watson (1994, 3rd edn, Garland Publishing, New York) or *Molecular cell biology* by H. Lodish , A. Berk, L. Zipursky , P. Matsudaira, D. Baltimore, and J. Darnell (2000, 4th edn, W. H. Freeman, New York). It is our hope that this book will take readers from the general textbooks to a level where they can enjoy the rapidly appearing mini reviews and the primary literature in this field. The chromatin field is rapidly expanding as it attracts new young scientists as well as interest from established scientists in related fields (e.g. DNA transcription, replication, repair). While progress in the past 5 years has been outstanding, new mysteries continue to unfold. We hope our readers will be intrigued and challenged by the problems posed here, and that some will be motivated to join us in our efforts to understand the regulation of gene expression established by nucleosomes and higher-order chromatin structure. We thank all the authors for their outstanding contributions in preparing the chapters, and Ms Lorene Stitzer for helping us to assemble the book. We hope that you enjoy and learn as much from these chapters as we have.

University Park, Pennsylvania   J.L.W.
St Louis, Missouri   S.C.R.E.
October 2000

Contents

2 DNA replication, nucleotide excision repair, and nucleosome assembly 24

PAUL D. KAUFMAN AND GENEVIÈVE ALMOUZNI

3 Chromatin structure and control of transcription *in vivo* 49

WOLFRAM HÖRZ AND SHARON ROTH

5   The SWI/SNF family of remodelling complexes   97

BRADLEY R. CAIRNS AND ROBERT E. KINGSTON

6   ATP-dependent chromatin remodelling by the ISWI complexes   114

CARL WU, PETER B. BECKER, AND TOSHIO TSUKIYAMA

7 Histone acetyltransferase/transcription co-activator complexes 135

SHELLEY L. BERGER, PATRICK A. GRANT, JERRY L. WORKMAN, AND
C. DAVID ALLIS

10 Chromatin contributions to epigenetic transcriptional states in yeast

LISA FREEMAN-COOK, ROHINTON KAMAKAKA, AND LORRAINE PILLUS

13 Chromatin boundaries

VICTOR G. CORCES AND GARY FELSENFELD

14 Linking large-scale chromatin structure with nuclear function

NICOLA L. MAHY, WENDY A. BICKMORE, TUDORITA TUMBAR, AND
ANDREW S. BELMONT

Contributors

C. DAVID ALLIS
Department of Biochemistry and Molecular Genetics, University of Virginia School of Medicine, Box 440, Jordan Hall, Charlottesville, VA 22908, USA

GENEVIÈVE ALMOUZNI
Nuclear Dynamics and Genome Plasticity, UMR218 CNRS/Institut Curie Research Section, 26 rue d'Ulm, 75248 Paris Cedex 05, France

MICHELLE CRAIG BARTON
Department of Molecular Genetics, Biochemistry and Microbiology, University of Cincinnati, 231 Bethesda Avenue, P.O. Box 670524, Cincinnati, OH 45267, USA

PETER B. BECKER
Adolf-Butenandt-Institüt-Molekularbiologie, Universität München, Schillerstr. 44, D-80336 München, Germany

ANDREW S. BELMONT
Department of Cell and Structural Biology, University of Illinois at Urbana-Champaign, B107 CLSL, 601 S. Goodwin Ave., Urbana, IL 61801, USA

SHELLEY L. BERGER
The Wistar Institute, 3601 Spruce Street, Philadelphia, PA 19104, USA

WENDY A. BICKMORE
MRC Human Genetics Unit, Western General Hospital, Crewe Road, Edinburgh EH4 2XU, UK

ADRIAN BIRD
Institute of Cell and Molecular Biology, University of Edinburgh, Swann Building, The King's Buildings, Mayfield Road, Edinburgh, EH9 3JR, UK

BRADLEY R. CAIRNS
Department of Oncological Sciences, Huntsman Cancer Institute, 2000 Circle of Hope, Salt Lake City, UT 84112, USA

VICTOR G. CORCES
Department of Biology, The Johns Hopkins University, 3400 N. Charles St., Baltimore, MD 21218, USA

JOEL C. EISSENBERG
Department of Biochemistry, St Louis University School of Medicine, 1402 Grand South Blvd., St Louis, MO 63104, USA

SARAH C. R. ELGIN
Department of Biology, CB-1229, Washington University, One Brookings Drive, St Louis, MO 63130, USA

GARY FELSENFELD
Laboratory of Molecular Biology, NIDDKD, Building 5, Room 212, National Institutes of Health, Bethesda, MD 20892, USA

ANDREW FREE
Institute of Cell and Molecular Biology, University of Edinburgh, Swann Building, The King's Buildings, Mayfield Road, Edinburgh, EH9 3JR, UK

LISA FREEMAN-COOK
Department of Genetics, Yale University School of Medicine, 333 Cedar Street, SHM I166, New Haven, CT 06510, USA

PATRICK A. GRANT
Department of Biochemistry and Molecular Genetics, University of Virginia Health Sciences Center, Jordan Hall, Box 800733, Charlottesville, VA 22908, USA

MICHAEL GRUNSTEIN
Department of Biological Chemistry, UCLA School of Medicine, University of California, Los Angeles, CA 90095, USA

WOLFRAM HÖRZ
Adolf-Butenandt-Institüt-Molekularbiologie, Universität München, Schillerstr. 44, D-80336 München, Germany

ROHINTON KAMAKAKA
Laboratory of Molecular Embryology, National Institutes of Child Health and Development, 18T 106, National Institutes of Health, Bethesda, MD 20892, USA

PAUL D. KAUFMAN
Lawrence Berkeley National Laboratory and Department of Molecular and Cell Biology, University of California-Berkeley, Berkeley, CA 94720, USA

ROBERT E. KINGSTON
Department of Molecular Biology, Massachusetts General Hospital, Harvard Medical School, Wellman 10, Fruit Street, Boston, MA 02114, USA

NICOLA L. MAHY
MRC Human Genetics Unit, Western General Hospital, Crewe Road, Edinburgh EH4 2XU, UK

RENATO PARO
Zentrum für Molekulare Biologie, Universität Heidelberg, Im Neuenheimer Feld 282, Heidelberg D-69120, Germany

LORRAINE PILLUS
Department of Biology, University of California-San Diego, Pacific Hall 0347, 9500 Gilman Drive, La Jolla, CA 92093, USA

TIMOTHY J. RICHMOND
Institut für Molekularbiologie und Biophysik, ETH-Hönggerberg, HPM F8 CH-8093 Zürich, Switzerland

SHARON ROTH
Department of Biochemistry and Molecular Biology, Program in Genes and Development, U.T. M.D. Anderson Cancer Center, Houston, TX 77030, USA

CHRISTOPHER J. SCHOENHERR
Princeton University, Lewis Thomas Lab, Washington Road, Princeton, NJ 08544, USA

M. MITCHELL SMITH
Department of Microbiology, University of Virginia, 1300 JPS Box 441, Charlottesville, VA 22908, USA

SHIRLEY M. TILGHMAN
Princeton University, Lewis Thomas Lab, Washington Road, Princeton, NJ 08544, USA

TOSHI TSUKIYAMA
Division of Basic Sciences, Fred Hutchinson Cancer Research Center, 1100 Fairview Avenue North, Mail Stop A-162, P.O. Box 19024, Seattle, WA 98109, USA

TUDORITA TUMBAR
Department of Cell and Structural Biology, University of Illinois at Urbana-Champaign, B107 CLSL, 601 S. Goodwin Ave., Urbana, IL 61801, USA

MARIA VOGELAUER
Department of Biological Chemistry, UCLA School of Medicine, University of California, Los Angeles, CA 90095, USA

JONATHAN WIDOM
Department of Biochemistry, Molecular Biology, and Cell Biology, Northwestern University, 2153 Sheridan Road, Evanston, Illinois 60208, USA

FRED WINSTON
Department of Genetics, Harvard University, 200 Longwood Avenue, Boston MA 02115, USA

ALAN P. WOLFFE
Sangamo Biosciences, Inc., Point Richmond Tech Center, 501 Canal Blvd, Suite A100, Richmond, CA 94804, USA

JERRY L. WORKMAN
Howard Hughes Medical Institute, Department of Biochemistry and Molecular Biology, The Pennsylvania State University, 306 Althouse Laboratory, University Park, PA 16802, USA

CARL WU
Laboratory of Molecular Cell Biology, National Cancer Institute, Building 37, Room 5E-26, National Institutes of Health, Bethesda, Maryland 20892, USA

Abbreviations

ABF	autonomously replicating sequence (ARS) binding factor
ACF	ATP-utilizing chromatin assembly and modifying factor
ACTR	co-activator of thyroid receptor
ACS	ARS consensus sequence
ADA	alteration/deficiency in activation
AFM	atomic force microscopy
AHC	Ada histone acetyltransferase complex component
ANT-C	*Antennapedia* gene complex in *Drosophila*
ARP	actin-related protein
ARS	autonomously replicating sequence
AS	Angelman syndrome
ASF	anti-silencing function
AS–IC	Angelman syndrome, imprinting centre
AT	acetyltransferase
ATM	ataxia telangiectasia mutated
BAF	BRG1-associated factor
BD	bromodomain
BEAD-1	blocking element alpha/delta 1
BEAF-32	boundary element associated factor of 32 kDa
bHLH	basic-helix–loop–helix
Brg1	brahma-related gene 1
BrUTP	5-bromouridine 5-triphosphate
BWS	Beckwith–Wiedemann syndrome
BX-C	bithorax complex
CAC	chromatin assembly complex
CAF	chromatin assembly factor
CBP	CREB-binding protein
CDK	cyclin dependent kinase
CHD	chromo domain–helicase–DNA-binding protein
CHIP	chromatin immunoprecipitation
CHRAC	chromatin accessibility complex
chromo	chromatin organization modifier
CLN	cyclin
CoA	coenzyme A
CpG	cytosine–guanine dinucleotide
CPSF	cleavage-polyadenylation specificity factor
CREB	cyclic AMP response element binding protein
CstF	cleavage stimulation factor

CTCF	CCTC-binding factor
CTD	carboxy-terminal domain of the large subunit of RNA polymerase II
DDM	decreased DNA methylation
DMR	differentially methylated region
Dnmt1	DNA methyltransferase 1
DPY	dumpy protein
DREF	DNA replication-related element-binding factor
EDTA	ethylenediaminetetraacetic acid
EGF	epidermal growth factor
EKLF	erythroid Kruppel-like factor
ELP	elongator protein
EM	electron microscopic, electron microscopy, or electron microscope
E-RC1	erythroid-remodelling complex 1
ESA	essential SAS-related acetyltransferase
ES cells	embryonic stem cells
E(var)	dominant enhancer of position effect variegation
FACT	facilitates activation of chromatin templates
FAT	factor acetyltransferase
FISH	fluorescence *in situ* hybridization
GAL	galactose non-utilizer
Gcn	general control non-derepressible
GFP	green fluorescent protein
GNAT	Gcn5-related *N*-acetyltransferase
GST	glutathione S-transferase
HAT	histone acetyltransferase
hBrm	human brahma
HDAC	histone deacetylase
Hdf	high-affinity DNA-binding factor
HHT	histone H3 genes
HIR	histone cell-cycle regulation defective
HIS	histidine requiring
HLH	helix–loop–helix
HMG	high-mobility group
HOS	HDA1 similarity
HP1	heterochromatin protein 1
HPC	histone periodic control
HRX	human trithorax protein
HS	hypersensitive (to nuclease digestion)
HSA	*Homo sapiens*
HSR	homogeneously staining region
HST	homologues of *SIR2*
HSV TK	herpes simplex virus thymidine kinase
HTA	histone H2A genes
HTB	histone H2B genes

HU	hydroxyurea
ICD	inter-chromosomal domain
ICR	imprinting control region
IFN	interferon
INO	inositol deficient
ISWI	imitation switch
LBR	lamin B receptor
LCR	locus control region
LTR	long terminal repeat
LYS	lysine requiring
MAR	matrix attachment region OR matrix associated region
MBC	methyl-benzimidazole-2yl carbamate
MBD	methyl-CpG-binding domain
MCM	minichromosome maintenance protein
MeCP2	methyl-CpG-binding protein 2
MEF	myocyte enhancer factor
MEL	mouse erythroleukaemia cell
MENT	myeloid and erythroid nuclear termination stage-specific protein
MHC	major histocompatability complex
MIX	mitosis- and X-associated protein
MMS	methyl methanesulphonate
MNase	micrococcal nuclease
Mot	modulator of transcription
MTA	metastasis-associated protein
NAD	nicotinamide–adenine dinucleotide
Nan	Net1-associated nucleolar protein
NAP	nucleosome assembly protein
NCoR	nuclear hormone receptor co-repressor OR negative co-repressor
NER	nucleotide excision repair
Net	nucleolar silencing establishing factor and telophase regulator
NMR	nuclear magnetic resonance
NOR	nucleolar organizer region
NuA3	nucleosomal acetyltransferase of histone H3
NuA4	nucleosomal acetyltransferase of histone H4
NuRD	nucleosome remodelling histone deacetylase
NURF	nucleosome remodelling factor
ORC	origin recognition complex
ori	origin of replication
PAH	paired amphipathic helix
PAL	palindromic binding factor
PC	Polycomb protein
PCAF	p300/CBP-associated factor
PcG	*Polycomb* group genes
PCNA	proliferating cell nuclear antigen

PCR	polymerase chain reaction
PEV	position effect variegation
PH	polyhomeotic protein
PHD	plant homeodomain
PHO	phosphatase deficient
PML	promyelocytic leukaemia
pol α/primase	DNA polymerase α/primase complex
pol δ	DNA polymerase δ
pol II	RNA polymerase II
PPase	inorganic pyrophosphatase
PRE	*Polycomb* response element
PSC	*Posterior sex combs* protein
PTS	promoter-targeting sequence
PWS	Prader–Willi syndrome
PWS–IC	Prader–Willi syndrome, imprinting centre
RAD	radiation (ultraviolet or ionizing) sensitive
RAP	repressor–activator protein
RbAp	retinoblastoma-associated protein
RCAF	replication-coupling assembly factor
rDNA	DNA encoding rRNA
REB1	rRNA enhancer-binding protein 1
RENT	regulator of nucleolar silencing and telophase exit
RFC	replication factor C
RNA Pol II	RNA polymerase II
RNP	ribonuclear protein, ribonucleoprotein
RO	repeat organizer of the *Xenopus* rRNA genes
RPA	replication protein A
RPD	reduced potassium deficiency
RSC	remodels the structure of chromatin
RSF	remodelling and spacing factor
RXR	retinoid-X receptor
SAR	scaffold attachment region
SAS	something about silencing
SBP	scs binding protein
SCII	scaffold protein II
scs, scs′	specialized chromatin structures (boundaries)
SET domain	suppressor of variegation, enhancer of *zeste*, *Trithorax*
SIN	switch independent
SIR	silent (mating type) information regulator
SMC	structural maintenance of chromosomes
SMRT	silencing mediator of retinoid and thyroid hormone receptors
SNF	sucrose non-fermenting
SPT	suppressors of TY transcription
SRB	suppressor of *rpb1* (RNA polymerase II subunit mutation)

SRC	steroid receptor co-activator
SSN6	suppressor of *snf1* (snf1 mutation)
STAGA	Spt3–TAF$_{II}$31–Gcn5L acetyltransferase
STAR	subtelomeric anti-silencing region
STH	SNF2 homologue
su(Hw)	suppressor of Hairy-wing (protein)
Su(var)	dominant suppressor of position effect variegation
SV40	simian virus 40
SV40 ori	simian virus 40 DNA replication origin
SWI	homothallic switching deficient
Swp	SWI/SNF complex protein
T$_3$	3, 5, 3'-triiodothyronine
T$_4$	3, 5, 3', 5'-tetraiodothyronine
TAF	TBP-associated factor
TAS	telomeric-associated sequence
Taz	telomere associated in *Schizosaccharomyces*
TBF	TTAGGG repeat-binding factor
TBP	TATA-binding protein
TBZ	thiabendazole
TEACl	triethanolamine hydrochloride
TEM	transmission electron microscopy
TFII	transcription factor for RNA polymerase II transcription
TFIII	transcription factor for RNA polymerase III transcription
TFTC	TBP-free TAF$_{II}$-containing
TH	thyroid hormone
TIF	transcriptional intermediary factor
TIP	Tat interacting protein
topo II	topoisomerase II
TR	thyroid hormone receptor
TRA1	yeast TRRAP homologue
TRD	transcriptional repression domain
TRE	thyroid response element
tRNA	transfer RNA
TRRAP	transcription/transformation domain associated protein
trxG	*trithorax* group genes
Ts	temperature sensitive
TSA	trichostatin A
TUP	deoxythymidine monophosphate uptake positive
UAS	upstream activating sequence
UME	unscheduled meiotic gene expression
VP16	herpes virus protein 16
WCRF	Williams syndrome transcription factor-related chromatin remodelling factor
WSTF	Williams syndrome transcription factor

Xa	active X chromosome
Xce	X-chromosome controlling element
Xi	inactive X chromosome
XIC	X-chromosome inactivation centre
XP	xeroderma pigmentosum
XPA–XPG	xeroderma pigmentosum complementation groups (proteins)

1 | Nucleosome and chromatin structure

TIMOTHY J. RICHMOND and JONATHAN WIDOM

1. Introduction: the essence of chromatin

Eukaryotic cells contain from 10 million to 100 billion DNA base pairs in a nucleus just a few microns in diameter. The DNA molecules that comprise the human genome would span almost 2 metres in length if they were laid end to end. These delicate threads encoding the blueprint of life would entangle and snap during cell division if not carefully organized. This vast quantity of DNA is packaged by the histone proteins into a hierarchical structure called chromatin. DNA in chromatin is compacted over 10 000-fold compared to its straight form. Importantly, the structure of chromatin permits localized decondensation and repackaging of DNA to facilitate the processes of replication, transcription, and repair. In contrast to the view that chromatin is an essential nuisance to DNA function, it and a host of enzymatic molecular machines have evolved to play a major role in the control of gene activation. The nucleosome is the fundamental repeating unit of chromatin, a consequence of the interaction between roughly equal proportions of histone proteins and DNA, and accounts for at least the first two levels of DNA organization in the cell nucleus.

2. Core histone proteins

2.1 Sequences

The histone proteins fall into five classes—H1, H2A, H2B, H3, and H4—derived from their amino-acid composition and sequence (1) (Fig. 1a). Each histone class includes some gene variants or subtypes which are likely to provide tissue-specific and developmental-stage-dependent variation of chromatin structure (2). The histones H2A, H2B, H3, and H4 have molecular masses of 10–14 kDa and are known as the 'core' histones, since they supercoil DNA around them to form the nucleosome core of chromatin (3, 4). The complete set of core histone proteins is essential for cell viability, as shown by genetic experiments using yeast (5). The core histone proteins are found in the same fixed equal molar stoichiometry, with histone H1 in half this amount, in all eukaryotic organisms (6). Their sequences are among the most highly conserved in evolution.

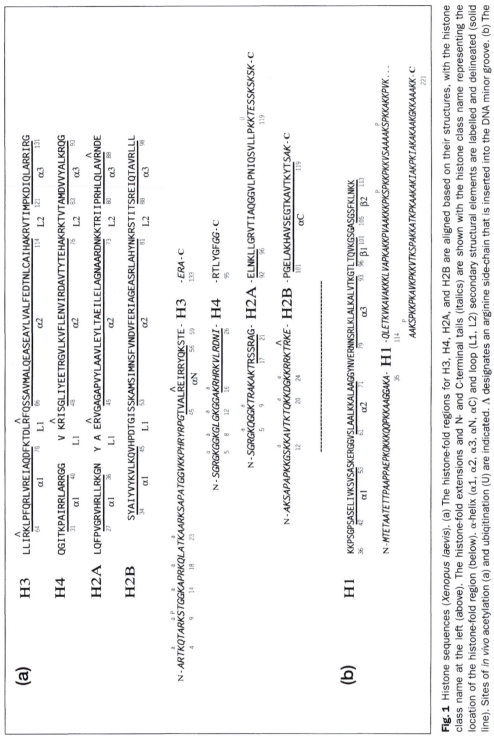

Fig. 1 Histone sequences (*Xenopus laevis*). (a) The histone-fold regions for H3, H4, H2A, and H2B are aligned based on their structures, with the histone class name at the left (above). The histone-fold extensions and N- and C-terminal tails (italics) are shown with the histone class name representing the location of the histone-fold region (below). α-helix (α1, α2, α3, αN, αC) and loop (L1, L2) secondary structural elements are labelled and delineated (solid line). Sites of *in vivo* acetylation (a) and ubiquitination (U) are indicated. Λ designates an arginine side-chain that is inserted into the DNA minor groove. (b) The winged helix domain of the histone H1 (above) and its N- and C-terminal tail regions (below) are shown. α-helix (α1, α2, α3) and β-strand (β1, β2) secondary structural elements are labelled and delineated (solid line). Sites of *in vivo* phosphorylation (P) are indicated.

2.2 Domain structure

The core histone proteins can be divided into three distinct types of structural domains, based on the occurrence of α-helices, β-strands, loops, and coil elements (7, 8) (Fig. 1a):

1. At the central region of the protein chain in each case, a region of approximately 70 amino acids, called the 'histone-fold', comprises in order a 3- to 4-turn α-helix (α1), a loop of 7–8 amino acids (L1), an 8-turn α-helix (α2), a loop of 6 amino acids, and a final 2- to 3-turn α-helix (α3).

2. Significantly less-uniform structural elements, called the 'histone-fold extensions', extend N-terminally from the histone-fold domains of H3 (αN) and H2A, and C-terminally from the histone-fold domain for H2A and H2B (αC). Each of these structural elements is packed against histone-fold regions or other extensions.

3. The N-termini of all core histones are random-coil elements, from 16 (H2A) to 44 (H3) amino acids in length, known as 'flexible tails' as they are largely unstructured, at least in the context of a single nucleosome core (8). By definition, they do not make contact with the histone-fold domains or their extensions within the same nucleosome core. H2A also has a significant 10 amino-acid C-terminal tail that is much longer in certain H2A variants.

Interestingly, both the tail and histone-fold domains have been added to the basic histone-fold domain in the course of evolution, as archaebacterial histones lack the tails and histone-fold extensions (9).

The histone tails contain the sites of post-translational modifications that include the acetylation, methylation, and ubiqitination of specific lysine amino acids, phosphorylation of serine, and poly(ADP-ribosylation) (10–14) (Fig. 1a). Specific patterns of tail acetylation have been shown to be highly correlated with histone deposition during replication and regulation of gene transcription (15). The mechanism by which these modifications are involved in these processes is not well understood, but under intensive investigation.

2.3 Heterodimeric histone pairs

The histone-fold domains combine to form H2A/H2B and H3/H4 heterodimeric pairs (16). The α1–L1–α2–L2–α3 regions of the histone proteins in a pair interact in an antiparallel orientation, predominantly through hydrophobic contacts made by the α1, α2, and α3 helices (8). The contact sites between the α2 helices are offset toward the helix N-terminus by one helical turn, which juxtaposes the α3 helices and the α2 helices and positions the α1 helices side by side (Plate 1a). There are ample differences between the packing interfaces of the actual heterodimers and the alternative pairings and homodimers to account for the specificity of dimerization. The histone-fold extensions in some cases also make hydrophobic interactions with the underlying histone-folds, contributing significantly to specificity. Alignment of the sequences of the histone-fold domains, based on the hydrophobic interaction sites governing dimerization, reveals that the histones H3 and H2B, and H4 and H2A are most similar to each other. The divergence of the core histones into four types presumably reflects the requirement to provide different stereospecificity of interaction with respect to the DNA entry and exit points around the circumference of the nucleosome core. The histone-fold domains appear in non-histone nuclear protein complexes as well, namely TFIID, a factor important for initiation of basal level transcription by RNA polymerase II, that contains 9–13 proteins (17–19).

2.4 Histone–DNA interactions

Two types of DNA-binding sites occur in a histone-fold heterodimer: (A) two L1L2 loop sites, and (B) one $\alpha 1\alpha 1$ site (8) (Plate 1a). The L1 and L2 loops and $\alpha 1$ helices are brought together because of the anti-parallel arrangement of histones. Each site binds to DNA centred on one of three adjacent minor grooves along approximately three double-helical turns, bending the DNA through 140°. Together, the four histone-fold domains in a nucleosome core account for the binding of 121 bp of DNA. The means by which the histones grip the DNA fall into three categories of roughly equal contribution:

1. Charge neutralization of the acidic DNA phosphate groups by lysine and arginine amino acids occurs either by direct or water-mediated hydrogen bonds between phosphate oxygen and side-chain amino and guanidinium groups. Importantly, an arginine side-chain penetrates the DNA minor groove once for each of the L1L2 binding sites and for the H3H4 $\alpha 1\alpha 1$ binding site. The minor groove sites not accommodated in this way by the histone-fold domains have an arginine inserted from another part of the structure. The inserted arginine guanidinium group is generally hydrogen bonded between an adjacent threonine and a water molecule linked to the N3 atom of the purine base in the closest base pair.

2. More irregularly, but possibly no less importantly, hydrophobic side-chains, primarily threonine, proline, valine, and isoleucine, interact with the deoxyribose moieties. In one case, a leucine reaches into the major groove to contact a thymidine methyl group.

3. Hydrogen bonds from the histone main-chain amide groups to the DNA are frequent, occurring with the phosphate oxygen atoms of both phosphodiester chains each time they face the histone core. About 50% of the direct hydrogen bonds from histone to DNA are between main-chain amide groups and phosphate oxygen atoms.

3. Nucleosome core

3.1 Histone and DNA organization

The universal repeating element of chromatin is the nucleosome core, containing 147 bp length of DNA and a histone octamer comprised of a pair of each of the core histone proteins (20) (Plate 2). The additional linker DNA that connects nucleosome cores is of variable length, depending on cell type and species (21). The nucleosome core, linker DNA, and histone H1 make up the complete nucleosome.

The histone heterodimers combine to form an octamer in the presence of DNA or in high salt conditions (e.g. monovalent NaCl greater than about 1.2 M) (22). The DNA within a nucleosome core is 147 bp in length and takes the form of a 1.65-turn left-handed superhelix (8). In the absence of DNA or salt, the stable histone oligomers are tetramers of two H3/H4 dimers and dimers of H2A and H2B. The α3 helices and C-terminal third of the α2 helices provide the interaction sites between histone pairs that generate the histone octamer containing two of each core histone protein. These helices form a four-helix bundle that binds the two H3 molecules together to make the histone tetramer (Plate 1b). The H3/H4 tetramer lies at the centre of the nucleosome core and therefore is centred on the 147 bp DNA within the nucleosome. The H4 molecules provide two analogous halves of a four-helix bundle, each combining with a counterpart half bundle in the H2B subunit of each H2A/H2B dimer to complete the histone octamer. The fourth half helix-bundle in each copy of H2A is buried in the nucleosome core and is thereby blocked from polymerization with further heterodimers. Together the four histone-heterodimeric pairs are arranged in the octamer to yield an overall molecular twofold axis of symmetry and bind 121 bp of the 147 bp nucleosome core DNA. The central base pair falls on the symmetry axis so that, as elsewhere in the structure, depending on the DNA sequence, the twofold symmetry is not perfect, but pseudo-twofold.

3.2 Histone-fold extensions

The histone-fold extensions play several obvious roles in the nucleosome core, but other functions may yet be discovered (8). The remaining 13 base pairs of DNA at either end of the DNA superhelix that complete the 147 bp are bound by the H3 αN helix, which precedes the histone-fold domain along the H3 sequence. The αN DNA-binding site appears to be the weakest, based on the number of interactions observed with the DNA, and, in addition, it may be flexibly associated with the histone core. Part of its association with the octamer is mediated through the C-terminal H2A coil extension. These features of the structure that provide the grip on the nucleosome core DNA at its entry and exit points may be important in allowing other proteins access to the DNA. They may also permit the linker DNA to adopt different traject-ories, depending on its length, that allow the nucleosome to fit into the next level or higher-order structure. The association of the H2A/H2B dimer with the H3/H4 tetramer not only depends on the four-helix bundle that occurs between H2B and H4, but also on the H2A 'docking' domain, which includes the α3 helix of the H2A histone-fold, a short α-helix C-terminal to α3, and a β-strand that combines with a β-strand extension that is the C-terminal stretch of H2A. The H2A protein also contains a one-turn α-helix that just precedes its α1 helix and may provide a contact point for a component of the chromatin remodelling machinery. The remaining histone-fold extension is the αC helix of H2B, which is the most protruding part of the histone octamer surface not contacted by DNA (excluding the tails). This α-helix is likely to be the binding site for a non-histone chromatin interaction assembly, or plays an important part in chromatin fibre formation.

4. Histone H1 and the nucleosome

4.1 Sequence and domain structure

Histone H1 is larger than the core histones, with a mass of 21 kDa, and is called the 'linker' histone because it associates with the DNA connecting the nucleosome cores (23) (Fig. 1b). Although it was originally thought that formation of the chromatin fibre depended critically on the presence of H1, in *in vitro* studies fibres have been shown to form in the absence of H1 (24). Indeed, chromatin can apparently fold into the highest levels of compaction even in the absence of histone H1 (25, 26), and yeast cells lacking the one H1 homologue present in the yeast genome live (27, 28). Taken together, these data suggest that H1 may have chiefly a regulatory rather than a structural role. In higher eukaryotes, H1 is present in approximately one copy per nucleosome, although it may not be associated with every nucleosome core. There is a much greater sequence diversity for H1 compared to core histones, and in lower eukaryotes (e.g. Protozoa, yeast), the candidate H1 molecules do not have the same domain organization as in higher eukaryotes (27, 29). The H1 molecule of higher eukaryotes is divided into a central, folded domain of about 80 amino acids flanked by a short N-terminal tail of about 20 amino acids and a C-terminal tail of approximately 100 amino acids. The C-terminal tail is highly positively charged and undergoes post-translational modification (29). The three-dimensional X-ray structure of the globular domain of histone H5, a variant of H1 found in avian erythrocytes, shows that it is a member of the winged-helix family of DNA-binding proteins (30, 31). The N- and C-terminal domains are essentially random coil in the free protein. H1 does not bind to the histone octamer or its subcomponents in the absence of DNA (23, 32).

4.2 Binding site in the nucleosome

The H1 globular domain is bound in the nucleosome in the vicinity of DNA entry and exit, and the C-terminal domain is most likely associated with the linker DNA. Although there is no direct imaging of H1 in the nucleosome or bound to DNA, several lines of evidence indicate that H1 binds near the centre of the nucleosomal DNA, at least for the bulk of chromatin, and makes contact with about 10–20 bp in one or both linker DNA segments just outside the core:

1. Electron microscopy of chromatin fibres shows that the linker entry and exit points around the circumference of the nucleosome are more highly fixed near each other when H1 is present (33).
2. Limited DNA digestion of whole chromatin releases a particle containing '166' bp of DNA and histone H1 prior to yielding a nucleosome core particle (34). Using

reconstitutes of the nucleosome containing only the globular domain of H1, the same protection of the linker DNA is found.

3. Site-specific, chemical cross-linking of the globular domain of the linker histone to DNA shows that it binds near the DNA centre (35, 36).

These findings suggest that histone H1 should have a second DNA-binding surface distinct from the homologous winged-helix DNA-binding motif. Mutagenesis and *in vitro* assembly experiments indicate that such a surface, containing a cluster of four basic amino acids, exists and can affect nucleosome formation (37).

4.3 Functional roles

The primary functional role of H1 in the nucleus has been assigned historically to establishment of the most compact form of the chromatin fibre (33, 38). At metaphase, chromatin compaction has been correlated, depending on the organism and cell type, with phosphorylation or dephosphorylation of multiple copies of S/TPXK sequences in the H1 C-terminal domain (39–41). Genetic 'knockout' experiments in higher eukaryotes have been complicated by the existence of compensatory factors, such as further H1 alleles or variants, and another class of similar proteins (i.e. HMG1/2) (36). Analogous experiments in lower eukaryotes may not be relevant to higher eukaryotes, since the domain structure of the closest homologues for H1 are not identical. For example, many of the studies yielding the most disparate results involve elimination of H1 from the macronucleus of *Tetrahymena*; however, this H1 molecule lacks the central, globular domain (29). Nevertheless, the presence or absence of H1 or its phosphorylation give rise to gene-specific effects, both positive and negative, on transcription (42). In *Tetrahymena*, constitutively dephosphorylated H1 is phenotypically similar to no histone H1 (41).

5. Chromatin fibre *in vitro* and *in situ*

5.1 Models for chromatin fibre structure

Throughout most of the cell cycle, eukaryotic chromatin is maintained in states that are more compact than a simple extended chain of nucleosomes. Often this chromatin is visible in electron micrographs of thin-sectioned nuclei as dispersed or clustered fibres, roughly 30 nm in diameter (30 nm fibre or chromatin fibre). In other cases— indeed, for all cell types as they enter mitosis or meiosis—the chromatin fibre undergoes additional levels of compaction, ultimately yielding the dense state seen at mitotic metaphase (43). Despite decades of intensive study, only limited information is available for the structure of the 30 nm fibre, and even less is certain for higher levels of chromatin compaction beyond the 30 nm fibre state.

Early electron microscopic (EM) studies of isolated chromatin fragments *in vitro* revealed that chromatin fibres would undergo a reversible cation-dependent compaction in response to changes in solution conditions (33) (Fig. 2). In dilute salt

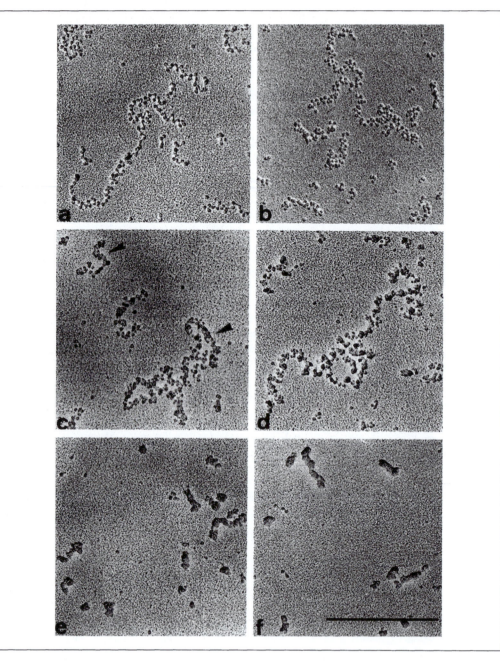

Fig. 2 Salt-dependent folding of chromatin *in vitro* observed by electron microscopy (from ref. 33). Chromatin fibres are examined after fixation (pH 7.0) in (a) 0.2 mM EDTA; (b) 5 mM triethanolamine hydrochloride (TEACl) + 0.2 mM EDTA; (c–f) as in (b) plus 10, 20, 40, and 60 mM NaCl. The bar indicates 500 nm. The chromatin fibre compacts gradually, from a zigzag fibre of distinct nucleosomes in low salt concentrations into shorter, ≈30 nm-wide fibres at higher salt concentrations. Reproduced from *The Journal of Cell Biology*, 1979, vol. 83, pp. 2107 by copyright permission of the Rockefeller University Press.

solutions (e.g. 1–10 mM monovalent ions), chromatin fibres appear as an extended chain of distinct nucleosomes, whereas in the presence of physiological ionic conditions (100–200 mM) the extended nucleosome filaments become compacted reversibly into shorter, wider fibres, approximately 30 nm in width. These fibres resemble roughly those visible in electron micrographs of chromatin *in situ*. Hydrodynamic studies of chromatin in solution reveal an equivalent cation-dependent increase in compaction, implying that the increasing compaction seen by EM is not an artefact of the microscopy (44). Many of the EM images reveal surface striations running oblique to the fibre axis with a pitch comparable to the 11 nm diameter of individual nucleosomes (45). The simplest interpretation of such images is that the chain of nucleosomes is compacting into the 30 nm fibre by a gentle helical coiling, with roughly six nucleosomes per helical turn, and with nucleosomes packed roughly side to side above each other, yielding a helical repeat comparable to the 11 nm nucleosome diameter (43) (Fig. 3a). This simple helical coiling resembles the coiling of wire in a solenoid; hence this model is frequently referred to as the solenoid model for 30 nm fibre structure (45).

X-ray and neutron scattering have provided useful information complementary to that obtained by electron microscopy. X-ray solution scattering reveals a set of features that correlate well with the folding of the extended nucleosome filament into 30 nm fibres (44). Importantly, X-ray scattering from chromatin fibres in isolated nuclei and even in intact living cells reveal this same set of characteristic features (46), suggesting that chromatin fibres have a similar structure *in vitro* and *in vivo*, notwithstanding the wide variability in EM images dependent on sample preparation methodologies (47). Studies on oriented samples of 30 nm chromatin fibres reveal how characteristic repeating distances within the structure are angularly distributed relative to each other and to the fibre axis, showing that the nucleosomes are packed side to side in the direction of the fibre axis and radially around it (48). These results are consistent with the solenoid model for 30 nm fibre structure (45) (Fig. 3a), as well as with subtle variants of it (Fig. 3b–d); they are likely inconsistent with models for which the nucleosomes are more loosely packed (Fig. 3e).

Neutron scattering has provided two important results bearing on chromatin fibre structure. First, the mass per unit length for chromatin in solution was determined directly. In dilute buffer, the results imply an extended nucleosome filament conformation, such as that seen by electron microscopy. As the cation concentration is increased, the mass density increases up to a value in physiological ionic conditions corresponding to roughly 6–8 nucleosomes per 11 nm along the fibre axis, consistent with models having the nucleosomes tightly packed (49) (Fig. 3a–d) but inconsistent with more open models (Fig. 3e). Secondly, deuterated H1 molecules could be distinguished from the other constituents of reconstituted chromatin fibres, and were found to lie toward the centre of the 30 nm fibre (50). Earlier EM studies had established that, in the presence of histone H1, the entry and exit points of DNA in the nucleosome are brought into near-juxtaposition, resulting in the zigzag appearance of the extended chromatin fibre (33) (Fig. 2). Additionally, chemical cross-linking data confirm that H1 sits near the point at which DNA enters/exits the nucleosome

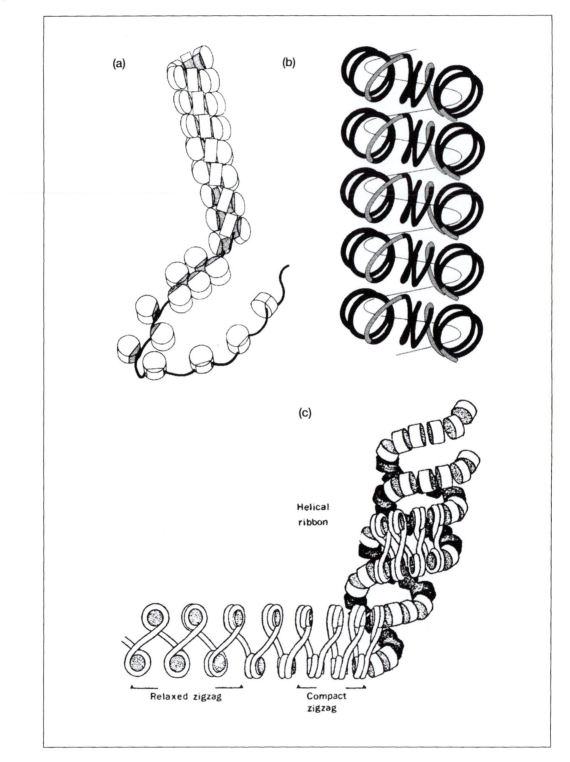

(a)

(b)

(c)

Helical
ribbon

Relaxed zigzag

Compact
zigzag

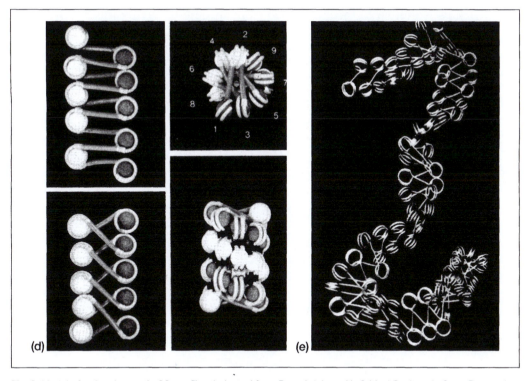

Fig. 3 Models for the chromatin 30 nm fibre (adapted from Ramakrishnan V, *Critical Reviews in Gene Expression*, 1997, vol. 7, pp. 215 by copyright permission of Begell House, Inc.). (a) The solenod model (33) reproduced by copyright permission of The Rockefeller University Press. The chain of nucleosomes is shown in increasing stages of compaction into the 30 nm fibre corresponding, for example, to increasing concentrations of monovalent salt (see text). (b) The continuous superhelix model (79) reproduced by copyright permission of Cell Press. (c) The twisted-ribbon model (80) reproduced by copyright permission of The Rockefeller University Press. (d) The cross-linker model (81) reproduced by copyright permission of The Biophysical Society. (e) Open chromatin fibre (extended, random-length linker DNA) (54) reproduced by copyright permission of The Rockefeller University Press.

(35). These approximate locations for H1 within the nucleosome and within the chromatin fibre favour the solenoid structure (Fig. 3a) over the uniform superhelical structure (Fig. 3b) because in this latter case, the nucleosomes, and therefore H1, would have no particular rotational setting around the axis of the nucleosome disc in the chromatin fibre.

Further data helping to distinguish between models for the 30 nm fibre structure comes from studies of the dependence of structural features on the length of linker DNA. While some models, such as the solenoid model, make no specific prediction, others make testable predictions. For example, the 'twisted-ribbon' model (Fig. 3c) implies that the axial striations and corresponding 11 nm X-ray scattering band should change with increased linker DNA length; however, this feature was found to be invariant to an increase of linker length of roughly 10 nm (44). In contrast to the predictions from 'crossed-linker' models (Fig. 3d), experiments on chromatin having an average linker length too short to cross the fibre centre reveal structures that are closely similar to those obtained from chromatin having more typical, longer linker lengths (51). Finally, the finding that some cell types have nucleosome repeat lengths as short as 156 bp constrains solenoidal models to a right-handed helical form (52) (the mirror-image of Fig. 3a).

5.2 Interactions stabilizing the chromatin fibre

A significant feature of the 30 nm fibre, as revealed from physical studies, is that it is not in a highly stabilized, inert state. Rather, in approximately physiological conditions, 30 nm fibres may be considered to be in dynamic equilibrium with more extended states (53). These open states may correspond to structures seen in three-dimensional tomographic images made by electron microscopy (54). Further stability may come through interactions with other proteins, such as the polycomb protein in heterochromatin (55).

It is of substantial interest to determine what interactions contribute to stabilizing the chromatin fibre. Regulation of these interactions would facilitate preferential activation by opening up, or repression by folding up, particular stretches of the genome. An especially intriguing finding is that the tail domain of histone H4 can make specific interactions with a conserved patch on the surface of a neighbouring nucleosome core (8). Earlier studies pointed to essential roles in higher-order chromatin folding for the core–histone tail domains collectively, and showed that acetylation of the tails influenced (albeit subtly) the energetics of chromatin folding (53). Taken together, these results suggest that the tail domains, particularly that of H4, may be primary determinants of chromatin higher-order structure, and that cells may modulate the structure or stability of the folded chromatin fibre through specific changes in the pattern of histone acetylation (56).

6. Nucleosome positioning

6.1 What is nucleosome positioning?

In vivo nucleosome mapping methods reveal that nucleosomes are often preferentially localized at specific genomic positions, often correlated with the underlying regulatory organization of genes (57). The preferred positions can be influenced by the presence of other proteins, but often are the same as those found when histones are reconstituted on the same naked DNA *in vitro* (58). Moreover, when nucleosomes are reconstituted *in vitro* with a unique but arbitrary DNA sequence and length longer than the 147 bp of the nucleosome core, it is often found that the histone octamer exhibits a pronounced preference for location at one or a few positions (59). Each position can be designated by the base pair that the pseudo-dyad (twofold) axis of the nucleosome core passes through.

'Translational' and 'rotational' positioning of the histone octamer on DNA have been distinguished from each other, although they are completely coupled (i.e. the DNA minor groove always faces outward in the nucleosome core at the central, dyad position—translation by one base pair requires a concomitant rotation by one base pair). Translational positioning refers to the extent to which a histone octamer selects a particular contiguous stretch of 147 bp of DNA in preference to other stretches of the same length. Rotational positioning is a degenerate form of translational positioning in which a set of discrete translational positions, differing by an integral multiple of the DNA helical repeat, is occupied in preference to the intervening bases in the repeat of the double helix. Rotational positioning can occur because a segment of DNA is more or less uniformly bent in the same direction along its length.

Even in the simplest case, that of preferential positioning of a histone octamer at a particular site along a single DNA sequence *in vitro*, nucleosome positioning must be viewed as a statistical phenomenon, not as 'precise' as it is often described. Standard protocols for nucleosome reconstitution involve equilibrium dialysis from concentrated salt solutions, in which histones and DNA have negligible affinity for each other, down to physiological or even lower salt concentrations, in which nucleosome formation is strongly favoured (60). Moreover, nucleosomes lacking histone H1 are mobile in physiological conditions (61–64). Thus, such procedures for nucleosome reconstitution are expected to lead to equilibrium distributions of nucleosome positions, implying that nucleosome positioning is a statistical property. Observations of apparent 'precise' positioning generally reflect insensitivity of mapping methods to lower levels of occupancy. More careful studies of positioning *in vitro* reveal occupancy of numerous translational positions, including those that are not related by the DNA helical twist (65).

Positioning *in vivo* will also be statistical, even though additional forces may contribute to establishing positional biases. The distribution of free energies for nucleosome core formation on different DNA sequences yields variations in stability of about tenfold, and, in exceptional cases, of no more than 100-fold (65). It is important to recognize this statistical property of positioning because it has substantial ramifications for mechanisms of gene regulation—essential DNA sequences may sometimes be buried when they need to be accessible, and vice versa. Mechanisms proposed for gene regulation must be robust with respect to statistical fluctuations in nucleosome positioning (66).

6.2 Mechanistic basis of nucleosome positioning

For *in vitro* systems containing only purified histone octamer and DNA, there are five distinct mechanisms through which a histone octamer may organize a particular stretch of DNA into a nucleosome in preference to other possible stretches of DNA in the same DNA molecule (67). These stem from considering free-energy changes in solution ongoing from a 147 bp defined-sequence, naked DNA segment, and free histone octamer to the nucleosome core, including the conformational differences on binding:

(1) the DNA may make more or better bonds with the histone octamer;

(2) the DNA may have static curvature that favours nucleosomal wrapping;

(3) the DNA might have increased bendability or flexibility—this property, as for

static bends as well, may extend over the entire sequence or occur in segments arranged to match the variation in requirements for bending along the surface of the histone octamer;

(4 & 5) the local twist of DNA in the nucleosome differs at most locations from the average helical twist of DNA in solution. It follows then the same arguments given under (2) and (3) for bending also apply to the twist parameter of the DNA double helix.

DNA determinants of nucleosome positioning have been discovered through analyses of DNA sequences present in isolated natural nucleosomes and of DNA sequences found by happenstance to be organized in preferentially positioned nucleosomes (68). Many dinucleotide and longer sequence motifs have been discovered that recur with a 10.2 bp periodicity. This periodicity matches the average DNA helical repeat seen in the core particle structure (8), implying that genomic DNA has evolved to contribute to its own nucleosomal packaging. It is noteworthy that DNA sequences that incorporate sequence motifs for static DNA bending spaced periodically commensurate with the DNA helical repeat, lead to rotationally positioned nucleosomes (69, 70).

Many additional forces may contribute to preferential positioning of nucleosomes *in vivo*. The population of DNA-binding proteins present in the cell at any moment may influence positioning, since protein binding to an accessible target site will almost certainly restrict the translational or rotational repositioning of that site in a nucleosome. These interactions by non-histone proteins would affect nucleosome positioning directly. Protein-dependent positioning could also be indirect; for example, such effects may arise when other DNA-binding proteins attract or exclude nucleosomes and so delineate a region to be filled in non-uniformly with nucleosome (71). Chromatin higher-order structure creates additional biases for the mutual positioning of arrays of nucleosomes because of the requirement for particular values of linker DNA lengths that permit formation of a compact fibre and especially because of likely direct nucleosome–nucleosome contacts in the 30 nm fibre (72).

7. Nucleosome dynamics

7.1 *In vitro* assays

Nucleosomes are remarkably dynamic structures, and these dynamics seem likely to be intimately involved in nucleosome function. Two forms of large-scale nucleosome dynamics are of much interest: these are 'site exposure', a non-dissociative, partial uncoiling of nucleosomal DNA from the histone octamer, starting from an end (73, 74), and 'nucleosome mobility', the ability of the histone octamer to translocate along DNA (75). Nucleosome mobility is a relatively slow process, typically requiring minutes to hours. A non-equilibrium distribution can generally be shifted to a different distribution by changing ionic strength, and elevated temperature will increase the rate of the redistribution.

Nucleosomal site exposure was first recognized as a process that occurred detectably near the ends of the nucleosomal DNA (76). However, recent studies have shown that it occurs at sites throughout the entire length of the nucleosome, and that it accounts quantitatively for the differences in binding phenomena for protein interactions with naked DNA as compared to the same DNA wrapped in nucleosomes (74). Importantly, site exposure is non-dissociative: access to sites anywhere within the nucleosomal DNA occurs without the DNA dissociating completely from the octamer surface. The available evidence suggests that access to sites internal to the nucleosomal DNA is achieved by progressive inward exposure, starting from an end. Equilibrium constants for site exposure (the fraction of time, on average, that buried sites are actually accessible) decrease progressively from values of $1\text{–}4 \times 10^{-2}$ near the ends of the nucleosomal DNA to $10^{-4}\text{–}10^{-5}$ for sites near the middle. Several mechanistically independent assays allow for quantitative analyses of site exposure equilibria. One approach measures the decrease in affinity of a site-specific DNA-binding protein for binding to naked DNA or for binding to DNA in a nucleosome; the equilibrium constant for exposure at that nucleosomal DNA site is given by the ratio of these affinities. An alternative approach measures the suppression of rates of restriction endonuclease cleavage for recognition sites contained within a nucleosome compared to the rate of cleavage of the same site as naked DNA; the equilibrium constant for a site is the ratio of the apparent rate constants for digestion. A third assay examines the cooperativity or synergy inherent in the site exposure model, for the binding of arbitrarily chosen pairs of DNA-binding proteins to DNA target sites contained within the same nucleosome (77).

7.2   Mechanisms of nucleosome site exposure and mobility

The mechanisms of nucleosomal site exposure and mobility are unknown; indeed, two plausible possibilities are mutually exclusive: mobility may itself be the mechanism of site exposure, or site exposure may be a required first step in nucleosome mobility. Two putative mechanisms can be suggested for site exposure:

(1) translocation of the histone octamer exposes a buried DNA target site, making possible subsequent site-specific binding of a DNA-binding protein (Fig. 4b); and

(2) site exposure at an internal site is achieved by progressive uncoiling of the DNA superhelix starting from an end (Fig. 4a).

As a matter of principle, both such processes may occur and be linked in equilibria. However, indirect evidence suggests that nucleosome mobility is not the basic mechanism of site exposure. Most importantly, kinetic studies reveal that site exposure can occur at all sites throughout the nucleosome on a timescale of seconds or less, even in conditions in which nucleosome mobility occurs on a timescale of hours (74).

The mechanism of nucleosome mobility is frequently referred to as 'sliding'. The structure of the nucleosome itself argues against a simple mechanism for sliding, however. Each of the 14 times that both DNA backbones contact the histone octamer surface, a number of energetically favourable interactions occur that, overall, over-

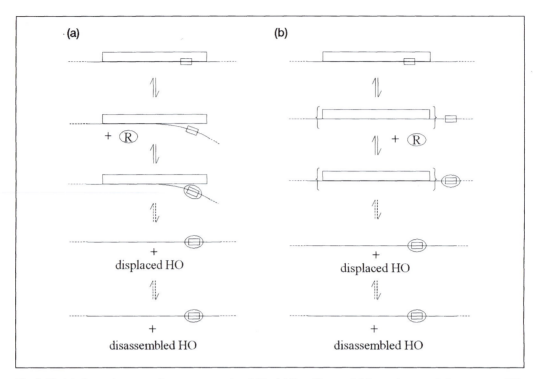

Fig. 4 Models for nucleosome site exposure and mobility. (a) Uncoiling model for nucleosomal site exposure (53, 77). Fully wrapped DNA is in rapid equilibrium with partially unwrapped states in which internal DNA target sites are now fully accessible, allowing binding of an arbitrary DNA binding protein 'R'. Binding of many such molecules, or processive translocation of one such molecule, may lead to displacement of the histone octamer, which in turn may lead to disassembly of the histone octamer. (b) Mobility model for site exposure. If sufficient naked DNA is available (or alternatively if concerted motions of several nucleosomes occur) then internal DNA target sites may be exposed with no need for net breakage of histone–DNA bonds, simply by translocating the histone octamer away from that target site.

come the resistance of DNA to undergo the bending and twisting distortions that accompany nucleosome formation. A literal sliding mechanism would involve prohibitively high energy barriers, since the steric interference due to the arginine side-chains and H3 and H2B tail segments penetrating the minor groove would not allow a simple rotation of the octamer within the gyres of the superhelix without breaking all bonds between protein and DNA simultaneously. Two possible mechanisms do exist, however, that would allow for mobility of the histone octamer along DNA while continuously maintaining contact with it:

1. The sliding mechanism is possible if local over- and under-twisting of the DNA is allowed, so that translation is coupled with rotation of the DNA: in effect the DNA acts as a flexible 'superhelical screw'. As seen in the X-ray structure of the nucleosome core particle, the DNA is sufficiently flexible to allow the sliding to occur, one site at a time, such that an energy penalty for simultaneous disruption of all sites

would not be necessary (8). The required DNA distortion actually occurs at one site in the crystal structure.

2. Partial DNA 'uncoiling' provides another mechanism for histone octamer translocation. In this case, the DNA unpeels from the nucleosome at an entry/exit site and rebinds at another site outside the original nucleosomal length of DNA. The rebinding forms a bulge in the DNA on the nucleosome that can migrate around the entire DNA superhelix by breaking bonds at a site on one side of the bulge and reforming them on the other side.

7.3 Site exposure and mobility in nuclear processes

Nucleosomal site exposure and mobility could account for the ability of site-specific, DNA-binding, gene regulatory proteins and enzymes involved in nuclear processes to function on nucleosomal DNA. In both cases, the dynamic properties of nucleosomes allow access to DNA that, in the time average, would ordinarily be considered inaccessible. They provide plausible mechanisms for the action of chromatin 'remodelling machines' or processive enzymes, such as RNA or DNA polymerases, to act on nucleosomal template DNA. For chromatin remodelling factors, the superhelical screw mechanism can well account for the primary experimentally determined effect of remodelling. Remodelled nucleosome templates show patterns of DNA cleavages virtually identical with those of free DNA, although with slower cleavage kinetics, when partially digested with the endonuclease DNAse I. The remodelling factor may possibly act by screwing the DNA along the surface of the histone octamer using ATP as an energy source, such that every phosphodiester linkage is available for cleavage by DNAse I at some time during the process.

The uncoiling mechanism has been proposed to explain the transcription of nucleosomal DNA by RNA and DNA polymerases without concomitant dissociation of histone octamer. In this model, the naked DNA accumulating behind the polymerase loops around and binds the exposed sites of the histone octamer. The DNA bulge formed accommodates the bound polymerase, and spontaneously migrates, or is actively driven by the polymerase, around the nucleosome, allowing stepwise advances of the polymerase. Continued forward motion of the polymerase would couple to propagation of the bulge around the remainder of the histone octamer, leading ultimately to a net displacement of the histone octamer in a direction opposite to the direction of transcription elongation, as suggested above in the mechanism of nucleosome mobility (78).

8. Summary and discussion

The structure of chromatin has at its foundation the nucleosome core, for which there is a high-resolution X-ray structure. The architecture of this elemental assembly is now well understood, and further details of DNA base pair and histone conformation are forthcoming. The crystal structure of further defined DNA sequences relevant to the question of nucleosome positioning along genomic DNA is an area of further study. Several allelic variants for each of the core histones are expressed at various stages in cellular differentiation (e.g. H2A.Z) and it is of substantial interest to determine whether or not their structures differ significantly from the most abundant versions. The histone H3 homologue, CENP-A, is found exclusively at chromosome centromeres and is apparently crucial to their formation. Determination of the structure of a CENP-A-containing nucleosome core may shed light on its importance there.

Mechanistic studies *in vitro* and *in vivo* concerning nucleosome positioning and dynamics, both for DNA uncoiling and histone octamer mobility, are essential to the understanding of regulation of gene expression. Positioning of any particular nucleosome biases regulatory protein binding and the preferred locations of adjacent nucleosomes, and these, in turn, bias the positioning of that particular nucleosome. Evidently, nucleosome positioning can only be understood as a large set of coupled chemical equilibria, the exact nature of which requires further investigation.

The perceived role of histone H1 has changed, from mainly facilitating the compaction of the chromatin fibre to one of importance in gene regulation. In higher eukaryotes, these issues are still to be resolved. Even questions concerning the exact way in which H1 binds in a nucleosome and its effect on nucleosome positioning and dynamics are without adequate answers.

The problem of obtaining atomic-level structural information about the chromatin fibre, even in its simplest form, is immense. However, it is this scale of chromatin structure that is likely to be most relevant to gene regulation. The occurrence of histone-tail acetylase and deacetylase complexes that are targeted site-specifically to gene promoter regions, together with the role of tail acetylation in chromatin fibre formation, indicate the connection. Currently, a definitive structure for the 30 nm fibre made by direct, high-resolution imaging methods does not exist. Taken together, the less direct data collected over more that two decades suggest that, at least as evaluated with regard to the models shown in Fig. 3, the solenoidal structure (Fig. 3a) is the most suitable model for the compact chromatin fibre. Further variations from this simple structure might be imagined. Despite the success of electron microscopy in reaching atomic-level detail for biological macromolecules in favourable cases, future progress in determining the structure of the chromatin fibre is likely to require additional, new approaches.

In discussion with other authors of this book, it was pointed out that recent AFM experimental observations, plus numerical simulations from the Woodcock and van Holde laboratories, have led to models of 30 nm fibre structure which do not involve nucleosome interactions but instead are dependent only on linker geometry and linker length (82). These images and simulations bear a striking similarity to images obtained from ultrastructural cryo-EM work on oligonucleosomes. These other authors suggest that the lack of nucleosome interactions may relate to the absence of divalent or polycations, and they suggest that perhaps under more physiological conditions, under which chromatin fibres will self-associate (and which cannot easily be studied by AFM or EM) one would also see higher-order fibres with internucleosome interactions. We agree that this is a likely possibility. Indeed, it is likely the case that chromatin fibres are evolved to undergo even higher levels of folding in physiological conditions, and that the conditions required to stabilize native-like chromatin structures are probably in fundamental conflict with the conditions that yield soluble isolated fibres that are optimal for structural studies. It is for this reason that we anticipate that new approaches will be required for further progress to be made on this problem.

References

1. Johns, E. W. (1967) The electrophoresis of histones in polyacrylamide gel and their quantitative determination. *Biochem. J.*, **104**, 78.
2. Franklin, S. G. and Zweidler, A. (1977) Non-allelic variants of histones 2a, 2b and 3 in mammals. *Nature*, **266**, 273.
3. Shaw, B. R., Herman, T. M., Kovacic, R. T., Beaudreau, G. S., and Van Holde, K. E. (1976) Analysis of subunit organization in chicken erythrocyte chromatin. *Proc. Natl Acad. Sci., USA*, **73**, 505.
4. Noll, M. and Kornberg, R. D. (1977) Action of micrococcal nuclease on chromatin and the location of histone H1. *J. Mol. Biol.*, **109**, 393.
5. Kim, U. J., Han, M., Kayne, P., and Grunstein, M. (1988) Effects of histone H4 depletion on the cell cycle and transcription of *Saccharomyces cerevisiae*. *EMBO J.*, **7**, 2211.
6. Hayashi, K., Hofstaetter, T., and Yakuwa, N. (1978) Asymmetry of chromatin subunits probed with histone H1 in an H1–DNA complex. *Biochemistry*, **17**, 1880.
7. Arents, G., Burlingame, R. W., Wang, B.-C., Love, W. E., and Moudrianakis, E. N. (1991) The nucleosomal core histone octamer at 3.1 Å resolution: A tripartite protein assembly and a left-handed superhelix. *Proc. Natl Acad. Sci., USA*, **88**, 10148.
8. Luger, K., Mader, A. W., Richmond, R. K., Sargent, D. F., and Richmond, T. J. (1997) Crystal structure of the nucleosome core particle at 2.8 A resolution. *Nature*, **389**, 251.
9. Sandman, K., Krzycki, J. A., Dobrinski, B., Lurz, R., and Reeve, J. N. (1990) HMf, a DNA-binding protein isolated from the hyperthermophilic archaeon *Methanothermus fervidus*, is most closely related to histones. *Proc. Natl Acad. Sci., USA*, **87**, 5788.
10. Vidali, G., Gershey, E. L., and Allfrey, V. G. (1968) Chemical studies of histone acetylation. The distribution of epsilon-N-acetyl lysine in calf thymus histones. *J. Biol. Chem.*, **243**, 6361.
11. Tidwell, T., Allfrey, V. G., and Mirsky, A. E. (1968) The methylation of histones during regeneration of the liver. *J. Biol. Chem.*, **243**, 707.
12. Hunt, L. T. and Dayhoff, M. O. (1977) Amino-terminal sequence identity of ubiquitin and the nonhistone component of nuclear protein A24. *Biochem. Biophys. Res. Commun.*, **74**, 650.
13. Jackson, V., Shires, A., Chalkley, R., and Granner, D. K. (1975) Studies on highly metabolically active acetylation and phosphorylation of histones. *J. Biol. Chem.*, **250**, 4856.
14. Ueda, K., Omachi, A., Kawaichi, M., and Hayaishi, O. (1975) Natural occurrence of poly(ADP-ribosyl) histones in rat liver. *Proc. Natl Acad. Sci., USA*, **72**, 205.
15. Kuo, M. H., Brownell, J. E., Sobel, R. E., Ranalli, T. A., Cook, R. G., Edmondson, D. G., Roth, S. Y., and Allis, C. D. (1996) Transcription-linked acetylation by Gcn5p of histones H3 and H4 at specific lysines. *Nature*, **383**, 269.
16. Kornberg, R. D. and Thomas, J. O. (1974) Chromatin structure; oligomers of the histones. *Science*, **184**, 865.

17. Xie, X., Kokubo, T., Cohen, S. L., Mirza, U. A., Hoffmann, A., Chait, B. T., Roeder, R. G., Nakatani, Y., and Burley, S. K. (1996) Structural similarity between TAFs and the hetero-tetrameric core of the histone octamer. *Nature*, **380**, 316.

18. Birck, C., Poch, O., Romier, C., Ruff, M., Mengus, G., Lavigne, A. C., Davidson, I., and Moras, D. (1998) Human TAF(II)28 and TAF(II)18 interact through a histone fold encoded by atypical evolutionary conserved motifs also found in the SPT3 family. *Cell*, **94**, 239.

19. Gangloff, Y. G., Werten, S., Romier, C., Carre, L., Poch, O., Moras, D., and Davidson, I. (2000) The human TFIID components TAF(II)135 and TAF(II)20 and the yeast SAGA components ADA1 and TAF(II)68 heterodimerize to form histone-like pairs. *Mol. Cell. Biol.*, **20**, 340.

20. Finch, J. T., Lutter, L. C., Rhodes, D., Brown, R. S., Rushton, B., Levitt, M., and Klug, A. (1977) Structure of nucleosome core particles of chromatin. *Nature*, **269**, 29.

21. McGhee, J. D. and Felsenfeld, G. (1980) Nucleosome structure. *Annu. Rev. Biochem.*, **49**, 1115.

22. Bode, J. and Wagner, K. G. (1980) Cooperative exposure of histone H3 thiols in core particles. *Int. J. Biol. Macromol.*, **2**, 129.

23. Allan, J., Hartman, P. G., Crane-Robinson, C., and Aviles, F. X. (1980) The structure of histone H1 and its location in chromatin. *Nature*, **288**, 675.

24. Carruthers, L. M., Bednar, J., Woodcock, C. L., and Hansen, J. C. (1998) Linker histones stabilize the intrinsic salt-dependent folding of nucleosomal arrays: mechanistic ramifications for higher-order chromatin folding. *Biochemistry*, **37**, 14776.

25. Ohsumi, K., Katagiri, C., and Kishimoto, T. (1993) Chromosome condensation in *Xenopus* mitotic extracts without histone H1. *Science*, **262**, 2033.

26. Dasso, M., Dimitrov, S., and Wolffe, A. P. (1994) Nuclear assembly is independent of linker histones. *Proc. Natl Acad. Sci., USA*, **91**, 12477.

27. Ushinsky, S. C., Bussey, H., Ahmed, A. A., Wang, Y., Friesen, J., Williams, B. A., and Storms, R. K. (1997) Histone H1 in *Saccharomyces cerevisiae*. *Yeast*, **13**, 151.

28. Patterton, H. G., Landel, C. C., Landsman, D., Peterson, C. L., and Simpson, R. T. (1998) The biochemical and phenotypic characterization of Hho1p, the putative linker histone H1 of *Saccharomyces cerevisiae*. *J. Biol. Chem.*, **273**, 7268.

29. Wu, M., Allis, C. D., Richman, R., Cook, R. G., and Gorovsky, M. A. (1986) An intervening sequence in an unusual histone H1 gene of *Tetrahymena thermophila*. *Proc. Natl Acad. Sci., USA*, **83**, 8674.

30. Ramakrishnan, V., Finch, J. T., Graziano, V., Lee, P. L., and Sweet, R. M. (1993) Crystal structure of globular domain of histone H5 and its implications for nucleosome binding. *Nature*, **362**, 219.

31. Gajiwala, K. S., Chen, H., Cornille, F., Roques, B. P., Reith, W., Mach, B., and Burley, S. K. (2000) Structure of the winged-helix protein hRFX1 reveals a new mode of DNA binding. *Nature*, **403**, 916.

32. Tiktopulo, E. I., Privalov, P. L., Odintsova, T. I., Ermokhina, T. M., Krasheninnikov, I. A., Aviles, F. X., Cary, P. D., and Crane-Robinson, C. (1982) The central tryptic fragment of histones H1 and H5 is a fully compacted domain and is the only folded region in the polypeptide chain. A thermodynamic study. *Eur. J. Biochem.*, **122**, 327.

33. Thoma, F., Koller, T., and Klug, A. (1979) Involvement of histone H1 in the organization of the nucleosome and of the salt-dependent superstructures of chromatin. *J. Cell Biol.*, **83**, 403.

34. Simpson, R. T. (1978) Structure of the chromatosome, a chromatin particle containing 160 base pairs of DNA and all the histones. *Biochemistry*, **17**, 5524.

35. Zhou, Y. B., Gerchman, S. E., Ramakrishnan, V., Travers, A., and Muyldermans, S. (1998) Position and orientation of the globular domain of linker histone H5 on the nucleosome. *Nature*, **395**, 402.

36. Thomas, J. O. (1999) Histone H1: location and role. *Curr. Opin. Cell Biol.*, **11**, 312.

37. Goytisolo, F. A., Gerchman, S. E., Yu, X., Rees, C., Graziano, V., Ramakrishnan, V., and Thomas, J. O. (1996) Identification of two DNA-binding sites on the globular domain of histone H5. *EMBO J.*, **15**, 3421.

38. Shen, X., Yu, L., Weir, J. W., and Gorovsky, M. A. (1995) Linker histones are not essential and affect chromatin condensation *in vivo*. *Cell*, **82**, 47.

39. Suzuki, M. (1989) SPKK, a new nucleic acid-binding unit of protein found in histone. *EMBO J.*, **8**, 797.

40. Roth, S. Y. and Allis, C. D. (1992) Chromatin condensation: does histone H1 dephosphorylation play a role? *Trends Biochem. Sci.*, **17**, 93.

41. Dou, Y., Mizzen, C. A., Abrams, M., Allis, C. D., and Gorovsky, M. A. (1999) Phosphorylation of linker histone H1 regulates gene expression *in vivo* by mimicking H1 removal. *Mol. Cell*, **4**, 641.

42. Shen, X. and Gorovsky, M. A. (1996) Linker histone H1 regulates specific gene expression but not global transcription *in vivo*. *Cell*, **86**, 475.

43. Adolph, K. W. (1981) A serial sectioning study of the structure of human mitotic chromosomes. *Eur. J. Cell Biol.*, **24**, 146.

44. Widom, J. (1986) Physicochemical studies of the folding of the 100 Å nucleosome filament into the 300 Å filament. Cation dependence. *J. Mol. Biol.*, **190**, 411.

45. Finch, J. T. and Klug, A. (1976) Solenoidal model for superstructure in chromatin. *Proc. Natl Acad. Sci., USA*, **73**, 1897.

46. Langmore, J. P. and Paulson, J. R. (1983) Low angle X-ray diffraction studies of chromatin structure *in vivo* and in isolated nuclei and metaphase chromosomes. *J. Cell Biol.*, **96**, 1120.

47. Giannasca, P. J., Horowitz, R. A., and Woodcock, C. L. (1993) Transitions between *in situ* and isolated chromatin. *J. Cell Sci.*, **105**, 551.

48. Widom, J. and Klug, A. (1985) Structure of the 300 Å chromatin filament: X-ray diffraction from oriented samples. *Cell*, **43**, 207.

49. Gerchman, S. E. and Ramakrishnan, V. (1987) Chromatin higher-order structure studied by neutron scattering and scanning transmission electron microscopy. *Proc. Natl Acad. Sci., USA*, **84**, 7802.

50. Graziano, V., Gerchman, S. E., Schneider, D. K., and Ramakrishnan, V. (1994) Histone H1 is located in the interior of the chromatin 30-nm filament. *Nature*, **368**, 351.

51. Lowary, P. T. and Widom, J. (1989) Higher-order structure of *Saccharomyces cerevisiae* chromatin. *Proc. Natl Acad. Sci., USA*, **86**, 8266.

52. Godde, J. S. and Widom, J. (1992) Chromatin structure of *Schizosaccharomyces pombe*. A nucleosome repeat length that is shorter than the chromatosomal DNA length. *J. Mol. Biol.*, **226**, 1009.

53. Widom, J. (1998) Structure, dynamics, and function of chromatin *in vitro*. *Annu. Rev. Biophys. Biomol. Struct.*, **27**, 285.

54. Horowitz, R. A., Agard, D. A., Sedat, J. W., and Woodcock, C. L. (1994) The three-dimensional architecture of chromatin *in situ*: electron tomography reveals fibers composed of a continuously variable zig-zag nucleosomal ribbon. *J. Cell Biol.*, **125**, 1.

55. Breiling, A., Bonte, E., Ferrari, S., Becker, P. B., and Paro, R. (1999) The *Drosophila* polycomb protein interacts with nucleosomal core particles *in vitro* via its repression domain. *Mol. Cell. Biol.*, **19**, 8451.

56. Brownell, J. E., Zhou, J., Ranalli, T., Kobayashi, R., Edmondson, D. G., Roth, S. Y., and Allis, C. D. (1996) *Tetrahymena* histone acetyltransferase A: a homolog to yeast Gcn5p linking histone acetylation to gene activation. *Cell*, **84**, 843.

57. Kornberg, R. D. and Lorch, Y. (1992) Chromatin structure and transcription. *Annu. Rev. Cell Biol.*, **8**, 563.

58. Shimizu, M., Roth, S. Y., Szent-Gyorgyi, C., and Simpson, R. T. (1991) Nucleosomes are positioned with base pair precision adjacent to the alpha 2 operator in *Saccharomyces cerevisiae*. *EMBO J.*, **10**, 3033.

59. Flaus, A. and Richmond, T. J. (1998) Positioning and stability of nucleosomes on MMTV 3′LTR sequences. *J. Mol. Biol.*, **275**, 427.

60. Lowary, P. T. and Widom, J. (1997) Nucleosome packaging and nucleosome positioning of genomic DNA. *Proc. Natl Acad. Sci., USA*, **94**, 1183.

61. Pennings, S., Meersseman, G., and Bradbury, E. M. (1994) Linker histones H1 and H5 prevent the mobility of positioned nucleosomes. *Proc. Natl Acad. Sci., USA*, **91**, 10275.

62. Meersseman, G., Pennings, S., and Bradbury, E. M. (1992) Mobile nucleosomes—a general behavior. *EMBO J.*, **11**, 2951.

63. Varga-Weisz, P. D., Blank, T. A., and Becker, P. B. (1995) Energy-dependent chromatin accessibility and nucleosome mobility in a cell-free system. *EMBO J.*, **14**, 2209.

64. Ura, K., Hayes, J. J., and Wolffe, A. P. (1995) A positive role for nucleosome mobility in the transcriptional activity of chromatin templates: restriction by linker histones. *EMBO J.*, **14**, 3752.

65. Roberts, M. S., Fragoso, G., and Hager, G. L. (1995) Nucleosomes reconstituted *in vitro* on mouse mammary tumor virus B region DNA occupy multiple translational and rotational frames. *Biochemistry*, **34**, 12470.

66. Panetta, G., Buttinelli, M., Flaus, A., Richmond, T. J., and Rhodes, D. (1998) Differential nucleosome positioning on *Xenopus oocyte* and somatic 5S RNA genes determines both TFIIIA and H1 binding: a mechanism for selective H1 repression. *J. Mol. Biol.*, **282**, 683.

67. Lowary, P. T. and Widom, J. (1998) New DNA sequence rules for high affinity binding to histone octamer and sequence-directed nucleosome positioning. *J. Mol. Biol.*, **276**, 19.

68. Widom, J. (1996) Short-range order in two eukaryotic genomes: relation to chromosome structure. *J. Mol. Biol.*, **259**, 579.

69. Satchwell, S. C., Drew, H. R., and Travers, A. A. (1986) Sequence periodicities in chicken nucleosome core DNA. *J. Mol. Biol.*, **191**, 659.

70. Shrader, T. E. and Crothers, D. M. (1990) Effects of DNA sequence and histone–histone interactions on nucleosome placement. *J. Mol. Biol.*, **216**, 69.

71. Kornberg, R. D. and Stryer, L. (1988) Statistical distributions of nucleosomes: nonrandom locations by a stochastic mechanism. *Nucleic Acids Res.*, **16**, 6677.

72. Yao, J., Lowary, P. T., and Widom, J. (1993) Twist constraints on linker DNA in the 30-nm chromatin fiber: implications for nucleosome phasing. *Proc. Natl Acad. Sci., USA*, **90**, 9364.

73. Polach, K. J. and Widom, J. (1995) Mechanism of protein access to specific DNA sequences in chromatin: a dynamic equilibrium model for gene regulation. *J. Mol. Biol.*, **254**, 130.

74. Protacio, R. U., Polach, K. J., and Widom, J. (1997) Coupled-enzymatic assays for the rate and mechanism of DNA site exposure in a nucleosome. *J. Mol. Biol.*, **274**, 708.

75. Weischet, W. O. (1979) On the *de novo* formation of compact oligonucleosomes at high ionic strength. Evidence for nucleosomal sliding in high salt. *Nucleic Acids Res.*, **7**, 291.

76. Linxweiler, W. and Horz, W. (1984) Reconstitution of mononucleosomes: characterization of distinct particles that differ in the position of the histone core. *Nucleic Acids Res.*, **12**, 9395.

77. Polach, K. J. and Widom, J. (1996) A model for the cooperative binding of eukaryotic regulatory proteins to nucleosomal target sites. *J. Mol. Biol.*, **258**, 800.

78. Studitsky, V. M., Clark, D. J., and Felsenfeld, G. (1995) Overcoming a nucleosomal barrier to transcription. *Cell*, **83**, 19.

79. McGhee, J. D., Nickol, J. M., Felsenfeld, G., and Rau, D. C. (1983) Higher order structure of chromatin: orientation of nucleosomes within the 30 nm chromatin solenoid is independent of species and spacer length. *Cell*, **33**, 831.

80. Woodcock, C. L., Frado, L. L., and Rattner, J. B. (1984) The higher-order structure of chromatin: evidence for a helical ribbon arrangement. *J. Cell Biol.*, **99**, 42.

81. Williams, S. P., Athey, B. D., Muglia, L. J., Schappe, R. S., Gough, A. H., and Langmore, J. P. (1986) Chromatin fibers are left-handed double helices with diameter and mass per unit length that depend on linker length. *Biophys. J.*, **49**, 233.

82. Zlatanoua, J., Leuba, S. H., and van Holde, K. (1998) Chromatin fiber structure: morphology, molecular determinants, structural transitions. *Biophysical Journal* **74**, 2554.

2 | DNA replication, nucleotide excision repair, and nucleosome assembly

PAUL D. KAUFMAN and GENEVIÈVE ALMOUZNI

1. Introduction

Chromatin assembly is essential for viability and the maintenance of epigenetic states in eukaryotic cells. Formation of the fundamental chromatin unit, the nucleosome, is initiated immediately on both the leading and lagging strands after DNA replication fork passage. During nucleotide excision repair (NER), the best-characterized DNA repair mechanism, nucleosome formation is also coordinated with DNA synthesis. This chapter begins with an introduction to the protein machineries involved in DNA replication and NER, emphasizing components common to both processes. Next, the universal two-step pathway of nucleosome formation is presented, and nucleosome assembly factors linked or unlinked to DNA synthesis are reviewed. Finally, this biochemical information is integrated with *in vivo* data concerning histone deposition proteins, to highlight forthcoming important questions in this area of research.

2. Overview of eukaryotic DNA replication

In eukaryotic cells, DNA replication and NER are both coordinated with nucleosome assembly. Therefore, a brief description of DNA synthesis proteins is crucial for our later consideration of the formation of chromatin structures.

The development of mammalian cell extracts that support the complete replication of plasmid DNA containing the simian virus 40 DNA origin (SV40 ori) has allowed the identification of the proteins that duplicate cellular DNA (1). The SV40 replication reaction occurs in several steps (Fig. 1a):

1. The virally encoded T antigen protein serves as an origin-specific DNA unwinding protein.
2. In the presence of unwound DNA coated by the single-strand binding protein replication protein A (RPA), DNA polymerase α/primase complex (pol α/primase) synthesizes a hybrid RNA/DNA primer on each strand.

3. After dissociation of the non-processive pol α/primase complex, replication factor C (RFC) binds to the single-strand/double-strand junction at the 3' end of the primer. RFC catalyses ATP-dependent loading of the DNA polymerase process-ivity factor, proliferating cell nuclear antigen (PCNA), onto the template. PCNA stimulates coordinated synthesis of both leading- and lagging-strand DNA by DNA polymerase δ (pol δ).

4. Removal of RNA primers and sealing of lagging-strand Okazaki fragments is accomplished by the combined action of endonuclease FEN1, RNAase H, and DNA ligase.

Additionally, either topoisomerase I or II is required to relieve the superhelical strain caused by DNA unwinding; topoisomerase II also accomplishes decatenation of com-pleted daughter molecules. The specific roles of each protein are considered in the next two sections.

2.1 Primer synthesis

After recognition of an origin of DNA replication, proteins required for *de novo* primer synthesis form a complex known as the primosome. Primosome assembly during SV40 DNA replication comprises:

(1) DNA unwinding at the origin of replication; followed by

(2) the loading of an enzyme capable of initiating DNA synthesis on both the leading and lagging strands.

DNA unwinding requires two functions: a helicase and a single-strand binding pro-tein. During SV40 DNA replication, the DNA helicase T antigen binds to the origin of replication site-specifically and promotes local unwinding (2). T antigen assembles as a pair of homohexameric rings at the SV40 origin which encircle the DNA. This is likely to be a common theme among replicative helicases. Although the cellular heli-case for chromosomal DNA replication has not been identified positively, mini-chromosome maintenance proteins (MCMs) are possible candidates (1).

The unwound DNA strands generated by the DNA helicase activity are then stabilized by the single-strand binding protein, RPA (3). RPA is a heterotrimeric complex of approximately 70, 34, and 11 kDa subunits in all eukaryotic cells studied to date. These subunits contain four single-stranded DNA-binding domains (4). Indeed, RPA appears to be required for all reactions that involve transient single-stranded intermediates, including DNA recombination and repair.

After DNA unwinding at the origin, the RPA-bound single-stranded region then recruits proteins capable of initiating DNA synthesis. Processive leading-strand DNA polymerases are unable to generate their own primers, so the first nucleic acid syn-thesized must be an RNA primer made by the polymerase α/primase complex. Following DNA unwinding, pol α/primase is recruited via protein–protein inter-actions among RPA, pol α/primase, and T antigen (1). Pol α/primase then initiates primer synthesis. It first synthesizes an approximately 10 nt RNA primer and then

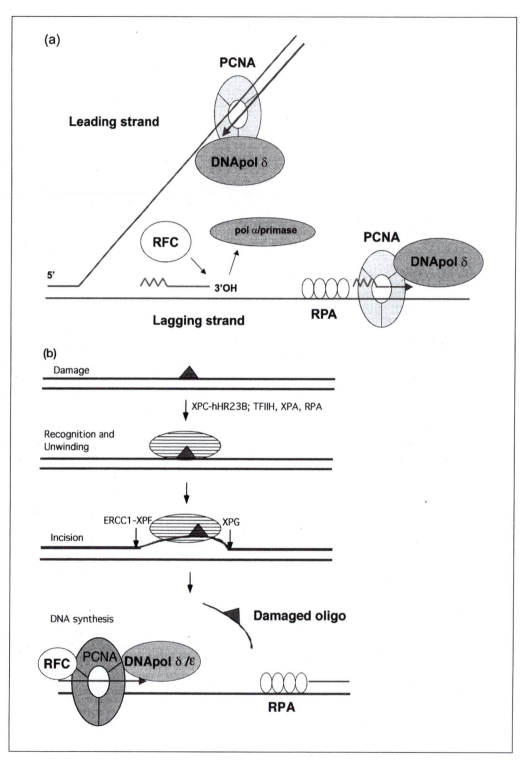

Fig. 1 (a) Bidirectional DNA replication. Key events described in the text are illustrated. On the leading strand, continuous DNA synthesis is catalysed by the processive DNA polymerase δ/PCNA complex. On the lagging strand, after the synthesis of each RNA–DNA primer by polymerase α/primase, RFC binds to the 3′OH terminus of the primer and loads a PCNA trimer around the template. The RFC/PCNA complex then recruits pol δ for synthesis of each Okazaki fragment. Not depicted are the initial unwinding events at the replication origin, the interactions between leading-strand and lagging-strand synthesis proteins, or the maturation of Okazaki fragments. (b) Nucleotide excision repair. A bulky DNA adduct, represented by a triangle, is first recognized by XPC-hHR23B. DNA unwinding is then catalysed by the XPB and XPD helicase subunits of TFIIH in the presence of XPA and RPA proteins. The recognition and unwinding proteins are depicted schematically as a striped oval. The unwound DNA is then cleaved by the ERCC1-XPF and XPG proteins, releasing the damaged oligonucleotide. The gap-filling DNA synthesis step involves DNA pol δ/ε, PCNA, and RFC and RPA. See text for details.

incorporates deoxynucleotides to produce a RNA–DNA hybrid primer approximately 40 nt long. These primers are then extended by DNA polymerase δ processively and continuously on the leading strand or discontinuously on the lagging strand (Fig. 1a). Pol α/primase itself is incapable of processive elongation, and instead dissociates from the template after primer synthesis.

2.2 Polymerase switching and maturation of Okazaki fragments

Because of the low processivity of pol α/primase, DNA polymerase δ is required for synthesis of both leading- and lagging-strand DNA once primers are formed. Pol δ is required for SV40 DNA synthesis *in vitro* and the corresponding yeast genes are essential *in vivo* (1, 5). The processivity of pol δ requires PCNA, a homotrimeric protein which encircles the DNA and serves as a sliding clamp, reducing the frequency of polymerase dissociation (6, 7). Loading of PCNA onto the DNA template requires the five-subunit ATPase complex termed RFC. RFC binds tightly to single-strand–double-strand junctions and nicks in DNA, and can load and unload PCNA at these sites.

The switch to pol δ occurs in a similar way on both strands, i.e. for continuous leading-strand synthesis and for discontinuous synthesis of Okazaki fragments on the lagging strand (Fig. 1a; reviewed in 1). After synthesis of the RNA–DNA primer, RFC binds to the 3′ end of the primer, displacing the non-processive pol α/primase. RFC then loads a PCNA trimer around the DNA, stimulating subsequent recruitment of pol δ. On the leading strand, DNA synthesis by the pol δ/PCNA complex is then processive and continuous over 5–10 kb. On the lagging strand, DNA synthesis continues until pol δ/PCNA encounters the next Okazaki fragment. Experiments using purified proteins have revealed how short Okazaki fragments are converted into long ungapped DNA products (1). Two nucleases, RNAase HI and FEN1, are involved in the removal of the RNA primer. At this point, gap filling by PCNA-stimulated DNA polymerase brings the 3′OH adjacent to the 5′ end to allow for ligation by DNA ligase.

Another processive DNA polymerase, pol ε, is not necessary for SV40 replication *in vitro*, but is essential for chromosomal replication *in vivo* in yeast (8). However, catalytically inactive alleles of the yeast pol ε gene support viability (9). The essential function of this protein involves S-phase checkpoint control (10), presumably as a sensor of damaged DNA or incomplete replication.

We wish to close this section by stressing some interesting properties of PCNA. The low conservation of the primary sequence of PCNA contrasts with the uniformity of its three-dimensional shape as a stable trimeric ring that encircles duplex DNA. The three-dimensional structures of *Saccharomyces cerevisiae* and human PCNA are super-imposable, even though these proteins share only 35% amino-acid identity (11, 12). PCNA binds to a variety of factors involved in different aspects of DNA metabolism (reviewed in 1), including the cell-cycle inhibitor p21, FEN1, DNA ligase 1, the nucleo-tide excision repair protein XPG, DNA(cytosine-5) methyltransferase, mismatch repair proteins MLH1 and MLH2, cyclin D, and the largest subunit of chromatin assembly factor-I (Section 6.2[4]). Thus, PCNA is predicted to coordinate DNA synthesis with DNA methylation, repair, epigenetic inheritance, and cell-cycle control. In mammalian cells, PCNA co-localizes with replication sites throughout S phase and becomes resistant to extraction with non-ionic detergents during this period, con-sistent with a stable association with chromatin (13). Outside of S phase, recruitment of PCNA to chromatin occurs upon DNA damage (14) (see Plate 3), reflecting the dual role of PCNA in replication and repair. Indeed, PCNA participates in several types of repair processes: NER, base excision repair, mismatch repair, postreplication repair, and double-strand break repair (6, 7). PCNA in *Drosophila melanogaster* is encoded by the gene *mus209*. Temperature-sensitive alleles of *mus209* were isolated which display enhanced sensitivity to DNA-damaging agents and also reduced gene silencing at a heterochromatic locus (15). These observations suggest a role of PCNA in the assembly or maintenance of chromatin structures, which is discussed further in Section 6.2.

3. Overview of nucleotide excision repair

Nucleotide excision repair (NER) is the main pathway used by mammalian cells to remove helix-distorting DNA lesions produced by UV irradiation or chemical muta-gens (16, 17). This evolutionary conserved pathway is very important for genome integrity and is the most thoroughly studied example of how DNA damage removal is coupled to DNA synthesis and nucleosome assembly. Defects in NER in humans are associated with the disease xeroderma pigmentosum (XP), the main character-istics of which are high UV-sensitivity and an increased incidence of sunlight-induced skin cancers. The seven complementation groups identified (XPA to XPG) correspond to proteins that act together in the initial steps of NER to recognize and remove damaged DNA. All core components of eukaryotic NER have been identified and the repair reaction has been reconstituted on naked DNA using purified proteins (17). In light of these experiments, two major steps can be distinguished: (1) recog-nition and incision–excision of the DNA strand containing the lesion, and (2) repli-cative DNA synthesis, restoring the original, undamaged DNA.

3.1 Specialized proteins for recognition and excision of damaged DNA

The enzymes involved in the first phase of NER are highly conserved from humans to *S. cerevisiae* (16, 17) (Fig. 1b):

1. For lesions on non-transcribed DNA strands, the XPC-hHR23B complex is likely to act as the initial recognition factor. On transcribed strands, initial recognition of damage is instead believed to involve stalling of RNA polymerase II.

2. Following recognition, an opened complex is formed in an ATP-dependent manner and DNA around the lesion site is unwound asymmetrically. This unwinding is achieved by the TFIIH complex, which contains the two DNA helicase activities, XPB (3′–5′) and XPD (5′–3′).

3. At this stage both XPA and the single-strand-binding protein RPA (described above) are also required to provide the substrate for cleavage by two specific endonucleases, XPG (3′ incision) and ERCC1-XPF(5′ incision), which cut near the single-stranded–double-stranded DNA junction. A 24–32 nt patch containing the lesion is thus released and the resulting gapped DNA is then available for replicative synthesis.

3.2 Recruitment of general replication proteins to seal gaps caused by repair

In the second step of NER, gapped DNA formed by removal of the damaged oligo-nucleotide is a substrate for replicative synthesis. *In vitro*, this has been recapitulated using DNA polymerase δ (or ε) holoenzymes together with the replication factors RPA, PCNA, and RFC (17) (Fig. 1b). The sugar–phosphate backbone is then sealed by a DNA ligase, most probably DNA ligase I (18). Thus, common factors are used for gap-sealing during both DNA replication and NER. These factors also provide an interface for coordination with chromatin assembly via the direct interaction between PCNA and histone deposition proteins (19, 20; and see Section 6.2).

The evidence that PCNA participates in mammalian NER *in vivo* is compelling. UV irradiation of non-S-phase human cells in culture leads to a relocation of PCNA, which becomes tightly bound to chromatin. This association of PCNA with chromatin after irradiation is absent in UV-irradiated quiescent XPA cells (21).

3.3 Restoration of chromatin organization after DNA repair

Immediately after repair synthesis, repair patches in the cell are sensitive to micro-coccal nuclease and do not reveal the 10 bp periodicity of DNAase I digestion typical of canonical nucleosomal DNA. Most repair patches become nuclease resistant over the course of several hours, similar to the time course of repair synthesis (22). These results suggest that nucleosomes are disrupted or mobilized during repair of UV photoproducts and that nucleosome assembly, repositioning, and packaging into higher-order structure are likely to be required steps after repair synthesis (23, 24). In many ways, the nuclease sensitivity of newly repaired DNA is reminiscent of that of newly replicated DNA following the passage of the replication fork *in vivo* (25; and see Section 4.2). We will compare nucleosome assembly during replication and repair again in Section 6, after first considering the basic aspects of nucleosome formation.

4. Histones, nucleosome formation, and chromatin maturation

4.1 *In vitro* reconstitution of nucleosomal core particles

Different experimental approaches, both *in vivo* and *in vitro*, have come to the uniform conclusion that nucleosomal core particles are formed in two steps. First, a tetramer of histones H3 and H4 (i.e. $(H3/H4)_2$) is deposited onto DNA; secondly, the deposited tetramer then serves as a binding site for two histone H2A and H2B dimers to complete the particle. Two-step nucleosome assembly occurs during *in vitro* experiments in which core histones are deposited onto DNA by slow dialysis from high salt (2 M NaCl) to physiological conditions (26). Therefore, histones and DNA are capable of self-assembly under artificial conditions. However, simultaneous mixing of all four core histones with DNA under physiological conditions does not result in efficient nucleosome formation; rather, aggregation and precipitation result unless low concentrations of histones are added very slowly to DNA and then subjected to long incubations (27). Thus, control of the potentially promiscuous electrostatic interactions between histones and DNA is essential for proper formation of a core particle. This theme will appear again during our consideration of histone chaperones (Sections 5 and 6). We also note that core particles deposited by dialysis are not arrayed with the highly regular spacing observed in cellular chromatin. Such post-deposition spacing adjustments require the action of ATP-dependent remodelling factors (see Chapter 6).

4.2 Coordination of histone deposition with DNA synthesis *in vivo*

The temporal order of histone deposition during DNA replication *in vivo* has been demonstrated by a variety of pulse–chase experiments. Worcel and colleagues treated *Drosophila* tissue culture cells with very short pulses of radioactive amino acids, followed by density gradient isolation of protein–DNA complexes. Newly synthesized histones H3 and H4 were already detected on nascent DNA after a 2 minute pulse, newly synthesized histones H2A and H2B associated with the nascent DNA 2–10 minutes later, and linker histone H1 was deposited 10–20 minutes later (28) (Fig. 2). Non-radioactive chase incubations after pulse labelling demonstrated that during these assembly steps nascent chromatin changes from being highly nuclease-accessible to a more resistant form. Subsequent work showed that this chromatin 'maturation' involves formation of regularly spaced nucleosomal ladders (25).

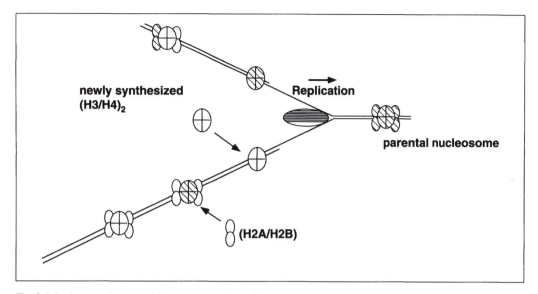

Fig. 2 Inheritance of parental histones and deposition of newly synthesized histones at a replication fork. Replication-fork machinery is depicted as a striped oval. Parental histones are hatched, newly synthesized histones are not. As described in the text, nucleosomes are assembled in two steps, with the $(H3/H4)_2$ tetramer deposited prior to (H2A/H2B) dimer deposition. Parental tetramers and dimers are inherited separately. Tetramers are divided onto daughter DNA duplexes in a non-conservative fashion; dimers are segregated independently, suggesting dissolution of parental octamers upon passage of the fork.

How is deposition of new histones onto DNA coordinated with inheritance of histones from the parental DNA duplex? In a series of elegant cross-linking experiments, Vaughan Jackson demonstrated that histones H3 and H4 are inherited following replication-fork passage as $(H3/H4)_2$ tetramers, separately from H2A/H2B dimers (29) (Fig. 2). These results predict that $(H3/H4)_2$ tetramers and dimers will require separate assembly factors to deliver them to DNA. Other experiments showed that the incorporation of newly synthesized histones into one octamer did not dictate whether the neighbouring octamer was labelled (30). In other words, segregation of parental histones was not conservative with respect to the two daughter DNA strands (Fig. 2).

Grunstein and colleagues performed experiments in the budding yeast *S. cerevisiae* which demonstrated the requirement for nucleosome formation during DNA replication *in vivo*. To do this, they generated cells expressing either histone H2A or H4 under the transcriptional control of a glucose-repressible promoter (31, 32). Release of such cells into S phase in the absence of histone gene expression allowed for duplication of the DNA as measured by flow cytometric analyses, but the cells arrested irrevocably in the following G_2 phase even if the depleted histone was supplied later. Thus, nucleosome formation must be concerted with DNA replication *in vivo* to ensure formation of chromosomes that allow completion of the cell cycle.

4.3 Synthesis and modification of histones prior to assembly

Synthesis of the four core histones is regulated during the eukaryotic cell cycle (33). In proliferating cells, the bulk of histone synthesis occurs during DNA replication because histone gene transcription is derepressed at the G_1/S phase transition of the cell cycle. This ensures that adequate histones are present when required for formation of new nucleosomes. Exceptions to this rule (34) include a subset of histone genes that are constitutively transcribed at low levels, presumably to ensure the presence of histones for DNA repair outside of S phase. Another exception occurs during the early development of *Xenopus* and *Drosophila* embryos, which contain a large pool of maternal histones.

In all species that have been examined, newly synthesized histones undergo reversible post-translation modifications distinct from those that occur on chromatin-deposited molecules. Early work demonstrated that newly synthesized histones H3 and H4 are acetylated on N-terminal residues and that these initial modifications are lost upon incorporation into chromatin; transient H4 phosphorylation was also observed (34). Development of techniques for isolation and Edman degradation sequencing of newly synthesized histones in the Allis laboratory revealed a non-random pattern of acetylated lysines in newly synthesized histones (35–37). Newly synthesized histone H4 is acetylated on lysines 5 and 12 (or on the analogous residues 4 and 11 in *Tetrahymena*). This pattern is observed in all organisms examined, and is distinct from the acetylation associated with transcriptionally active chromatin (38) accomplished by proteins such as Gcn5p in yeast (see Chapter 7). In contrast to histone H4, acetylation of newly synthesized histone H3 molecules occurs on different lysines in different species (37).

What are the roles of these transient modifications? It is not presently clear whether they are required for deposition by nucleosome assembly factors, for an upstream process such as cellular trafficking of newly synthesized histones, or for an event subsequent to deposition (see Section 6.2). In budding yeast, the *HAT1* gene encodes an enzyme responsible for the majority of the histone H4 lysine 12-directed acetyltransferase activity. However, deletion of this gene produces no phenotype other than a reduction in lysine 12-acetyltransferase activity observed in extracts (39, 40). Thus, functionally overlapping enzymes may compensate for loss of Hat1p *in vivo*. Furthermore, at least in budding yeast, either of the histone H3 or H4 N-termini are sufficient for viability, whereas deletion of both tails is lethal (41), suggesting some redundancy among the various N-terminal modifications. Experiments designed to determine the minimal set of N-terminal residues in yeast (42) will be discussed further in Section 6.2.

5. *In vitro* nucleosome assembly systems unlinked to DNA synthesis

Given that salt dialysis is able to mediate formation of nucleosomes (see Chapter 1 and Section 4.1), it is not surprising that anionic polymers able to neutralize the high positive charge of the histone proteins are also able to act as nucleosome assembly factors. Specifically, prebinding of core histones to polyglutamic acid (43) or RNA (44) creates complexes that form nucleosomes upon subsequent addition of DNA. As observed in dialysis experiments, nucleosomes formed with these polymers do not display the regular spacing observed in cellular chromatin. However, many of the factors isolated in biochemical assays for nucleosome assembly have tracts of acidic amino acids in common (45–48), suggesting that charge neutralization by histone binding is a key aspect of their functionality.

The initial development of crude extracts derived from *Xenopus* eggs and *Drosophila* embryos which support nucleosome assembly onto exogenously added DNA in the absence of replication provided the first systems for identification of critical components in these reactions. Such extracts contain a large maternal store of histones bound to acidic 'chaperone' proteins thought to allow for multiple rounds of cell division prior to the onset of transcription. Variation in the preparation and use of extracts proved very useful for revealing the importance of ATP in the production of long arrays of nucleosomes (49).

The nucleoplasmin and N1/N2 proteins from *Xenopus* eggs responsible for formation of core particles *in vitro* have been purified and cloned (34, 49). In *Xenopus* eggs, nucleoplasmin is a H2A/H2B-binding protein, while N1/N2 binds to histones H3 and H4; together, these complexes allow for core particle formation at physiological conditions by providing charge neutralization to the histones (50). Such results are consistent with the generality of the two-step nucleosome assembly pathway. However, nucleoplasmin and N1/N2 do not require DNA synthesis to assemble nucleosomes, and they assemble nucleosomes *in vitro* significantly more slowly than the rapid cell divisions that occur during early embryogenesis. Also, nucleosome assembly with these components does not produce regular nucleosomal arrays like those obtained using crude extracts.

In *Drosophila* embryo extracts, histone-binding proteins involved in nucleosome formation were identified, including the *Drosophila* homologue of NAP-1 (47), an acidic protein previously isolated from mammalian cells and yeast that can bind all four core histones. Another protein complex isolated from the *Drosophila* embryo extract, termed ACF (ATP-utilizing chromatin assembly and modifying factor), rapidly generates very long, spaced nucleosomal arrays in the presence of core histones and a histone-binding protein (51). ACF is required in catalytic amounts in this reaction, and has been shown to cooperate in this assay with a variety of histone-binding proteins which must be present in amounts stoichiometric with the histones. Notably, one subunit of ACF is ISWI, an ATPase present in a number of other protein complexes implicated in the energy-dependent mobilization of nucleosomes (see Chapter 6). An important question remains as to how these DNA-replication-independent reactions are related to nucleosome assembly during S phase *in vivo*.

6. *In vitro* histone deposition systems linked to DNA synthesis

6.1 Replication- and repair-linked nucleosome assembly by cellular extracts

As described in Section 4, the majority of nucleosome assembly *in vivo* occurs during DNA synthesis. Biochemical assays were therefore developed to gain insight into the link between replication fork passage and nucleosome assembly.

As noted above, nucleosome assembly on double-stranded templates in *Xenopus* egg extracts leads to slow formation of nucleosomes. However, addition of single-stranded templates results in rapid DNA replication, presumably using lagging-strand replication proteins (see Section 2.1). This single- to double-strand replication is accompanied by equally rapid formation of nucleosomes, either on replicated templates *in vitro* (52) or upon injection into *Xenopus* oocytes (53). Consistent with data from other systems (see Section 4), histones H3 and H4 were deposited onto DNA prior to H2A/H2B (54). Similar reactions could be reproduced in *Drosophila* extracts (55, 56) or in a human cell-free system (57).

Bruce Stillman approached this problem using the *in vitro* assay for SV40 DNA replication (see Section 2). After dounce homogenization of human cells in very low ionic strength buffer, nuclei were removed by low-speed centrifugation. The 'cytosolic' supernatant contained sufficient DNA polymerases and their cofactors, which had leaked from the nuclei, to support bidirectional replication (58). DNA replication in this extract yielded topologically relaxed products. However, addition of a salt-extracted nuclear fraction from the same cells caused the products to become negatively supercoiled due to formation of nucleosomes (59).

Similarly, *in vitro* assays using extracts from *Xenopus* eggs, human cells, and *Drosophila* embryos were developed which perform nucleosome assembly linked to NER (60, 61). In the *Drosophila* embryo extracts (20, 61) the assembly reaction was triggered from a single lesion site, and nucleosome formation propagated bidirectionally from this site. These results make sense in the light of experiments identifying PCNA as the trigger for DNA synthesis-linked nucleosome formation (see Section 6.2).

6.2 Chromatin assembly factor-I (CAF-I): biochemical functions

The SV40-based assay was used to isolate the DNA replication-linked histone deposition factor from human cell nuclear extracts. A three-subunit protein complex was purified and named chromatin assembly factor-I (CAF-I) (62). Purification of CAF-I to homogeneity demonstrated that the histones deposited in these experiments were provided by the 'cytosolic' replication extract rather than the CAF-I-containing nuclear extract. This suggested that the histones deposited were newly synthesized molecules not previously incorporated into chromatin. This notion was supported by detection of newly synthesized, acetylated H3 and H4 molecules in the cytosolic extract (63). By chromatographically depleting histones H2A and H2B from the replication extract, Smith and Stillman demonstrated that CAF-I performs the first step of nucleosome assembly, bringing histones H3 and H4 to replicated DNA. Subsequently, histones H2A/H2B bind to this 'chromatin precursor', completing the histone octamer (63). Thus, CAF-I assembles nucleosomes *in vitro* by the same two-step mechanism observed *in vivo* (Section 4.2), and it does so in a manner linked to DNA synthesis.

Later experiments demonstrated that CAF-I is bound to newly synthesized histones H3 and H4 in cellular extracts (45, 64), and that these histones display an acetylation pattern similar to that of newly synthesized molecules. A subfraction of the histone H3 bound to CAF-I displayed reduced mobility on triton–acid–urea gels, which likely represents either monoacetylation or phosphorylation. Also, antibodies which specifically recognize individual acetylated lysine residues detected acetylation on up to three histone H4 residues: lysines 5, 8, and 12 (64). At cellular sites of DNA replication within heterochromatin, CAF-1 co-localizes with histone H4 acetylated at lysines 5 or 12, suggesting functional relevance for this complex (65; see also Section 7.1).

The acetylation pattern of histone H4 bound to CAF-I may explain why any one of H4 lysines 5, 8, or 12 is required for viability in the absence of the H3 N-terminus in yeast (42). One hypothesis that would unify these data is that a single acetylatable residue on either H3 or H4 is sufficient for delivery of newly synthesized histones to CAF-I or other assembly factors. Alternatively, the observed requirement for one acetylatable lysine could instead reflect structural requirements for a functional chromosome. However, we note that H3/H4 N-termini are not required for binding to the p48 subunit of CAF-I (66; see Section 6.4). Furthermore, N-terminal residues of histones H3/H4 are not required for histone deposition by CAF-I (66a). N-terminal modifications are therefore dispensable for nucleosome core formation by CAF-I and their exact role is still unclear (see Section 8.2).

Although CAF-I functions when nucleosome assembly is concomitant with DNA synthesis, *in vitro* experiments using the DNA polymerase inhibitor aphidicolin demonstrated that the two processes can be separated temporally (67). In these experiments, DNA replication reactions performed in the absence of CAF-I were treated with aphidicolin; subsequent addition of CAF-I was still able to cause nucleosome formation preferentially on molecules that had been replicated prior to CAF-I addition. Importantly, this order of events was not reversible, because addition of aphidicolin at the start of the reaction prevented both DNA synthesis and nucleosome assembly. Also, the ability of the replicated template to support nucleosome assembly by CAF-I was temporally labile, decaying during incubation in the replication extract. These experiments suggested that a mark left by the replication process serves as a specificity factor for later recognition by the CAF-I/H3/H4 complex, and that this mark decays with time.

Recent experiments demonstrated that CAF-I deposits histones specifically onto replicated DNA molecules via binding of the CAF-I p150 subunit to PCNA (19). In this work, DNA templates replicated in the absence of CAF-I were separated by size exclusion chromatography away from the bulk of proteins in the replication extract. The replicated templates were shown to contain PCNA molecules loaded by RFC (see Section 2). These PCNA-containing plasmids were functional substrates for nucleosome assembly by CAF-I in the absence of ongoing replication, although additional, as yet unidentified molecules from cellular extract were also required. Importantly, incubation of the PCNA-containing templates with RFC and ATP prevented subsequent CAF-I activity, because RFC removed the PCNA trimers from the templates. Other experiments in the context of DNA repair have also demonstrated that extensive DNA synthesis is not required for CAF-I activity (20, 61). In this work, a single nick on a DNA template was sufficient to trigger CAF-I-dependent chromatin assembly, even in the presence of aphidicolin, presumably due to RFC-mediated loading of PCNA. Also, incubation of DNA molecules with human cell extracts led to recruitment of PCNA together with CAF-I p150 and p60 subunits in an ATP-dependent and DNA nick-dependent manner (20). Together, these data support a role for PCNA as a link between CAF-I and both DNA replication and DNA repair.

6.3   Primary structure and conservation of CAF-I subunits

Human CAF-I is comprised of subunits termed p150, p60, and p48, and is likely to be a heterotrimeric complex containing one of each subunit (62, 64). p150 is an acidic protein with several highly charged regions (Fig. 3); the other subunits, p60 and p48, contain WD repeats (45, 64), a prevalent repeating motif in many eukaryotic proteins (68). The subunit structure and biochemical function of CAF-I is conserved from budding yeast to humans (46). CAF-I activity has also been identified in *Drosophila*

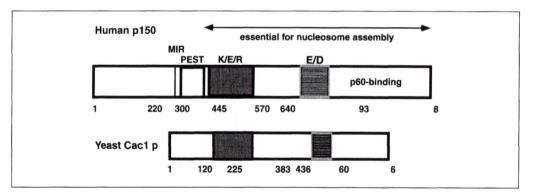

Fig. 3 Structure and conservation of the large subunit of chromatin assembly factor-I. The human p150 and budding yeast Cac1p proteins are shown. MIR indicates the MOD1-interacting region (amino acids PXVXL) required for localization to heterochromatin; MOD1 is a heterochromatin protein 1 (HP1) family member. PEST indicates a putative destruction box. K/E/R indicates the region enriched in lysine, glutamic acid, and arginine residues predicted to form a coiled-coil. E/D indicates a region enriched in acidic residues. The regions of human p150 essential for *in vitro* nucleosome assembly activity and for binding the p60 subunit are indicated. Amino-acid numbers are indicated below each gene diagram. See text for details.

and *Xenopus* extracts by cross-species complementation of the human SV40 replication and NER assays (60, 67, 69).

The large subunit of CAF-I is the least well conserved in evolution (Fig. 3), although homologues from fission yeast and the plant *Arabidopsis* can be found in databases. The human p150 protein and its budding yeast homologue Cac1p are approximately 25% identical at the amino-acid level, sharing a large region comprised mostly of the charged amino acids lysine, glutamic acid, and arginine (46). This region is predicted to form a coiled-coil (70) and is referred to as the K/E/R domain (Fig. 3). Deletion of this region in human p150 abolishes activity in the *in vitro* nucleosome assembly assay (45). Human p150 and yeast Cac1p also share a region enriched in aspartic and glutamic acid residues (E/D in Fig. 3). This domain is also essential for p150 function (45). Notably, the PEST region in human p150, perhaps a proteolytic degradation signal (71) is absent in Cac1p. Finally, the N-terminal 296 amino acids of human p150 are not required for nucleosome activity *in vitro* (45), but are required for heterochromatic localization of CAF-I outside of S phase due to a specific interaction between residues near the p150 N-terminus and HP1 (heterochromatin protein 1) family members (72; see sections 6.4 and 7.1). In contrast, the C-terminal deletions of p150 are inactive, presumably due to loss of the p60 binding site (45).

The other two CAF-I subunits are more highly conserved. Like human p60, the yeast Cac2p subunit is a WD-repeat protein. However, like Cac1p, Cac2p lacks the PEST region in its human counterpart. Including conservative substitutions, the human and yeast proteins share 47% homology within the WD-repeat region. More significantly, alignment of the two proteins shows that the homology extends over each of the seven WD repeats and includes residues other than those that comprise the canonical amino acids that define the repeat itself (46, 73). Examination of database entries showed that the yeast protein most similar to Cac2p is Hir1p, which contains WD motifs in its N-terminal half (74) as does its human homologue, the HIRA protein (75). Comparison of *CAC2*, *HIR1*, and their human homologues showed that they encode a distinct subfamily of WD proteins (73), postulated to contain histone-binding activity (76, 77). Functions of the Hir proteins with regard to CAF-I will be discussed in Section 7.2.

The p48 subunit of CAF-I is a member of a highly conserved subfamily of WD proteins which does not overlap with the p60/Cac2p/Hir1p subfamily. The p48 subfamily members are histone-binding proteins which are subunits of a variety of multisubunit complexes contributing to multiple aspects of histone metabolism, notably nucleosome assembly and histone acetylation and deacetylation (78; see also Chapters 7 and 8). Thus, these proteins are likely to coordinate interactions between these complexes and their histone substrates.

6.4 CAF-I interacting proteins: histones H3/H4, PCNA, and MOD1

Several recent studies have demonstrated direct protein–protein interactions between CAF-I subunits and other proteins implicated in CAF-I function. First of all, the human p48 subunit and its very close (>90% identical) human homologue, p46, were shown to bind directly histones H2A and H4 (66). Interestingly, the binding sites on these two histones were mapped and shown to include a similar amino-acid sequence within the α1 helix of both histones; interaction did not require the amino termini. These data are consistent with the role of CAF-I in deposition of newly synthesized histones, as these α-helices become far less accessible when incorporated into nucleosomes (79). CAF-I binds and deposits newly synthesized histones H3 and H4 (45, 63, 64), but is not known to contact directly histones H2A and H2B. The interaction between p48 and H2A may therefore be more relevant for other p48-containing complexes which are involved in histone modification.

Mammalian CAF-I p150 subunits bind directly heterochromatin protein 1 (HP1; see Figure 3 and Chapter 11). This interaction requires a PXVXL motif near the N-terminus of the human and mouse p150 proteins (Fig. 3) (72). Notably, this motif is present in the human and mouse p150 proteins (45, 72) but is absent in the corresponding proteins of budding and fission yeasts and *Arabidopsis*, consistent with the highly divergent sets of heterochromatin proteins in these different organisms (see also Chapter 10).

Because PCNA recruits CAF-I to replicated templates via direct binding to p150 (19), it is of interest to understand the molecular details of this interaction. Surprisingly, recent pull-down experiments using glutathione S-transferase fusion proteins demonstrated an interaction between PCNA and residues 1–31 of human p150 (20). The significance of these data is unclear because amino acids 1–296 are not required for nucleosome assembly *in vitro* (Fig. 3) (45). Therefore, future experiments will be required to determine whether additional PCNA binding sites exist or if PCNA-independent modes of action exist for CAF-I. An interesting possibility, although unexplored yet, is that targeting of CAF-1 to replication or repair sites may be regulated by different sets of protein–protein interactions.

6.5   Stimulation of CAF-I activity by ASF1

The budding yeast *ASF1* gene was identified in two independent genetic screens for high-dosage disrupters of position-dependent gene silencing (80, 81). Recently, the *Drosophila* homologue of *ASF1* was identified as a factor that stimulates the nucleosome assembly activity of CAF-I *in vitro* (82). In this assay, DNA replication-linked assembly assays were performed with an increased number of larger DNA templates. CAF-I acts as a stoichiometric histone donor *in vitro* (45, 62), and no significant change in the topology of replicated molecules occurred under these modified conditions. *Drosophila* embryo extracts were added to this assay, and an activity which increased the supercoiling of replicated products was detected. This assay was then used to purify a complex termed RCAF (replication-coupling assembly factor), consisting of three polypeptides: dASF1, the *Drosophila* homologue of yeast Asf1p, and acetylated forms of histones H3 and H4. Interestingly, the histone H3 was acetylated at lysine 14, but not at positions 4 or 9, and histone H4 was acetylated on lysine 5 and 12; these are the modification patterns of newly synthesized *Drosophila* histones (37). Addition of RCAF to CAF-I synergistically increased supercoiling of replicated products, although larger amounts of RCAF alone were able to supercoil unreplicated molecules. The supercoiling results are consistent with increased nucleosome formation in the presence of both RCAF and CAF-I, although the products were not analysed by micrococcal nuclease digestion to confirm this. Both dASF1 and CAF-I are histone H3/H4-binding proteins, so one hypothesis based on these data is that ASF1 may be a general donor of newly synthesized histones to other protein complexes, including CAF-I, and that CAF-I becomes able to deposit multiple rounds of histones in the presence of ASF1. Genetic data concerning the role of the yeast *ASF1* gene will be considered in Section 7.2.

7. *In vivo* experiments with histone deposition proteins

7.1 Regulation and modification of CAF-I during the cell cycle

In mammalian cells, cellular DNA synthesis takes place at discrete granular sites. These foci correspond to clustered replication forks which vary in number, location, and size at different times during S phase (83, 84). Many of the replication proteins identified biochemically have been located at these sites during S phase. These include the p150 and p60 subunits of CAF-I (85), suggesting that CAF-I assembles nucleosomes within these foci. The extent of CAF-I co-localization with early and late replication foci is similar (65) (Plate 3). However, CAF-1 is maintained at late replication foci for 20 minutes after detectable nucleotide incorporation (65). In other experiments, re-localization of mouse-cell CAF-I to heterochromatin outside of S phase was shown to require HP1 (heterochromatin protein 1) family members, as it is abolished in CAF-I p150 mutants lacking the HP1-interacting region (Fig. 3) (72). Together, these data suggest a role for CAF-I in the maintenance of heterochromatic structures in the absence of DNA synthesis. Similar conclusions have been reached based on experiments in yeast (86; see also Section 7.2), suggesting that this is a conserved function.

The p150 and p60 subunits of human CAF-I are phosphoproteins (87), and p60 phosphorylation is increased dramatically during mitosis (88), at which point CAF-I largely dissociates from chromatin (89). Also, CAF-I-containing extracts from mitotic cells are deficient in nucleosome assembly activity (89), suggesting that mitotic p60 hyperphosphorylation is either directly inhibitory or prevents productive interaction with replicated DNA templates. Upon ultraviolet irradiation of cells, CAF-I localizes to PCNA-containing sites of DNA repair, and p60 also becomes phosphorylated, but with a smaller decrease in electrophoretic mobility than seen in mitotic cells (90). These data are consistent with the biochemical data implicating PCNA as the link between CAF-I activity during both DNA replication and nucleotide excision repair (Section 6.2).

7.2 Silencing and growth phenotypes of yeast cells lacking CAF-I, Hir proteins, and Asf1

Budding yeast cells lacking any or all of the three genes encoding CAF-I subunits (*CAC1*, *CAC2*, and *CAC3*) display reduced position-dependent gene silencing at telomeres (46, 70, 91), the silent *HM* mating loci (73, 86), and at ribosomal DNA (92). Gene silencing at these loci is a chromatin-mediated process analogous to hetero-chromatic silencing in multicellular organisms: mutation of histone termini, or histone-binding silencing proteins can abolish silencing (93; see also Chapter 10). Therefore, the silencing defects of *cac* mutants are consistent with the biochemical and cell biological analyses of mammalian CAF-I (Sections 6.2 and 7.1). We also note that chromatin assembly during DNA replication has been hypothesized to be a window of opportunity for remodelling epigenetic states, particularly in yeast (see Chapter 10).

Yeast *cac* mutants are defective in maintenance of the silenced state. This was demonstrated by analysis of *cac* cells arrested by the mating pheromone alpha-factor (86). Normally, alpha-factor arrests yeast cells of mating type **a** during G_1 phase of the cell cycle, requiring the cell to prepare for mating rather than continuing cell-cycle progression. However, this arrest is dependent on continued silencing of the *HM* loci; loss of *HM* silencing allows co-expression of the cryptic **a** and α genes which specify cell type, generating a resistance to the mating pheromone. In *cac* mutant strains, alpha-factor-mediated arrest is temporally more labile in all members of the population. In contrast, cells of *sir1* strains exist in two epigenetic states, either fully arrested or fully resistant to pheromone arrest (94). These results suggest that Sir1p is important of establishing of the silenced state, whereas CAF-I appears to be more important for its maintenance. Similar conclusions regarding CAF-I have been reached by analysis of the stability of silencing at telomeres (91).

Although *cac* mutants have chromatin-related phenotypes, they are viable and have wild-type growth rates. Thus, other factors which contribute to histone deposition *in vivo* have been looked for. Recent work has implicated the histone regulation (*HIR*) genes, (*HIR1*, *HIR2*, *HIR3* and *HPC2*) (95, 96). *HIR* genes are required for cell-cycle-dependent repression of histone gene transcription outside of the G_1/S transition. Mutation of *HIR* genes has minor effects on silencing (73), and growth rates are not affected. However, *cacΔ hirΔ* double mutant strains display synergistic reduction of position-dependent gene silencing at both telomeres and the silent mating loci, exhibit slow growth after germination, increased sensitivity to methyl methane-sulphonate (MMS), and increased Ty element transposition (73, 97). Thus, Hir proteins become important for normal growth and silencing in the absence of CAF-I. One hypothesis to explain this synergy is that Hir proteins contribute to an alternative nucleosome formation pathway which becomes more important for silencing in *cacΔ* cells. Consistent with this idea, genetic evidence suggests that Hir proteins bind histones (76, 98), as does biochemical analysis of a mammalian *HIR* homologue (77).

The yeast *ASF1* gene was identified in two independent genetic screens for high-dosage disrupters of position-dependent gene silencing (80, 81). Because *Drosophila* ASF1 is a histone H3/H4-binding protein, the overexpression phenotype of yeast

ASF1 likely results from sequestration of newly synthesized histones. Consistent with the biochemical data, *cac1Δ asf1Δ* double mutants display synergistic decreases in telomeric and *HM* silencing, slower growth rates, and accumulation of cells with 2C DNA content (82). However, these double mutants are viable; an important question remaining is whether other histone deposition factors remain unidentified, or whether a degree of histone deposition can occur in an unchaperoned manner *in vivo* (see Section 8.1).

7.3 DNA repair defects of yeast cells lacking CAF-1 and Asf1

Deletion of any of the three *CAC* genes confers a mild increase in sensitivity to killing by ultraviolet (UV) radiation (46). Furthermore, deletion of the *CAC1* gene increases the UV sensitivity of cells mutant in genes from each of the known DNA repair epistasis groups (99). For example, double mutants involving *cac1Δ* and NER gene deletions *rad1Δ* or *rad14Δ* showed increased UV sensitivity, as did double mutants involving *cac1Δ* and deletions of members of the recombinational repair group. *cac1Δ* also increased the UV sensitivity of strains with defects in both the error-prone and error-free branches of post-replicative DNA repair pathway. Together, these data suggest that CAF-I has a role in multiple damage repair pathways. All these repair pathways utilize PCNA, consistent with the idea that the contribution of CAF-I to DNA repair is via nucleosome assembly triggered via direct interaction (Section 6.4) with PCNA.

asf1Δ mutants are sensitive to treatment with ultraviolet light, hydroxyurea (HU), bleomycin and MMS, and are more slow-growing than the wild type (80–82). UV, MMS, and HU sensitivity of *asf1Δ* cells is exacerbated upon deletion of *CAC* genes (82). *cacΔ hirΔ* double mutants also display synergistically increased sensitivity to MMS (82, 97), but not to UV. These data suggest that Asf1p and Hir proteins contribute to DNA repair in a manner which at least partially relies on CAF-I function.

8. Final comments

8.1 Important questions

Mechanistic similarities between DNA replication and NER provide a framework for understanding how both these processes can be linked to nucleosome assembly. Several nucleosome assembly factors identified biochemically have been validated by genetic and cell biological means. However, none of the histone deposition factors identified to date are essential for viability in yeast. This could be explained either by functional overlap among these factors, or by the hypothesis that histone deposition can occur in a largely unchaperoned manner. We disfavour the latter hypothesis as it contradicts a wealth of biochemical data from many organisms which demonstrate that newly synthesized histones are always bound to other proteins prior to deposition. It will, however, be important to assess the contributions of these factors in multicellular organisms with more complex chromatin organization.

An important issue is therefore how different nucleosome assembly factors cooperate, e.g. how CAF-I and ASF1 proteins mechanistically achieve synergy. It is equally important to understand their relative contributions to chromosome formation *in vivo* at specialized regions of the genome. For example, specialized histone H3-like proteins (CENP-A in mammals, Cse4p in yeast, see Chapter 4) are specifically incorporated into centromeric chromatin. Does this occur via a specialized, dedicated chaperone, or through regulated use of one of the assembly factors we have discussed? In mammalian and plant cells, the inheritance of chromatin states also needs to take into account the mode of inheritance of DNA methylation patterns. Thanks to recent identification of several chromatin-associated DNA methylase complexes, much progress is expected in this field. Finally, an important challenge will be to move from nucleosome assembly to higher levels of nuclear organization. For example, how are domains bearing different classes of proteins (e.g. HP1, Polycomb group proteins, etc.) established and maintained, and how are these structures modulated during differentiation and development?

8.2 Discussion

Although synthesis-related N-terminal modifications of histones H3/H4 are found throughout eukaryotic organisms (37), it is still unclear what functional significance they have. Nucleosomes lacking tail domains of all four core histones can be assembled *in vitro* by dialysis (100) and, as described in Section 6.2, the N-termini of histones H3/H4 are dispensable for deposition by CAF-I. Thus, the biological role(s) of the H3/H4 modifications could be:

(1) upstream of deposition by CAF-1,

(2) downstream in a later step in chromatin maturation;

(3) used as a signal to monitor progression of the assembly process; or

(4) important for activity of a different assembly factor.

In the first case, proper trafficking of newly synthesized molecules into the nucleus, or loading onto CAF-I, could be the important steps. However, our knowledge about the kinetics and possible regulation of histone transport and loading is minimal, and this is an area ripe for future exploration. For example, embryonic cells that contain large histone pools presumably regulate both the amount and timing of histone availability for nucleosome assembly. However, we do not presently have a good estimate of the concentrations of histones available for assembly and how these may vary during the cell cycle, differentiation, etc. Such data will be crucial for understanding how nucleosome assembly may compete with transcription factor binding, which may be especially important at heterochromatic loci.

In the second case, chromatin maturation or higher-order folding events may be affected by histone modification. However, these modifications are removed rapidly after deposition *in vivo* (101), and it is not clear how these labile groups would affect later events. The third case brings up the intriguing possibility that a checkpoint exists which monitors defects in chromatin structure. Consistent with this idea, mutation of acetylatable lysines in histone H4 (102), or the acetyltransferases Gcn5 (103) or Esa1 (104) cause delays during the G_2/M phase of the cell cycle which are relieved by mutation of *RAD9*, a component of the DNA damage checkpoint.

Acknowledgements

The authors would like to thank J. Workman, J. Widom, and S. Roth for insightful comments on the manuscript. PDK has been supported by grants from the NIH, NSF, the California Breast Cancer Research Program, and by Department of Energy funds administered through the Lawrence Berkeley National Laboratory. GA has been supported by the CNRS, the Curie Institute, and European networking programmes.

References

1. Waga, S. and Stillman, B. (1998) The DNA replication fork in eukaryotic cells. *Annu. Rev. Biochem.*, **67**, 721.
2. Fanning, E. and Knippers, R. (1992) Structure and function of simian virus 40 large tumor antigen. *Annu. Rev. Biochem.*, **61**, 55.
3. Wold, M. S. (1997) Replication protein A: a heterotrimeric, single-stranded DNA-binding protein required for eukaryotic DNA metabolism. *Annu. Rev. Biochem.*, **66**, 61.
4. Brill, S. J. and Bastin-Shanower, S. (1998) Identification and characterization of the fourth single-stranded-DNA binding domain of replication protein A. *Mol. Cell. Biol.*, **18**, 7225.
5. Burgers, P. M. (1998) Eukaryotic DNA polymerases in DNA replication and DNA repair. *Chromosoma*, **107**, 218.
6. Kelman, Z. (1997) PCNA: structure, functions and interactions. *Oncogene*, **14**, 629.
7. Tsurimoto, T. (1998) A multifunctional ring on DNA. *Biochim. Biophys. Acta*, **1443**, 23.
8. Morrison, A., Araki, H., Clark, B., Hamatake, R. K. J., and Sugino, A. (1990) A third essential DNA polymerase in *S. cerevisiae*. *Cell*, **62**, 1143.
9. Kesti, T., Flick, K., Keranen, S., Syvaoja, J. E., and Wittenberg, C. (1999) DNA polymerase epsilon catalytic domains are dispensable for DNA replication, DNA repair, and cell viability. *Mol. Cell*, **3**, 679.
10. Navas, T. A., Zhou, Z., and Elledge, S. J. (1995) DNA polymerase epsilon links the DNA replication machinery to the S phase checkpoint. *Cell*, **80**, 29.
11. Krishna, T. S. R., Kong, P.-P., Gary, S., Burgers, P. M. J., and Kurijan, J. (1994) Crystal structure of the eukaryotic DNA polymerase processivity factor PCNA. *Cell*, **79**, 1233.
12. Gulbis, J. M., Kelman, Z., Hurwitz, J., O'Donnell, M., and Kurijan, J. (1996) Structure of the C-terminal region of p21 complexed with PCNA. *Cell*, **87**, 297.
13. Bravo, R. and MacDonald-Bravo, H. (1987) Existence of two populations of cyclin/proliferating cell nuclear antigen during the cell cycle: association with DNA replication sites. *J. Cell. Biol.*, **105**, 1549.
14. Toshi, L. and Bravo, R. (1988) Changes in PCNA distribution during DNA repair synthesis. *J. Cell. Biol.*, **107**, 1623.
15. Henderson, D. S., Banga, S. S., Grigliatti, T. A., and Boyd, J. B. (1994) Mutagen sensitivity and suppression of position-effect variegation result from mutations in *mus209*, the *Drosophila* gene encoding PCNA. *EMBO J.*, **13**, 1450.

16. Friedberg, E. C., Walker, G. C., and Siede, W. (ed.) (1995) *DNA repair and mutagenesis.* ASM Press, Washington, DC.

17. Araujo, S. J. and Wood, R. D. (2000) Protein complexes in nucleotide excision repair. *Mutat. Res.,* **435**, 233.

18. Nocentini, S. (1999) Rejoining kinetics of DNA single- and double-strand breaks in normal and DNA ligase-deficient cells after exposure to ultraviolet C and gamma radiation: an evaluation of ligating activities involved in different DNA repair processes. *Radiat. Res.,* **151**, 423.

19. Shibahara, K. and Stillman, B. (1999) Replication-dependent marking of DNA by PCNA facilitates CAF-1-coupled inheritance of chromatin. *Cell,* **96**, 575.

20. Moggs, J. G., Grandi, P., Quivy, J. P., Jonsson, Z. O., Hubscher, U., Becker, P. B., and Almouzni, G. (2000) A CAF-1-PCNA-mediated chromatin assembly pathway triggered by sensing DNA damage. *Mol. Cell. Biol.,* **20**, 1206.

21. Miura, M., Domon, M., Sasaki, T., Kondo, S., and Takasaki, Y. (1992) Two types of PCNA complex formation in quiescent normal and XPA fibroblasts following UV irradiation. *Exp. Cell Res.,* **201**, 541.

22. Smerdon, M. J. and Lieberman, M.W. (1978) Nucleosome rearrangement in human chromatin during UV-induced DNA-repair synthesis. *Proc. Natl Acad. Sci., USA,* **75**, 4238.

23. Smerdon, M. and Thoma, F. (1998) Modulations in chromatin structure during DNA damage formation and DNA repair. In *DNA damage and repair: molecular and cell biology,* (ed. M.F. Hoekstra and J.A. Nickoloff), Vol. 2. p. 199. Humana Press, Totowa, NJ,

24. Moggs, J. G. and Almouzni, G. (1999) Chromatin rearrangements during nucleotide excision repair. *Biochimie,* **81**, 45.

25. Smith, P. A., Jackson, V., and Chalkley , R. (1984) Two-stage maturation process for newly replicated chromatin. *Biochemistry,* **23**, 1576.

26. Hansen, J. C., van Holde, K. E., and Lohr, D. (1991) The mechanism of nucleosome assembly onto oligomers of the sea urchin 5S DNA positioning sequence. *J. Biol. Chem.,* **266**, 4276.

27. Ruiz-Carrillo, A., Jorcano, J. L., Eder, G., and Lurz, R. (1979) *In vitro* core particle and nucleosome assembly at physiological ionic strength. *Proc. Nat. Acad. Sci., USA,* **76**, 3284.

28. Worcel, A., Han, S., and Wong, M. L. (1978) Assembly of newly replicated chromatin. *Cell,* **15**, 969.

29. Jackson, V. (1987) Deposition of newly synthesized histones: new histones H2A and H2B do not deposit in the same nucleosome with new histones H3 and H4. *Biochemistry,* **26**, 2315.

30. Jackson, V. (1988) Deposition of newly synthesized histones: hybrid nucleosomes are not tandemly arranged on daughter DNA strands. *Biochemistry,* **27**, 2109.

31. Han, M., Chang, M., Kim, U.-J., and Grunstein, M. (1987) Histone H2B repression causes cell-cycle-specific arrest in yeast: effects on chromosomal segregation, replication, and transcription. *Cell,* **48**, 589.

32. Kim, U.-J., Han, M., Kayne, P., and Grunstein, M. (1988) Effects of histone H4 depletion on the cell cycle and transcription of *S. cerevisiae. EMBO J.,* **7**, 2211.

33. Osley, M. A. (1991) The regulation of histone synthesis in the cell cycle. *Annu. Rev. Biochem.,* **60**, 827.

34. van Holde, K. E. (1989) *Chromatin.* Springer Series in Molecular Biology, (ed. A. Rich). Springer-Verlag, Berlin.

35. Chicoine, L. G., Schulman, I. G., Richman, R., Cook, R. G., and Allis, C. D. (1986) Non-random utilization of acetylation sites in histones isolated from *Tetrahymena.* Evidence for functionally distinct H4 acetylation sites. *J. Biol. Chem.,* **261**, 1071.

36. Sobel, R. E., Cook, R. G., and Allis, C. D. (1994) Non-random acetylation of histone H4 by a cytoplasmic histone acetyltransferase as determined by novel methodology. *J. Biol. Chem.*, **269**, 18576.

37. Sobel, R. E., Cook, R. G., Perry, C. A., Annunziato, A. T., and Allis, C. D. (1995) Conservation of deposition-related acetylation sites in newly synthesized histones H3 and H4. *Proc. Natl Acad. Sci., USA*, **92**, 1237.

38. Berger, S. L. (1999) Gene activation by histone and factor acetyltransferases. *Curr. Opin. Cell Biol.*, **11**, 336.

39. Kleff, S., Andruis, E. D., Anderson, C. W., and Sternglanz, R. (1995) Identification of a gene encoding a yeast histone H4 acetyltransferase. *J. Biol. Chem.*, **270**, 24674.

40. Parthun, M., Widom, J., and Gottschling, D. E. (1996) The major cytoplasmic histone acetyltransferase in yeast: links to chromatin replication and histone metabolism. *Cell*, **87**, 85.

41. Ling, X., Harkness, T. A. A., Schultz, M. C., Fisher-Adams, G., and Grunstein, M. (1996) Yeast histone H3 and H4 amino termini are important for nucleosome assembly *in vivo* and *in vitro*: redundant and position-independent functions in assembly but not in gene regulation. *Genes Dev.*, **10**, 686.

42. Ma, X. J., Wu, J., Altheim, B. A., Schultz, M. C., and Grunstein, M. (1998) Deposition-related sites K5/K12 in histone H4 are not required for nucleosome deposition in yeast. *Proc. Natl Acad. Sci., USA*, **95**, 6693.

43. Stein, A. (1989) Reconstitution of chromatin from purified components. *Meth. Enzymol.*, **170**, 585.

44. Nelson, T., Wiegand, R., and Brutlag, D. (1981) Ribonucleic acid and other polyanions facilitate chromatin assembly *in vitro*. *Biochemistry*, **16**, 1490.

45. Kaufman, P. D., Kobayashi, R., Kessler, N., and Stillman, B. (1995) The p150 and p60 subunits of chromatin assembly factor 1: a molecular link between newly synthesized histones and DNA replication. *Cell*, **81**, 1105.

46. Kaufman, P. D., Kobayashi, R., and Stillman, B. (1997) Ultraviolet radiation sensitivity and reduction of telomeric silencing in *Saccharomyces cerevisiae* cell lacking chromatin assembly factor-I. *Genes Dev.*, **11**, 345.

47. Ito, T., Bulger, M., Kobayashi, R., and Kadonaga, J.T. (1996) *Drosophila* NAP-1 is a core histone chaperone that functions in ATP-facilitated assembly of regularly spaced nucleosomal arrays. *Mol. Cell. Biol.*, **16**, 3112.

48. Ito, T., Tyler, J. K., Bulger, M., Kobayashi, R., and Kadonaga, J. T. (1996) ATP-facilitated chromatin assembly with a nucleoplasmin-like protein from *Drosophila melanogaster*. *J. Biol. Chem.*, **271**, 25041.

49. Wolffe, A. (ed.) (1998) *Chromatin: Structure and Function*. Academic Press, San Diego, California.

50. Kleinschmidt, J. A., Seiter, A., and Zentgraf, H. (1990) Nucleosome assembly *in vitro*: separate histone transfer and synergistic interaction of native histone complexes purified from nuclei of *Xenopus laevis* oocytes. *EMBO J.*, **9**, 1309.

51. Ito, T., Bulger, M., Pazin, M. J., Kobayashi, R., and Kadonaga, J. T. (1997) ACF, an ISWI-containing and ATP-utilizing chromatin assembly and remodeling factor. *Cell*, **90**, 145.

52. Almouzni, G. and Mechali, M. (1988) Assembly of spaced chromatin promoted by DNA synthesis in extracts from *Xenopus* eggs. *EMBO J.*, **7**, 665.

53. Almouzni, G. and Wolffe, A. P. (1993) Replication-coupled chromatin assembly is required for the repression of basal transcription *in vivo*. *Genes Dev.*, **7**, 2033.

54. Almouzni, G., Clark, D. J., Mechali, M., and Wolffe, A. P. (1990) Chromatin assembly on replicating DNA in vitro. *Nucleic Acids Res.*, **18**, 5767.

55. Becker, P. B. and Wu, C. (1992) Cell-free system for assembly of transcriptionally repressed chromatin from *Drosophila* embryos. *Mol. Cell. Biol.*, **12**, 2241.

56. Kamakaka, R. T., Bulger, M., and Kadonaga, J. T. (1993) Potentiation of RNA polymerase II transcription by Gal4-VP16 during but not after DNA replication and chromatin assembly. *Genes Dev.*, **7**, 1779.

57. Krude, T. and Knippers, R. (1993) Nucleosome assembly during complementary DNA strand synthesis in extracts from mammalian cells. *J. Biol. Chem.*, **268**, 14432.

58. Stillman, B. W. and Gluzman, Y. (1985) Replication and supercoiling of simian virus 40 DNA in cell extracts from human cells. *Mol. Cell. Biol.*, **5**, 2051.

59. Stillman, B. (1986) Chromatin assembly during SV40 DNA replication *in vitro*. *Cell*, **45**, 555.

60. Gaillard, P.-H. L., Martini, E. M.-D., Kaufman, P. D., Stillman, B., Moustacchi, E., and Almouzni, G. (1996) Chromatin assembly coupled to DNA repair: a new role for Chromatin Assembly Factor-I. *Cell*, **86**, 887.

61. Gaillard, P. H., Moggs, J. G., Roche, D. M., Quivy, J. P., Becker, P. B., Wood, R. D., and Almouzni, G. (1997) Initiation and bidirectional propagation of chromatin assembly from a target site for nucleotide excision repair. *EMBO J.*, **16**, 6281.

62. Smith, S. and Stillman, B. (1989) Purification and characterization of CAF-I, a human cell factor required for chromatin assembly during DNA replication. *Cell*, **58**, 15.

63. Smith, S. and Stillman, B. (1991) Stepwise assembly of chromatin during DNA replication *in vitro*. *EMBO J.*, **10**, 971.

64. Verreault, A., Kaufman, P. D., Kobayashi, R., and Stillman, B. (1996) Nucleosome assembly by a complex of CAF-I and acetylated histones H3/H4. *Cell*, **87**, 95.

65. Taddei, A., Roche, D., Sibarita, J. B., Turner, B. M., and Almouzni, G. (1999) Duplication and maintenance of heterochromatin domains. *J. Cell Biol.*, **147**, 1153.

66. Verreault, A., Kaufman, P. D., Kobayashi, R., and Stillman, B. (1998) Nucleosomal DNA regulates the core-histone-binding subunit of the human Hat1 acetyltransferase. *Curr. Biol.*, **8**, 96.

66a. Shibahara, K., Verreault, A., and Stillman, B. (2000) The N-terminal domains of histones H3 and H4 are not necessary for Chromatin Assembly Factor-I-mediated nucleosome assembly onto replicated DNA *in vitro*. *Proc. Natl Acad. Sci., USA*, **97**, 7766.

67. Kamakaka, R. T., Bulger, M., Kaufman, P. D., Stillman, B., and Kadonaga, J. T. (1996) Post-replicative chromatin assembly by *Drosophila* and human Chromatin Assembly Factor-I. *Mol. Cell. Biol.*, **16**, 810.

68. Neer, E. J., Schmidt, C. J., Nambudripad, R., and Smith, T. F. (1994) The ancient regulatory-protein family of WD-repeat proteins. *Nature*, **371**, 297.

69. Tyler, J. K., Bulger, M., Kamakaka, R. T., Kobayashi, R., and Kadonaga, J. T. (1996) The p55 subunit of *Drosophila* Chromatin Assembly Factor-I is homologous to a histone deacetylase-associated protein. *Mol. Cell. Biol.*, **16**, 6149.

70. Enomoto, S., McCune-Zierath, P. D., Gerami-Nejad, M., Sanders, M., and Berman, J. (1997) *RLF2*, a subunit of yeast chromatin assembly factor I, is required for telomeric chromatin function *in vivo*. *Genes Dev.*, **11**, 358.

71. Rogers, S., Wells, R., and Rechstiener, M. (1986) Amino acid sequences common to rapidly degraded proteins: The PEST hypothesis. *Science*, **234**, 364.

72. Murzina, N., Verreault, A., Laue, E., and Stillman, B. (1999) Heterochromatin dynamics in mouse cells: interaction between chromatin assembly factor 1 and HP1 proteins. *Mol. Cell*, **4**, 529.

73. Kaufman, P. D., Cohen, J. L., and Osley, M. A. (1998) Hir proteins are required for position-dependent gene silencing in *Saccharomyces cerevisiae* in the absence of Chromatin Assembly Factor I. *Mol. Cell. Biol.*, **18**, 4793.

74. Sherwood, P. W., Tsang, S. V., and Osley, M. A. (1993) Characterization of *HIR1* and *HIR2*, two genes required for regulation of histone gene transcription in *Saccharomyces cerevisiae*. *Mol. Cell. Biol.*, **13**, 28.

75. Lamour, V., Lecluse, Y., Desmaze, C., Spector, M., Bodescot, M., Aurias, A., Osley, M. A., and Lipinski, M. (1995) A human homolog of the *S. cerevisiae* HIR1 and HIR2 transcriptional repressors cloned from the DiGeorge syndrome critical region. *Hum. Mol. Genet.*, **4**, 791.

76. Recht, J., Dunn, B., Raff, A., and Osley, M. A. (1996) Functional analysis of histones H2A and H2B in transcriptional repression in *Saccharomyces cerevisiae*. *Mol. Cell. Biol.*, **16**, 2545.

77. Lorain, S., Quivy, J. P., Monier-Gavelle, F., Scamps, C., Lecluse, Y., Almouzni, G., and Lipinski, M. (1998) Core histones and HIRIP3, a novel histone-binding protein, directly interact with WD repeat protein HIRA. *Mol. Cell. Biol.*, **18**, 5546.

78. Roth, S. Y. and Allis, C. D. (1996) Histone acetylation and chromatin assembly: a single escort, multiple dances? *Cell*, **87**, 5.

79. Luger, K., Mäder, A. W., Richmond, R. K., Sargent, D.F., and Richmond, T. J. (1997) Crystal structure of the nucleosome core particle at 2.8Å resolution. *Nature*, **389**, 251.

80. Le, S., Davis, C., Konopka, J. B., and Sternglanz, R. (1997) Two new S-phase-specific genes from *Saccharomyces cerevisiae*. *Yeast*, **13**, 1029.

81. Singer, M. S., Kahana, A., Wolf, A. J., Meisinger, L. L., Peterson, S. E., Goggin, C., Mahowald, M., and Gottschling, D. E. (1998) Identification of high-copy disruptors of telomeric silencing in *Saccharomyces cerevisiae*. *Genetics*, **150**, 613.

82. Tyler, J. K., Adams, C. R., Chen, S. R., Kobayashi, R., Kamakaka, R. T., and Kadonaga, J. T. (1999) The RCAF complex mediates chromatin assembly during DNA replication and repair. *Nature*, **402**, 555.

83. Spector, D. L. (1993) Macromolecular domains within the cell nucleus. *Annu. Rev. Cell. Biol.*, **9**, 265.

84. Leonhardt, H. and Cardoso, M. C. (1995) Targeting and association of proteins with functional domains in the mammalian nucleus the insoluble solution. *Int. Rev. Cytol.*, **162B**, 303.

85. Krude, T. (1995) Chromatin assembly factor 1 (CAF-1) colocalizes with replication foci in HeLa cell nuclei. *Exp. Cell Res.*, **220**, 304.

86. Enomoto, S. and Berman, J. (1998) Chromatin assembly factor I contributes to the maintenance, but not the reestablishment, of silencing at the yeast silent mating loci. *Genes Dev.*, **12**, 219.

87. Smith, S. and Stillman, B. (1991) Immunological characterization of chromatin assembly factor-I, a human cell factor required for chromatin assembly during DNA replication *in vitro*. *J. Biol. Chem.*, **266**, 12041.

88. Matsumoto-Taniura, N., Pirollet, F., Monroe, R., Gerace, L., and Westendorf, J. M. (1996) Identification of novel M phase phosphoproteins by expression cloning. *Mol. Biol. Cell*, **7**, 1455.

89. Marheineke, K. and Krude, T. (1998) Nucleosome assembly activity and intracellular localization of human CAF-1 changes during the cell division cycle. *J. Biol. Chem.*, **273**, 15279.

90. Martini, E., Roche, D. M., Marheineke, K., Verreault, A., and Almouzni, G. (1998) Recruitment of phosphorylated chromatin assembly factor 1 to chromatin after UV irradiation of human cells. *J. Cell Biol.*, **143**, 563.

91. Monson, E. K., de Bruin, D., and Zakian, V.A. (1997) The yeast Cac1 protein is required for the stable inheritance of transcriptionally repressed chromatin at telomeres. *Proc. Natl Acad. Sci., USA*, **94**, 13081.

92. Smith, J., Caputo, E., and Boeke, J. (1999) A genetic screen for ribosomal DNA silencing defects identifies multiple DNA replication and chromatin-modulating factors. *Mol. Cell. Biol.*, **19**, 3184.

93. Loo, S. and Rine, J. (1995) Silencing and heritable domains of gene expression. *Annu. Rev. Cell Dev. Biol.*, **11**, 519.

94. Pillus, L. and Rine, J. (1989) Epigenetic inheritance of transcriptional states in *S. cerevisiae*. *Cell*, **59**, 637.

95. Osley, M. A. and Lycan, D. (1987) Trans-acting regulatory mutations that alter transcription of *Saccharomyces cerevisiae* histone genes. *Mol. Cell. Biol.*, **7**, 4204.

96. Xu, H., Kim, U. J., Schuster, T., and Grunstein, M. (1992) Identification of a new set of cell cycle-regulatory genes that regulate S-phase transcription of histone genes in *Saccharomyces cerevisiae*. *Mol. Cell. Biol.*, **12**, 5249.

97. Qian, Z., Huang, H., Hong, J. Y., Burck, C. L., Johnston, S. D., Berman, J., Carol, A., and Liebman, S. W. (1998) Yeast Ty1 retrotransposition is stimulated by a synergistic interaction between mutations in Chromatin Assembly Factor-I and histone regulatory (Hir) proteins. *Mol. Cell. Biol.*, **17**, 4783.

98. Moran, L., Norris, D., and Osley, M. A. (1990) A yeast H2A–H2B promoter can be regulated by changes in histone gene copy number. *Genes Dev.*, **4**, 752.

99. Game, J. and Kaufman, P. D. (1999) Role of *Saccharomyces cerevisiae* Chromatin Assembly Factor-I in repair of ultraviolet radiation damage *in vivo*. *Genetics*, **151**, 485.

100. Vitolo, J. M., Thiriet, C., and Hayes, J. J. (2000) The H3–H4 N-terminal tail domains are the primary mediators of transcription factor IIIA access to 5S DNA within a nucleosome. *Mol. Cell. Biol.*, **20**, 2167.

101. Annunziato, A. T. and Seale, R. L. (1983) Histone deacetylation is required for the maturation of newly replicated chromatin. *J. Biol. Chem.*, **258**, 12675.

102. Megee, P. C., Morgan, B. A., and Smith, M. M. (1995) Histone H4 and the maintenance of genome integrity. *Genes Dev.*, **9**, 1716.

103. Zhang, W., Bone, J. R., Edmondson, D. G., Turner, B. M., and Roth, S. Y. (1998) Essential and redundant functions of histone acetylation revealed by mutation of target lysines and loss of the Gcn5p acetyltransferase. *EMBO J.*, **17**, 3155.

104. Clarke, A. S., Lowell, J. E., Jacobson, S. J., and Pillus, L. (1999) Esa1p is an essential histone acetyltransferase required for cell cycle progression. *Mol. Cell. Biol.*, **19**, 2515.

3 | Chromatin structure and control of transcription *in vivo*

WOLFRAM HÖRZ and SHARON ROTH

1. Introduction

Chapter 1 described in detail the structure of the nucleosome as well as higher-order levels of chromatin folding. Chapter 2 explained how these structures are assembled onto newly replicated DNA and how histones are inherited from parental to daughter DNA strands. The next few chapters will explore the dynamic nature of chromatin structures, with an emphasis on how the packaging of the DNA into chromatin affects gene expression, and the mechanisms by which the cell can remodel chromatin to regulate transcription. In this chapter, some basic concepts concerning the process of transcription will be introduced. Hallmark changes in chromatin structure associated with transcriptional competence will be described. Alterations in chromatin structures associated with activation and repression of specific genes in yeast will then be summarized to illustrate these concepts.

2. Chromatin and transcription

2.1 Transcription by RNA polymerase II

The transcription of DNA into RNA requires an orchestration of basal and regulatory transcription factors (1, 2). Like bacterial RNA polymerases, eukaryotic RNA polymerase II (pol II) is a multisubunit enzyme that is capable of synthesizing RNA from a DNA template but is not capable of recognizing promoters on its own (3). Promoter recognition requires several general transcription factors, including TFIID, TFIIB, TFIIA, TFIIF, and TFIIH. *In vitro*, these factors assemble onto promoters in a stepwise fashion to form a preinitiation complex:

(1) binding of TFIID to the promoter initiates assembly of the preinitiation complex. TFIID contains the TATA-binding protein (TBP) and several TBP-associated factors (TAFs);

(2) TFIIB and TFIIA bind to TFIID, and these DNA-bound complexes recruit TFIIF, which is stably associated with Pol II;

(3) a specific promoter architecture is created by the binding of TFIID, TFIIB, and TFIIA, which positions the polymerase at the transcription start site;

(4) additional factors, such as TFIIH, are then recruited to complete preinitiation complex assembly, promoter melting, transcription initiation, and elongation.

In vivo, many of the general transcription factors (as well as other proteins, including the mediator complex; see refs 3 and 4) appear to be associated with the polymerase in the form of a holoenzyme, which likely increases the efficiency of transcription initiation and which provides an obvious target for regulatory proteins (3, 4).

In vitro, a basal level of transcription can be achieved with the polymerase and the general transcription factors alone. Transcription is activated beyond the basal level by the addition of *trans*-acting factors (transactivators) that bind to specific sequences (enhancers) upstream or downstream of the promoter. These *trans*-acting factors often stabilize the binding of TBP (and TFIID) to the promoter through protein–protein interactions, thereby facilitating preinitiation complex formation (5–7).

Conversely, transcription can be inhibited by blocking the binding of TBP to the promoter (7). Specific repressors (such as NC2, Not1, or Mot1) act *in vitro* to limit or reverse TBP binding to TATA sequences (8). *In vivo*, the binding of TBP is limited further by the assembly of the promoter into nucleosomes (9–11). Nucleosomes can also occlude DNA-binding sites for transactivators. Moreover, packaging of the DNA into chromatin alters the topological state of the DNA, which likely limits formation of promoter architectures conducive to transcription (3). As will be seen throughout this text, both repression and activation of transcription are influenced directly by particular chromatin structures, and the overall level of transcription at any given time, in any given cell, likely reflects a flux between active and inactive chromatin states.

2.2 Alterations in chromatin structure associated with transcription

The first hints that alterations in chromatin structure accompany changes in gene expression came from now classic nuclease digestion studies (reviewed in 12). These investigations were predicated upon findings that chromatin could be cleaved into mono- and oligonucleosome-sized particles by micrococcal nuclease (MNase), which preferentially cleaves linker DNA between nucleosomes (13). Subsequently, it became clear that genes were more sensitive to digestion by MNase, DNAase I, or DNAase II in tissues where they were transcribed than in tissues where they were not transcribed, and that, within a cell, transcribed genes were more sensitive to digestion than were non-transcribed genes (14–17). The globin gene cluster, for example, is present in a nuclease-sensitive region in chick erythrocyte nuclei but is nuclease resistant in other tissues, such as liver or oviduct, where globin is not expressed (18,

19). Similarly, the chick lysozyme gene is nuclease sensitive in oviduct, where it is highly expressed, but not in red blood cells or liver, where it is not expressed (20–22). Often, the region of enhanced sensitivity extends both upstream and downstream of the coding region of the gene but has relatively sharp boundaries. A 24 kb domain surrounding and including the lysozyme gene, for instance, exhibits increased nuclease sensitivity in the oviduct relative to other tissues (23). These nuclease-sensitive regions are punctuated by more localized, hypersensitive sites that span several hundred base pairs and that often reflect the locations of important *cis*-acting regulatory sequences (12, 20). Increased nuclease sensitivity indicates that active genes have an altered chromatin structure. The localized hypersensitive sites reflect regions where nucleosomes have a substantially altered conformation that renders the associated DNA vulnerable to nuclease digestion. How the extent of open, active domains is delimited is not clear and is still the subject of much study (see Chapter 13). Importantly, nuclease sensitivity studies can not determine whether the opening of chromatin structure is a requisite to transcriptional activation or is a downstream result of the transcription process. Thus, these studies can not distinguish cause versus effect of changes in chromatin structures and transcriptional competence.

The subsequent development of *in vitro* transcription systems allowed the effects of chromatin on transcription to be tested directly. Hallmark studies using purified systems demonstrated that promoter regions assembled into nucleosomes proved refractory to transcription (9,10, 24, 25). In contrast, prior assembly of promoters into RNA pol II preinitiation complexes prevented nucleosome formation (9, 25, 26). These findings indicated that nucleosomes and transcription factors might compete for promoter occupancy *in vivo*, providing further evidence that chromatin modulations might regulate promoter accessibility, and thereby, transcriptional activity.

Perhaps some of the clearest data indicating that chromatin is important to the regulation of transcription came from studies of the effects of particular histone mutations in yeast (reviewed in 27). Unlike most organisms, which contain hundreds of copies and multiple variants of the histone genes, *Saccharomyces cerevisiae* only contains two copies of each of the core histone genes. This simplicity, together with the high degree of homologous recombination that occurs in yeast, facilitates gene replacement studies to probe the functions of particular histone domains. The first evidence that implicated histones in gene regulation came from studies in which histone gene dosage was altered (28) or expression of histone H4 was repressed (29). Three laboratories subsequently determined that the highly conserved amino-terminal tail domain of histone H4 is required for transcriptional silencing of the yeast *HM* mating loci (30–32) (see Chapter 10). These loci are normally never transcribed and serve only as donors in recombination events that lead to mating type switching. The loss of silencing in the presence of the H4 mutations clearly indicated that silencing is dependent on a specialized chromatin structure. Similar approaches have been used to probe the functions of all four core histones in various cellular processes, and many studies confirm the fundamental contributions of chromatin to transcriptional regulation (see below and Chapter 4) (33, 34). One caveat to these approaches, however, is that the histone mutations affect all the nucleosomes in the cell and often lead to many pleiotrophic phenotypes. Thus, it is difficult to determine which of these phenotypes reflect direct versus indirect effects on gene expression.

2.3 Hallmark features of transcriptionally active chromatin

Many insights as to how chromatin is remodelled to reprogramme gene expression have come from careful comparisons of the biochemical properties of 'active' versus 'inactive' chromatin (35–37). Such comparisons indicate that the active state is associated with one or more of the following:

- depletion and/or phosphorylation of H1-type linker histones;
- increased core histone acetylation;
- increased content of specific high-mobility group (HMG) proteins;
- increased incorporation of specific histone variants.

Specialized linker histones (such as B4 or H5) are often associated with large-scale changes in developmental transcription programmes, such as occurs prior to the activation of zygotic transcription in frog embryos or the inactivation of nuclei in avian erythrocytes (reviewed in 38). Phosphorylation of H1 accompanies chromosome condensation at mitosis, but also accompanies specific changes in gene expression (37). H1 phosphorylation might cause depletion of this histone from chromatin, as has been observed upon activation of the MMTV promoter (39). Alternatively, H1 phosphorylation might facilitate the binding of other regulatory factors to the DNA by loosening H1–DNA contacts (37).

Similarly, acetylation of lysine residues in the core histones neutralizes positive charge associated with the ε-amino group of these amino acids, altering DNA–histone interactions (see Chapters 1, 7, and 8). Histone acetylation also influences nucleosome–nucleosome interactions involved in higher-order chromatin folding, as well as the association of non-histone, regulatory proteins with chromatin (see below). The incorporation of special, variant forms of the histones or association of the HMGs may directly alter nucleosome structure, higher-order folding, and the association of non-histone regulatory factors as well (40, 41). A conserved, specialized form of H3 (Cse4 or CENP-A), for example, is associated with the specialized structures of centromeres in yeast and man (see Chapter 4) (42).

Inactive genes exhibit properties that are largely reversals of the above, including association or dephosphorylation of H1, deacetylation of the core histones (see Chapter 8), or diminishment of HMGs and histone variants. In addition, in some cases (see below), the locations of nucleosomes are fixed in promoter regions to obscure particular promoter elements under conditions of repression. The subnuclear location of a gene can also be altered upon repression (see Chapters 10, 11, and 14). Heterochromatin, for example, is generally found near the nuclear periphery and is usually associated with gene silencing, an extreme form of repression. Silenced genes are associated with specific silencing factors that organize heterochromatic structures (Chapter 10 and 11) and which may dictate subnuclear locations. In mammalian cells, DNA methylation often accompanies gene silencing, particularly in parentally imprinted genes (Chapter 12). DNA methylation may be related directly to decreased histone acetylation, since methyl-binding proteins appear to recruit histone de-acetylase complexes (43, 44). Indeed, the identification of histone acetyltransferases (HATs; Chapter 7) and histone deacetylases (HDACs; Chapter 8) in the past few years has provided perhaps the most clear molecular links between chromatin re-modelling and transcriptional regulation. Moreover, physical connections between HATs, HDACs, and DNA-bound activators or repressors have yielded important clues as to how chromatin modifications might be targeted to the right gene, at the right time, in the right cells (7, 45).

Below we will describe in detail two well-characterized systems in yeast that high-light the regulatory potential of chromatin, as well as the types of questions still open for the future.

3. Mechanism of chromatin remodelling at the *PHO5* and *PHO8* promoters

The PHO regulon in yeast is composed of a number of structural and regulatory genes (46). The structural genes correspond to phosphatases and the regulatory genes pro-vide the cell with the ability to respond to phosphate starvation. Phosphate depletion in the growth medium results in at least a 50-fold increased production of acid phos-phatase, which is composed of three isozymes produced from the *PHO5*, *PHO10*, and *PHO11* genes (47, 48). The *PHO5* gene product contributes more than 90% of the acid phosphatase enzymatic activity. In addition, an alkaline phosphatase encoded by the *PHO8* gene is similarly regulated by phosphate levels (49).

3.1 Transcription factors involved in *PHO5* regulation

Two transcription factors are responsible for the strong activation of *PHO5*. One is Pho4, a basic-helix–loop–helix (bHLH) transactivator, and the other is Pho2, which is a homeobox DNA-binding protein. Pho4 binds to two upstream activating sequence (UAS) elements in the *PHO5* promoter (UASp1 and UASp2) (Fig. 1) (50) which are critically required for promoter activation. Pho2 binds to multiple sites that are clustered in the vicinity of the two UAS elements at the *PHO5* promoter (51). Pho4 and Pho2 bind highly cooperatively to DNA, and this cooperativity also utilizes protein–protein interactions between Pho4 and Pho2 (52). The primary role of Pho2 seems to be to help Pho4 bind to its target sites.

In phosphate-containing media, i.e. conditions under which *PHO5* is repressed, Pho4 is phosphorylated through the action of Pho80/Pho85, a cyclin/cyclin dependent kinase (CDK) pair (53). Phosphorylation of Pho4 negatively affects transcriptional activation at three different levels: it prevents import of Pho4 into the nucleus, pro-motes its export from the nucleus, and prevents Pho4/Pho2 interaction (54).

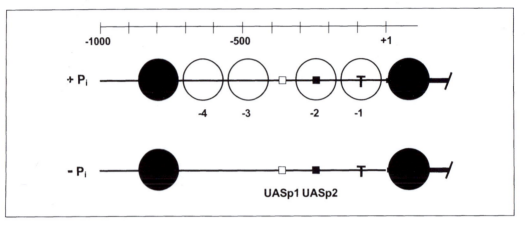

Fig. 1 Chromatin structure at the *PHO5* promoter. Nucleosome organization under repressed (+Pi) and induced (−Pi) conditions is shown. The nucleosomes shown in white (−1, −2, −3, and −4) are remodelled upon activation. The small squares mark UASp1 (open) and UASp2 (solid), which are Pho4-binding sites found by *in vitro* and *in vivo* footprinting experiments. The coding sequence is shown in solid black. T denotes the TATA-box.

3.2   The *PHO5* chromatin transition

A remarkable feature of the *PHO5* promoter is its chromatin organization, which undergoes a dramatic change from the repressed to the activated state. The repressed *PHO5* gene is packaged in a positioned array of nucleosomes, that is interrupted only in the promoter region by a short 80 bp region which is hypersensitive to nucleases (55) (Fig. 1). This hypersensitive site contains one of the two Pho4-binding sites (UASp1) while the second (UASp2) is localized near the centre of nucleosome −2. Upon *PHO5* activation by phosphate starvation, a 600 bp region of the *PHO5* promoter becomes hypersensitive to nucleases (see Fig. 1), indicating that the structure of four nucleosomes is profoundly altered in a way that renders the promoter accessible to nucleases and presumably to components of the transcriptional machinery (Fig. 1) (55).

The chromatin transition seems to be an essential step in the activation of the promoter. Although it is generally difficult to establish unambiguous cause and effect relationships, several findings indicate that chromatin remodelling at the *PHO5* promoter is a prerequisite for activation. Key observations include:

- among the large number of promoter variants generated by either modifying *cis*-acting elements or transacting factors, none was ever found that was fully active and that did not also undergo the complete chromatin transition;

- insertion into the *PHO5* promoter of a 150 bp DNA fragment from African green monkey α-satellite DNA that forms a stable nucleosome *in vitro* (56) and *in vivo* (57) abolishes its ability to be activated (58);

- conversely, inserting a prokaryotic DNA fragment of the same size which associates more weakly with histones maintains the promoter in an inducible state.

Altogether, these results argue that the quality of the histone–DNA interactions at the promoter is a determinant of *PHO5* activation and that chromatin remodelling is necessary for full activity.

It should be noted, however, that open chromatin is not by itself sufficient for full activation. Interfering with histone biosynthesis, which leads to derepression of many genes, activates *PHO5* to a level which is only about 15% of the fully induced level (29). For this activation the UAS elements are not required, demonstrating that it is due to increased accessibility of the proximal promoter. Along the same line, several mutations in components of the transcriptional machinery substantially lower activity of the promoter without affecting chromatin opening at all. For example, disruption of the *GAL11* gene, which codes for a component of the RNA polymerase holoenzyme, leads to a fourfold decrease in *PHO5* inducibility, yet chromatin opens up completely (59). In summary, chromatin remodelling appears to be necessary but not sufficient for full activity.

3.3   The Pho4 activation domain is required for chromatin remodelling at the *PHO5* promoter

The unique chromatin remodelling at the *PHO5* promoter makes this system well suited for studying the requirements of the remodelling step and the underlying mechanism. In the following, the relevant information available at present is summarized. It would be conceivable that chromatin remodelling is limited to the time of replication, when existing structures are, by necessity, disassembled and reassembled onto the newly replicated DNA. However, use of a temperature-sensitive allele of the negative regulator Pho80 made it possible to activate the system even at high phosphate in the absence of replication by a simple temperature shift. It turned out that replication was not required for chromatin opening (60). Alternatively, the act of transcription through the gene might be required for chromatin remodelling, but analysis of a non-transcribed promoter variant with a deleted TATA-box eliminated this possibility (61). Finally, binding of the transacting regulatory factors, Pho2 or Pho4, might trigger nucleosome movements. Indeed, Pho4 is critically required for remodelling, whereas Pho2 is dispensable, provided that Pho4 is overexpressed at the same time (62). This is consistent with a primary function of Pho2 in assisting Pho4 in DNA binding as described above. The absolute Pho4 requirement for chromatin disruption raised the question of whether a separate domain exists in Pho4 capable of chromatin opening. A systematic dissection of Pho4 revealed, however, a close correlation with the domain required for transcriptional activation. So far it has not been possible to separate the two Pho4 activities (63, 64).

The availability of Pho4 derivatives retaining DNA-binding activity yet unable to disrupt chromatin has made it possible to address directly the consequences of a binding site being within a nucleosome or not. Such Pho4 variants are able to bind UASp2 when it is situated within the hypersensitive site, but not when it is in its normal location within nucleosome –2 (50).

3.4   SAGA and SWI/SNF are not needed for chromatin opening at the *PHO5* promoter

The central question arising form the aforementioned results is whether the Pho4 activation domain accomplishes chromatin disruption by recruiting auxiliary activities, and if so, what kind? Obvious candidates are chromatin-remodelling complexes and histone-modifying activities. This possibility has become even more attractive in view of the recent demonstration that transcriptional activation domains can indeed recruit such activities (see Chapters 5 and 7). Up to now, the requirements for Gcn5 and Snf2, the catalytic subunits of the Spf-Ada-Gcn5 acetyltransferase (SAGA) and the SWI/SNF complex, respectively, that are essential for activity, have been analysed. It turned out that neither of them was required for chromatin remodelling at the *PHO5* promoter following phosphate starvation. The activity of the promoter in both a *gcn5* and a *snf2* strain reached 70–80% of wild-type level, and nucleosome disruption attained levels close, although not completely identical, to the wild-type levels (59, 65).

These findings leave open the possibility that still other activities redundant with SAGA and/or SWI/SNF are critical at the *PHO5* promoter. Alternatively, the possibility that interactions with the basal transcription machinery are sufficient must be considered. That such interactions might actually lead to disruption has been suggested by the demonstration that fusions of Gal11, a component of the RNA polymerase holoenzyme, with the Pho4 DNA-binding domain are able to disrupt nucleosomes at the *PHO5* promoter in exactly the same way as wild-type Pho4. This is true even for the TATA-delete construct described before (59). Although the actual results of these experiments are unambiguous, their interpretation is by no means straightforward. At one extreme, no extra function might be required for nucleosome remodelling when the total energy of the protein–DNA interactions is sufficiently high. Alternatively, factors that specifically labilize histone–DNA interactions might be always required. Such factors may turn out to be components of the holoenzyme or may be associated with it transiently in a way that is functionally significant.

Interestingly, conditions actually do exist under which the *PHO5* promoter requires Gcn5 for proper chromatin remodelling. As mentioned above, Pho80 and Pho85 are negative regulators of Pho4, and in their absence the *PHO5* promoter is remodelled even at high phosphate, i.e. non-inducing conditions. Promoter activity is only about 30% of the maximal level at high phosphate in a *pho80* or *pho85* strain, however, and reaches 100% only after phosphate starvation. The mechanism underlying this additional, Pho80-independent activation step is currently unknown. It turns out that in a *pho80* background, Gcn5 is required for complete remodelling at high phosphate. In its absence, a structure results consistent with the presence of nucleosomes occupying random positions across the promoter, in contrast to both the positioned nucleosomal array of the repressed promoter, and the disrupted state that is characteristic of the active state. Furthermore, the loss of the acetylation function via the targeted substitution of amino acids critical for catalytic Gcn5 HAT activity is sufficient to generate this novel chromatin structure. This intermediate state in the opening of chromatin at the promoter gives way to the fully open hypersensitive structure under conditions of maximal induction (phosphate starvation), even in the absence of Gcn5 (65).

An involvement of histone acetylation in *PHO5* regulation is also indicated by an investigation of the role of Rpd3, a histone deacetylase, in gene regulation. It was reported that disruption of *RPD3* leads to slight derepression of *PHO5* at high phosphate, while at the same time activation upon phosphate starvation was incomplete (66). Such a behaviour would be consistent with a disorganized chromatin configuration which cannot properly repress a promoter and at the same time resists full remodelling under inducing conditions.

3.5 The related *PHO8* promoter requires Gcn5 and Snf2 for remodelling

Recent work on the *PHO8* promoter has generated unexpected results that complement and extend our understanding of the interplay between chromatin and transcription in the PHO system. Chromatin at the *PHO8* promoter also undergoes remodelling upon activation (Fig. 2). However, there are important differences to the *PHO5* promoter:

(1) both Pho4-binding sites are accessible in the repressed promoter;

(2) the Pho4-dependent chromatin remodelling is partial, even under fully activating conditions (67) (see Fig. 2).

It turns out that absence of SWI/SNF components, or mutation of the ATPase activity of the SWI/SNF complex, results in the complete loss of chromatin remodelling

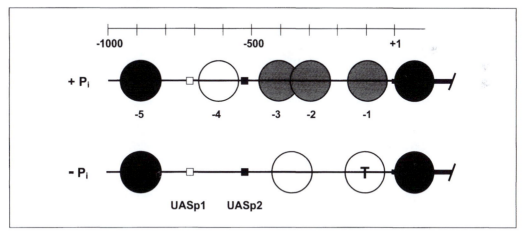

Fig. 2 Chromatin structure at the *PHO8* promoter. Nucleosome organization under repressed (+Pi) and induced (−Pi) conditions is shown. The shadowing of the nucleosomes reflects the level of DNA protection against restriction nuclease digestion: black ≈ 95%, grey ≈ 80%, white ≈ 50% protection (67). The small squares mark UASp1 (open) and UASp2 (solid), which are Pho4-binding sites found by *in vitro* footprinting experiments. The coding sequence is shown in solid black. T denotes the TATA-box.

at the *PHO8* promoter under inducing conditions. Gcn5 is also required for full remodelling and transcriptional activation of the *PHO8* promoter, and it appears to be the histone acetyltransferase activity that is required, since mutations that abolish Gcn5 HAT activity result in the same defect in activation and chromatin remodelling. *In vivo* DMS footprinting experiments demonstrated Pho4 DNA binding to the essential UASp2 element of the *PHO8* promoter in both *Δgcn5* and *Δsnf2* strains (68). Thus SAGA and SWI/SNF must act at a chromatin-remodelling step downstream of activator binding. It has been suggested previously that many SAGA-dependent genes also require the SWI/SNF complex for activation (69). This observation holds true for the *PHO8* promoter, although the features of this promoter that confer such double dependency are currently unknown. It is also not yet clear, what features of the *PHO8* promoter make it dependent on the SWI/SNF and SAGA complex and what features make *PHO5* independent, although the same activator is responsible for remodelling in both cases. The current working hypothesis is that it is the quality of the histone–DNA interactions that makes the difference.

4. Repression by Ssn6–Tup1

In contrast to Pho4, which initiates chromatin alterations associated with gene activation, the Ssn6–Tup1 complex appears to orchestrate repressive chromatin remodelling events. *SSN6* and *TUP1* were identified independently in multiple genetic screens for transcriptional regulators in yeast, and these proteins are required for the repression of many diverse genes (70, 71). Indeed, microarray analysis indicates that expression of as much as 2% of the yeast genome is increased fivefold or more upon loss of either *SSN6* or *TUP1* (72). The concurrence of the phenotypes of *SSN6* and *TUP1* mutations was explained by later studies which demonstrated that these proteins form a complex (73, 74). This complex does not bind to DNA directly but is recruited to individual promoters through interactions with DNA-bound repressors, leading to the designation of Ssn6 and Tup1 as co-repressors. Once recruited, the Ssn6–Tup1complex interferes with transcription through a process that may involve interactions both with the transcription machinery (71) and with chromatin (75) (Fig. 3).

One set of well-characterized genes subject to Ssn6–Tup1 repression is the **a**-cell-specific genes. Yeast exist in two haploid mating types, termed **a** and α, and these conjugate together to form **a**/α diploid cells (reviewed in 76). Mating type is determined by two alleles of the MAT locus, MATα and MAT**a**. In α cells, MATα encodes two transcriptional regulators, α1 and α2, which activate α-cell-specific and repress **a**-cell-specific genes, respectively. In **a** cells, MATα is replaced with the MAT**a** allele, so α1 and α2 are not expressed. Thus, α-cell-specific genes are not activated and **a**-cell-specific genes are not repressed. The repression of the **a**-cell-specific genes in α cells requires Ssn6–Tup1, which is recruited through interactions with the α2 repressor protein, which binds together with Mcm1, to a 32 bp operator located upstream of the **a**-cell-specific genes (77). Both Ssn6 and Tup1 interact directly with α2 (78, 79). Artificial recruitment of Tup1 through fusion to a heterologous, *E. coli* lexA DNA-

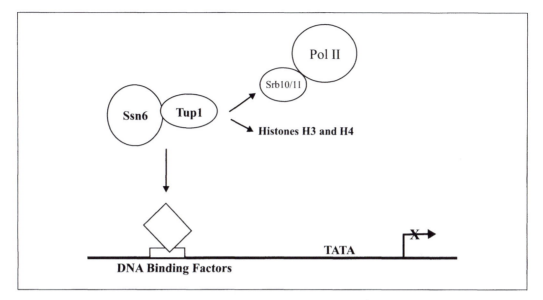

Fig. 3 Model for Ssn6–Tup1 repression in yeast. The Ssn6–Tup1 complex is recruited to promoter regions through interactions with sequence-specific DNA-binding proteins. Once recruited, the complex confers repression through interactions with components of the basal transcription machinery and/or histones H3 and H4.

binding domain bypasses the need for α2, resulting in repression of lexA operator (binding site)-containing promoters (80). Recruitment of Ssn6–lexA fusion proteins also confers repression, but this repression requires expression of *TUP1* (81), whereas Tup1–lexA-mediated repression does not require *SSN6* (80). Together these experiments indicate that Tup1 confers repressor function to the complex. Ssn6p may stabilize interactions between the complex and DNA-binding proteins such as α2.

4.1 Histones H3 and H4 are required for repression by Ssn6–Tup1

The chromatin structures of minichromosomes carrying the α2/Mcm1p binding site differ significantly in **a** and α cells, as do the structures of bona fide **a**-cell-specific genes such as *STE6* and *BAR1* (82, 83). In all cases, highly ordered arrays of positioned nucleosomes are observed in α cells (repressive conditions) but not in **a** cells (active conditions). Positioning of these nucleosomes requires α2, Ssn6p, and Tup1, as well as amino acids 4–19 in histone H4 (84, 85). Loss of nucleosome positioning correlates with loss of repression. Subsequent studies revealed that this degree of organization is not observed at all genes regulated by Tup1. None the less, in all cases examined to date chromatin appears to play an important role in Tup1 functions, since histone mutations cause a loss of repression (86, 87).

Further insights into this mechanism came through the finding that Tup1 interacts directly with the amino-terminal tails of histones H3 and H4 *in vitro* (86). These tail domains are both necessary and sufficient for Tup1 binding. Moreover, the region of Tup1 that interacts with the histones closely coincides with a domain that can confer repression independently when fused to lexA (80), further indicating that Tup1–histone interactions are important to Tup1 functions *in vivo*. Individual mutations in H3 or H4 alone that disrupt Tup1 binding have modest effects on repression. However, combined mutations in H3 and H4 synergistically compromise repression of at least three separate classes of Tup1-regulated genes, including the a-cell-specific genes described above, as well as haploid-specific genes and DNA-damage-inducible genes (86, 87). Thus, these histone domains may serve redundant roles in Tup1-mediated repression but are centrally important to Tup1 functions. Indeed, a recent report indicates that up to two Tup1p molecules may interact with each nucleosome of a Tup1p repressed reporter gene (88). The importance of Tup1–histone interactions is further supported by the finding that Tup1 homologues from an unrelated yeast, *Saccharomyces pombe*, also bind to H3 and H4 (89). Similarly, Tup1 analogues in *Drosophila* (groucho; 90) and mammalian cells (the TLE proteins; 91) bind to histones and serve as transcriptional repressors. Tup1 histone-binding functions, then, appear to be conserved across evolution, further supporting the importance of these interactions to Tup1 functions.

4.2 Acetylation of H3 and H4 antagonizes Ssn6–Tup1 functions

The amino-termini of H3 and H4 can be acetylated *in vivo*, as described above (also see Chapter 7), and acetylation affects the ability of Tup1 to interact with these histones (86). Tup1 binds poorly to highly acetylated forms of H3 and H4 *in vitro* but interacts very well with underacetylated isoforms of these histones. Consistent with these *in vitro* findings, Tup1-regulated genes are associated with underacetylated isoforms of H3 and H4 *in vivo* under conditions of repression and with more highly acetylated forms of these histones upon activation. Further, Tup1 repression is lost in cells containing a combination of mutations in three specific genes encoding histone deacetylases (HDACs)—*RPD3*, *HOS1*, and *HOS2*—which cause a concomitant hyperacetylation of both H3 and H4. Taken together, these data again indicate that Tup1 interactions with chromatin are important for its function as a repressor, and, moreover, that histone acetylation modulates Tup1 functions by interfering with these interactions.

The interaction of Tup1 with less acetylated forms of H3 and H4 is in keeping with the long-standing correlation between histone deacetylation and decreased transcription. Usually this correlation is taken to reflect a specialized, folded, nuclease-resistant chromatin structure associated with the repressed state. The effects of Tup1 binding on higher-order chromatin structures or on single nucleosomes are currently unknown. The above studies, however, indicate that decreased histone acetylation may be generally important for the binding of chromatin-organizing proteins such as Tup1. In turn, such binding may help to maintain the underacetylated state of the histones, further facilitating chromatin folding.

4.3 Transcription factors also contribute to Ssn6–Tup1 repression

The interactions of Tup1 with the histones likely only represents part of the repression mechanism (Fig. 3). Indeed, mutations in components of the mediator complex associated with the C-terminal domain (CTD) of RNA polymerase II (pol II), including Sin4, Srb8, Srb10, and Srb11, cause a partial loss of Tup1-mediated repression at multiple promoters (71, 92). The effects on repression are similar in extent to those caused by the histone mutations described above. These findings suggest that Tup1 might somehow inactivate or sequester one or more of these components of the basal transcription machinery to inhibit preinitiation complex formation or clearance. In support of this idea, a direct interaction has been observed between Tup1 and Hrs1/Med3, another component of the mediator (93). A modest amount of repression can be achieved *in vitro* upon addition of Tup1 to basal transcription factors alone, in the absence of transcriptional activators, and likely in the absence of the mediator or histones, supporting a role for Tup1 in the inhibition of basal activities as well (73, 94).

Taken together, the data to date indicate that Ssn6–Tup1 likely function both through interactions with the transcription machinery and with chromatin. Perhaps Tup1 interactions with the Srb or basal proteins halt transcription, whereas interactions with the histones maintain the repressed state. Such duality in functions is recognized more and more frequently in regulatory proteins, both for repression and for gene activation. The SAGA complex, for example, modifies chromatin through the HAT activity of Gcn5 but also affects the transcription machinery through its Spt components (95) (see Chapter 7). Tup1 serves as a unique example of a factor that organizes euchromatic genes into an inactive state, and since its association with chromatin is sensitive to changes in histone acetylation levels, Tup1 also highlights an additional mechanism by which histone acetylation can affect gene expression.

5. Summary

5.1 Important questions

Currently, there is little doubt that chromatin plays an active part of in the regulation of gene expression. The challenge now is to understand what types of structures actually exist *in vivo*, how these structures are regulated, and how they affect the functions of the transcription machinery.

The nature of the remodelled nucleosomes at the *PHO5* promoter is not clear. At one extreme it is conceivable that nucleosomes are completely absent from the promoter in the active state. It is equally possible, however, that a full set of histones is present but that they are in a non-native conformation, leaving DNA much more exposed. The speed at which nucleosomes reform in the strain with the temperature-sensitive *PHO80* allele after bringing the temperature down again from 37 °C to 24 °C (much less than an hour, again in the absence of replication in stationary cells) argues against a complete loss of histones from the promoter. DNA–protein cross-linking experiments might provide a clue to answering this question in the future.

The differential Gcn5 and Swi/Snf requirement for chromatin remodelling at the *PHO5* and *PHO8* promoter is a surprising finding, since both genes respond to the same signal, phosphate starvation, and the same transcription factor. The different response of two co-regulated genes opens up the possibility of potentially inform-ative experiments. Is it the nature of the histone–DNA interactions that makes the difference, or the location of the Pho4-binding sites? *PHO5–PHO8* hybrid promoter constructs should provide answers to these questions.

It will be important to understand the nature of the histone-modification status at the two promoters. The current excitement about acetylation is probably just the beginning. Experiments addressing histone acetylation at the two promoters before and after activation are currently under way. Are other modifications involved? Does Pho4 recruit acetylase and remodelling activity at *PHO8*? If so, are the same activities also recruited to *PHO5* but not stringently required? One of the unexplained prop-erties of the *PHO5* promoter is the precise boundaries of the chromatin transition. What is different between nucleosome –4 and –5? Does the four-nucleosome-transition reflect features of the modification status of the histones of precisely these nucleosomes?

A significant question for the future is how the differential states of histone acetyl-ation are established and maintained. Normally histone acetylation/deacetylation is very dynamic, such that the average half-life of an acetyl moiety on the histones is only a few seconds to minutes. Thus, it would seem difficult to create stable chromatin domains in the absence of a mechanism to direct and stabilize particular histone-modification states. Using Tup1-mediated events as an example, at least three different mechanisms appear to be possible (Fig. 4). First, Tup1 might be directly associated with an HDAC activity (Fig. 4A). Indeed, decreased histone acetylation states are associated with the functions of other repressors, such as Ume6 in yeast (96) and the MAD complex in mammalian cells (97) (Chapter 8). In these cases, HDACs are targeted directly to promoter regions through interactions with the re-pressor complexes. Secondly, Tup1 might interfere with the activity of HATs, thereby preventing re-acetylation of nearby histones (Fig. 4B). Precedent for this comes from the Twist and E1A proteins, which regulate the activity of two HATs, p300 and PCAF (98). Finally, transiently underacetylated histones might be 'captured' by proteins such as Tup1, and may thereby be shielded from re-acetylation (Fig. 4C). Thus, even if Tup1 does not affect the location or the activity of HDACs or HATs, it might directly affect acetylation levels through steric hindrance of the binding of these enzymes to the histone tails. Such a mechanism might also be used by 'architectural' proteins such as the Polycomb group proteins in *Drosophila*, which interact with chromatin to maintain repressed states established earlier in development by homeotic proteins (99).

If histone modifications affect the interactions of proteins such as Tup1p with chromatin, and proteins such as Tup1 reinforce particular modification states, then there is potential for feedback regulation of chromatin domains. Moreover, given

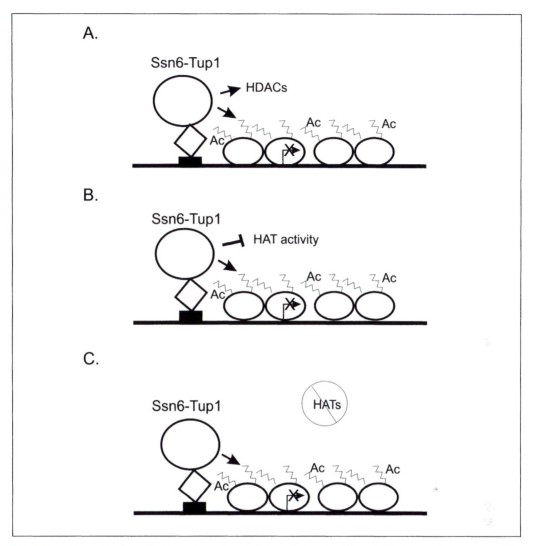

Fig. 4 Possible mechanisms to generate underacetylation of histones H3 and H4 at promoters repressed by Ssn6–Tup1. (A) The Ssn6–Tup1 complex may be associated with an HDAC activity. (B) Ssn6–Tup1 might inhibit HAT activities. (C) Binding of Tup1 to H3 and H4 might shield them from HATs, preventing their acetylation. For simplicity, the model drawn here illustrates the effects of the co-repressor only on promoter proximal nucleosomes, but these effects may extend into coding regions of the repressed genes as well.

that each histone is subject to variety of post-translational modifications in addition to acetylation (100), including phosphorylation, methylation, and ADP-ribosylation, different combinations and types of modifications on the histone tails may provide unique 'receptors' in chromatin for specific *trans*-acting regulatory factors. Indeed, many thousands of different combinations of histone modifications are possible, providing a plethora of regulatory potential. Understanding the nature and mechanism of cross-talk between histone-modifying activities is clearly an important endeavour for future research.

5.2   Discussion

Two important points were raised in discussion of the mechanism of Ssn6–Tup1 repression with other book authors. First the requirement of Srb proteins for repression implies that the RNA polymerase II holoenzyme may be held at the promoter of the repressed genes by the co-repressor complex. Such a 'docking' mechanism might occur after transcription initiation and might be entirely uncoupled from chromatin. Chromatin immunoprecipitation studies that address the presence of these complexes at repressed promoters would be of great interest. In addition, the identification of *tup1* mutants that can only repress by one of these two mechanisms would be of interest, as would the identification of mutants that alter the Srbs to render them resistant to the effects of Ssn6–Tup1. Alternatively, it may be that these two functions are linked and are not separable. For example, interactions between Ssn6–Tup1 and the Srbs might be transient, with the purpose of halting transcription momentarily in order to create a window of opportunity for establishment of the fully repressed chromatin state. Histone chromatin immunoprecipitation experiments have already demonstrated that nucleosomes are present in the promoter regions of Ssn6–Tup1-regulated genes, both when they are repressed and when they are active. Halting of transcription via short-term interaction of Ssn6–Tup1 with the Srbs might tip the balance away from maintenance of the active, acetylated state towards creation of the inactive, underacetylated state, which would then be further stabilized by Tup1–histone interactions.

The creation of this repressed state is relevant to the second point raised by other authors, which suggests parallels between the functions of Ssn6–Tup1 and the functions of the Sir proteins in silencing of the *HM* loci and of genes located near telomeres. Like Tup1, Sir3 and Sir4 interact with the amino-terminal tails of H3 and H4 (101), and silenced regions are enriched in hypoacetylated isoforms of H4 (102). Silencing, however, is an extreme, permanent mode of transcriptional repression, whereas many of the genes repressed by Ssn6–Tup1 are rapidly induced upon changing cellular conditions (such as DNA damage or a change in carbon source). Silenced structures, then, may more resemble heterochromatin (33) and may provide a stronger barrier to transcription than do the structures created by Ssn6–Tup1. Testing these ideas will require the development of *in vitro* systems that faithfully reconstruct these structures so that they can be analysed biophysically, or the development of *in vivo* assays for higher-order chromatin structures.

As pointed out in other discussions with book authors, the striking difference between *PHO5* and *PHO8* in their Swi/Snf requirement is especially interesting in light of the recent studies that have suggested that Swi/Snf is recruited by transcriptional activators (103, 104). It is possible that Pho4 generally recruits Swi/Snf to promoters even in cases such as *PHO5* where it is not needed. The reason why, at *PHO5*, Swi/Snf would not be needed would reside in a redundancy with other factors or mechanisms. This possibility should be testable in genetic screens looking for mutations that render *PHO5* Swi/Snf dependent. Such a screen might uncover other trans-acting complexes such as RSC (see Chapter 5) or, alternatively, *cis*-acting chromatin components important for the Snf/Swi independence (like, for example, other histone modifications).

It was also discussed whether the relative placement of the Pho4-binding sites relative to nucleosomes and the start site in *PHO5* and *PHO8* could account for the differential Gcn5 and Snf2 requirement. Assuming it might be the position of the Pho4-binding site relative to the nucleosomes, this would go against the generally held belief that Swi/Snf helps activators bind to a nucleosomal site. Both sites are non-nucleosomal in the (Snf2 dependent) *PHO8* promoter and Pho4 is actually bound in the absence of Snf2, while the promoter with an intranucleosomal site (*PHO5*) does not require Snf2.

When it comes to distance between UASp2 and TATA-box/start site in the two promoters, this stretch is considerably longer for *PHO8*, which could weaken the promoter and contribute to the dependence on the co-activators. However, in this scenario we would interpret the quality of the histone binding to the extra DNA being the source of the problem, and not the relative locations of the DNA elements themselves.

References

1. Roeder, R. G. (1998) Role of general and gene-specific cofactors in the regulation of eukaryotic transcription. *Cold Spring Harbor Symp. Quant. Biol.*, **63**, 201.
2. Reinberg, D., *et al.* (1998) The RNA polymerase II general transcriptin factors: past, present, and future. *Cold Spring Harbor Symp. Quant. Biol.*, **63**, 83.
3. Kornberg, R. D. (1998) Mechanism and regulation of yeast RNA polymerase II transcription. *Cold Spring Harbor Symp. Quant. Biol.*, **63**, 229.
4. Koleske, A. J. and Young, R. A. (1995) The RNA polymerase II holoenzyme and its implications for gene regulation. *Trends in Biochem. Sci.*, **20**, 113.
5. Buratowski, S. (1995) Mechanisms of gene activation. *Science*, **270**, 1773.
6. Lee, T. I. and Young, R. A. (1998) Regulation of gene expression by TBP associated proteins. *Genes and Dev.*, **12**, 1398.
7. Struhl, K., Kadosh, D., Keaveney, M., Kuras, L., and Moqtaderi, Z. (1998) Activation and repression mechanisms in yeast. *Cold Spring Harbor Symp. Quant. Biol.*, **63**, 413.
8. Maldonado, E., Hampsey, M., and Reinberg, D. (1999) Repression: targeting the heart of the matter. *Cell*, **99**, 455.
9. Knezetic, J. A. and Luse, D. A. (1986) The presence of nucleosomes on a DNA template prevents initiation by RNA polymerase II *in vitro*. *Cell*, **45**, 95.
10. Lorch, Y., LaPointe, J. W., and Kornberg, R. D. (1987) Nucleosomes inhibit the initiation of transcription but allow chain elongation with the displacement of histones. *Cell*, **49**, 203.
11. Kornberg, R. D. and Lorch, Y. (1991) Irresistible force meets immovable object: transcription and the nucleosome. *Cell*, **67**, 833.
12. Gross, D. S. and Garrard, W. T. (1988) Nuclease hypersensitive sites in chromatin. *Annu. Rev. Biochem.*, **57**, 159.

13. Noll, M. and Kornberg, R. D. (1977) Action of micrococcal nuclease on chromatin and the location of histone H1. *J. Mol. Biol.*, **109**, 393.

14. Levy, A. and Noll, M. (1981) Chromatin fine structure of active and repressed genes. *Nature*, **289**, 198.

15. Wu, C., Bingham, P. M., Livak, K. J., Holmgren, R., and Elgin, S. C. (1979) The chromatin structure of specific genes: I. Evidence for higher order domains of defined DNA sequence. *Cell*, **16**, 797.

16. Wu, C., Wong, Y. C., and Elgin, S. C. (1979) The chromatin structure of specific genes: II. Disruption of chromatin structure during gene activity. *Cell*, **16**, 807.

17. Keene, M. A. and Elgin, S. C. (1981) Micrococcal nuclease as a probe of DNA sequence organization and chromatin structure. *Cell*, **27**, 57.

18. Weintraub, H., Larsen, A., and Groudine, M. (1981) α-Globin gene switching during the development of chicken embryos: expression and chromosome structure. *Cell*, **24**, 333.

19. Stalder, J., Larsen, A., Engel, J. D., Dolan, M., Groudine, M., and Weintraub, H. (1980) Tissue-specific DNA cleavages in the globin chromatin domain introduced by DNAaseI. *Cell*, **20**, 451.

20. Sippel, A. E., Saueressig, H., Huber, M. C., Hoefer, H. C., Stief, A., Borgmeyer, U., and Bonifer, C. (1996) Identification of *cis*-acting elements as DNaseI hypersensitive sites in lysozyme gene chromatin. *Meth. Enzymol.*, **274**, 233.

21. Fritton, H. P., Igo-Kemenes, T., Nowock, J., Strech-Jurk, U., Theisen, M., and Sippel, A. E. (1984) Alternative sets of DNaseI hypersensitive sites characterize the various functional states of the chicken lysozyme gene. *Nature*, **311**, 163.

22. Fritton, H. P., Sippel, A. E., and Igo-Kemenes, T. (1983) Nuclease hypersensitive sites in the chromatin domain of the chicken lysozyme gene. *Nucleic Acids Res.*, **11**, 3467.

23. Jantzen, K., Fritton, H. P., and Igo-Kemenes, T. (1986) The DNase I sensitive domain of the chicken lysosyme gene spans 24 kb. *Nucleic Acids Res.*, **14**, 6085.

24. Matsui, T. (1987) Transcription of Adenovirus 2 major late and peptide IX genes under conditions of *in vitro* nucleosome assembly. *Mol. Cell. Biol.*, **7**, 1401.

25. Workman, J. L. and Roeder, R. G. (1987) Binding of transcription factor TFIID to the major late promoter during *in vitro* nucleosome assembly potentiates subsequent initiation by RNA polymerase II. *Cell*, **51**, 613.

26. Knezetic, J. A., Jacob, G. A., and Luse, D. S. (1988) Assembly of RNA polymerase II preinitiation complexes before assembly of nucleosomes allows efficient initiation of transcription on nucleosomal templates. *Mol. Cell. Biol.*, **8**, 3114.

27. Grunstein, M. (1990) Nucleosomes: regulators of transcription. *Trends Genet.*, **6**, 395.

28. Clark-Adams, C. D., Norris, D., Osley, M. A., Fassler, J. S., and Winston, F. (1988) Changes in histone gene dosage alter transcription in yeast. *Genes Develop.*, **2**, 150.

29. Han, M., Kim, U. J., Kayne, P., and Grunstein, M. (1988) Depletion of histone H4 and nucleosomes activates the PHO5 gene in *Saccharomyces cerevisiae*. *EMBO J.*, **7**, 2221.

30. Park, E. C. and Szostak, J. W. (1990) Point mutations in the yeast histone H4 gene prevent silencing of the silent mating type locus HML. *Mol. Cell. Biol.*, **10**, 4932.

31. Megee, P. C., Morgan, B. A., Mittman, B. A., and Smith, M. M. (1990) Genetic analysis of histone H4: essential role of lysines subject to reversible acetylation. *Science*, **247**, 841.

32. Kayne, P. S., Kim, U., Han, M., Mullen, J. R., Yoshizaki, F., and Grunstein, M. (1988) Extremely conserved histone H4 amino terminus is dispensable for growth but essential for repressing the silent mating loci in yeast. *Cell*, **55**, 27.

33. Grunstein, M. (1998) Yeast heterochromatin: regulation of its assembly and inheritance by histones. *Cell*, **93**, 325.

34. Smith, M. M. (1991) Histone structure and function. *Curr. Opin. Cell Biol.*, **3**, 429.

35. Weintraub, H. (1985) Assembly and propagation of repressed and derepressed chromosomal states. *Cell*, **42**, 705.

36. Kornberg, R. D. and Lorch, Y. (1992) Chromatin structure and transcription. *Annu. Rev. Cell Biol.*, **8**, 563.

37. Roth, S. Y. and Allis, C. D. (1992) Chromatin condensation: does histone H1 dephosphorylation play a role? *Trends Biochem. Sci.*, **17**, 93.

38. Wolffe, A. P. and Dimitrov, S. (1993) Histone-modulated gene activity: developmental implications. *Crit. Rev. Euk. Gene Express.*, **3**, 167.

39. Bresnick, E. H., Bustin, M., Marsaud, V., Richard-Foy, H., and Hager, G. L. (1992) The transcriptionally active MMTV promoter is depleted of histone H1. *Nucleic Acids Res.*, **20**, 273.

40. Wolffe, A. P. and Pruss, D. (1996) Deviant nucleosomes: the functional specialization of chromatin. *Trends Genet.*, **12**, 58.

41. Wolffe, A. P. (1999) Architectural regulations and HMG1. *Nature Genet.*, **22**, 215.

42. Wolffe, A. P. (1995) Centromeric chromatin. Histone deviants. *Curr. Biol.*, **5**, 452.

43. Wade, P. A., Jones, P. L., Vermaak, D., and Wolffe, A. P. (1998) A multiple subunit Mi-2 histone deacetylase from *Xenopus laevis* cofractionates with an assoicated Snf2 superfamily ATPase. *Curr. Biol.*, **8**, 843.

44. Zhang, Y., LeRoy, G., Seelig, H. P., Lane, W. S., and Reinberg, D. (1998) The dermatomyositis-specific autoantigen Mi2 is a component of a complex containing histone deacetylase and nucleosome remodeling activities. *Cell*, **95**, 279.

45. Wade, P. A., Pruss, D., and Wolffe, A. P. (1997) Histone acetylation: chromatin in action. *Trends Biochem. Sci.*, **22**, 128.

46. Oshima, Y. (1997) The phosphatase system in *Saccharomyces cerevisiae*. *Genes Genet. Syst.*, **72**, 323.

47. Vogel, K. and Hinnen, A. (1990) The yeast phosphatase system. *Mol. Microbiol.*, **4**, 2013.

48. Venter, U. and Hörz, W. (1989) The acid phosphatase genes PHO10 and PHO11 in *S. cerevisiae* are located at the telomeres of chromosomes VIII and I. *Nucleic Acids Res.*, **17**, 1353.

49. Kaneko, Y., Hayashi, N., Toh-e, A., Banno, I., and Oshima, Y. (1987) Structural characteristics of the PHO8 gene encoding repressible alkaline phosphatase in *Saccharomyces cerevisiae*. *Gene*, **58**, 137.

50. Venter, U., Svaren, J., Schmitz, J., Schmid, A., and Hörz, W. (1994) A nucleosome precludes binding of the transcription factor Pho4 *in vivo* to a critical target site in the PHO5 promoter. *EMBO J.*, **13**, 4848.

51. Barbaric, S., Münsterkötter, M., Svaren, J., and Hörz, W. (1996) The homeodomain protein Pho2 and the basic-helix–loop–helix protein Pho4 bind DNA cooperatively at the yeast PHO5 promoter. *Nucleic Acids Res.*, **24**, 4479.

52. Barbaric, S., Münsterkötter, M., Goding, C., and Hörz, W. (1998) Cooperative Pho2–Pho4 interactions at the PHO5 promoter are critical for binding of Pho4 to UASp1 and for efficient transactivation by Pho4 at UAS2p. *Mol. Cell. Biol.*, **18**, 2629.

53. Kaffman, A., Herskowitz, I., Tjian, R., and O'Shea, E. K. (1994) Phosphorylation of the transcription factor Pho4 by a cyclin CDK complex, Pho80–Pho85. *Science*, **263**, 1153.

54. Komeili, A. and O'Shea, E. K. (1999) Roles of phosphorylation sites in regulating activity of the transcription factor Pho4. *Science*, **284**, 977.

55. Almer, A., Rudolph, H., Hinnen, A., and Hörz, W. (1986) Removal of positioned nucleosomes from the yeast PHO5 promoter upon PHO5 induction releases additional upstream activating DNA elements. *EMBO J.*, **5**, 2689.

56. Neubauer, B., Linxweiler, W., and Hörz, W. (1986) DNA engineering shows that nucleo-

some phasing on the African green monkey alpha satellite is the result of multiple additivie histone–DNA interactions. *J. Mol. Biol.*, **190**, 639.

57. Zhang, X. Y., Fittler, F., and Hörz, W. (1983) Eight different highly specific nucleosome phases on alpha-satellite DNA in the African green monkey. *Nucleic Acids Res.*, **11**, 4287.

58. Straka, C. and Hörz, W. (1991) A functional role for nucleosomes in the repression of a yeast promoter. *EMBO J.*, **13**, 4856.

59. Gaudreau, L., Schmid, A., Blaschke, D., Ptashne, M., and Hörz, W. (1997) RNA polymerase II holoenzyme recruitment is sufficient to remodel chromatin at the yeast PHO5 promoter. *Cell*, **89**, 55.

60. Schmid, A., Fascher, K. D., and Hörz, W. (1992) Nucleosome disruption at the yeast PHO5 promoter upon induction occurs in the absence of DNA replication. *Cell*, **71**, 853.

61. Fascher, K. D., Schmitz, J., and Hörz, W. (1993) Structural and functional requirements for the chromatin transition at the PHO5 promoter in *Saccharamyces cerevisiae* upon PHO5 activation. *J. Mol. Biol.*, **231**, 658.

62. Fascher, K. D., Schmitz, J., and Hörz, W. (1990) Role of *trans*-activating proteins in the generation of active chromatin at the PHO5 promoter in *S. cerevisiae*. *EMBO J.*, **9**, 2523.

63. Svaren, J., Schmitz, J., and Hörz, W. (1994) The transactivation domain of Pho4 is required for nucleosome disruption at the PHO5 promoter. *EMBO J.*, **13**, 4856.

64. McAndrew, P. C., Svaren, J., Martin, S. R., Hörz, W., and Goding, G. R. (1998) Requirements from chromatin modulation and transcription activation by the Pho4 acidic activation domain. *Mol. Cell. Biol.*, **18**, 5818.

65. Gregory, P. D., Schmid, A., Zavari, M., Lui, L., Berger, S. L., and Hörz, W. (1998) Absence of Gcn5 HAT activity defines a novel state in the opening of chromatin at the PHO5 promoter in yeast. *Mol. Cell.*, **1**, 495.

66. Vidal, M. and Gaber, R. F. (1991) RPD3 encodes a second factor required to achieve maximum positive and negative transcription states in *Saccharomyces cerevisiae*. *Mol. Cell. Biol.*, **11**, 6317.

67. Barbaric, S., Fascher, K. D., and Hörz, W. (1992) Activation of the weakly regulated PHO8 promoter in *S. cerevisiae*: Chromatin transition and binding sites for the positive regulator protein Pho4. *Nucleic Acids Res.*, **20**, 1031.

68. Gregory, P. D., Schmid, A., Zavari, M., Münsterkötter, M., and Hörz, W. (1999) Chromatin remodeling at the PHO8 promoter requires SWI-SNF and SAGA at a step subsequent to activator binding. *EMBO J.*, **18**, 6407.

69. Pollard, K. J. and Peterson, C. L. (1998) Chromatin remodeling: a marriage between two families? *Bioessays*, **20**, 771.

70. Roth, S. Y. (1995) Chromatin-mediated transcriptional repression in yeast. *Curr. Opin. Genet. Develop.*, **5**, 168.

71. Wahi, M., Komachi, K., and Johnson, A. D. (1998) Gene regulation by the yeast Ssn6–Tup1 corepressor. *Cold Spring Harbor Symp. Quant. Biol.*, **63**, 447.

72. DeRisi, J. l., Iyer, V. R., and Brown, P. O. (1997) Exploring the metabolic and genetic control of gene expression on a genomic scale. *Science*, **278**, 680.

73. Redd, M. J., Arnaud, M. B., and Johnson, A. D. (1997) A complex composed of Tup1 and Ssn6 represses transcription *in vitro*. *J. Biol. Chem.*, **272**, 11193.

74. Varanasi, U. S., Klis, M., Mikesell, P. B., and Trumbly, R. J. (1996) The Cyc8 (Ssn6)–Tup1 corepressor complex is composed of one Cyc8 and four Tup1 subunits. *Mol. Cell. Biol.*, **16**, 6707.

75. Edmondson, D. G., Zhang, W., Watson, A., Xu, W., Bone, J. R., Yu, Y., Stillman, D., and Roth, S. Y. (1998) *In vivo* functions of histone acetylation/deacetylation in Tup1p repression and Gcn5p activation. *Cold Spring Harbor Symp. Quant. Biol.*, **63**, 459.

76. Herskowitz, I. (1989) A regulatory hierarchy for cell specialization in yeast. *Nature*, **342**, 749.
77. Keleher, C. A., Passmore, S., and Johnson, A. D. (1989) Yeast repressor alpha 2 binds to its operator cooperatively with yeast protein Mcm1. *Mol. Cell. Biol.*, **9**, 5228.
78. Komachi, K., Redd, M. J., and Johnson, A. D. (1994) The WD repeats of Tup1 interact with the homeo domain protein alpha 2. *Genes Develop.*, **8**, 2857.
79. Smith, R. L., Redd, M. J., and Johnson, A. D. (1995) The tetratricopeptide repeats of Ssn6 interact with the homeo domain of alpha2. *Genes and Develop.*, **9**, 2903.
80. Tzamarias, D. and Struhl, K. (1994) Functional dissection of the yeast Cyc8–Tup1 corepressor complex. *Nature*, **369**, 758.
81. Keleher, C. A., Redd, M. J., Schultz, J., Carlson, M., and Johnson, A. D. (1992) Ssn6–Tup1 is a general repressor of transcription in yeast. *Cell*, **68**, 709.
82. Roth, S. Y., Dean, A., and Simpson, R. T. (1990) Yeast alpha 2 repressor positions nucleosomes in TRP1/ARS1 chromatin. *Mol. Cell. Biol.*, **10**, 2247.
83. Shimizu, M., Roth, S. Y., Szent-Gyorgyi, C., and Simpson, R. T. (1991) Nucleosomes are positioned with base pair precision adjacent to the alpha 2 operator in *Saccharomyces cerevisiae*. *EMBO J.*, **10**, 3033.
84. Roth, S. Y., Shimizu, M., Johnson, L., Grunstein, M., and Simpson, R. T. (1992) Stable nucleosome positioning and complete repression by the yeast alpha 2 repressor are disrupted by amino-terminal mutations in histone H4. *Genes Develop.*, **6**, 411.
85. Cooper, J. P., Roth, S. Y., and Simpson, R. T. (1994) The global transcriptional regulators, SSN6 and TUP1, play distinct roles in the establishment of a repressive chromatin structure. *Genes Develop.*, **8**, 1400.
86. Edmondson, D. G., Smith, M. M., and Roth, S. Y. (1996) Repression domain of the yeast global repressor Tup1 interacts directly with histones H3 and H4. *Genes Develop.*, **10**, 1247.
87. Huang, L., Zhang, W., and Roth, S. Y. (1997) Amino termini of histones H3 and H4 are required for a1/α2 repression in yeast. *Mol. Cell. Biol.*, **17**, 6555.
88. Ducker, C. E. and Simpson, R. T. (2000) The organized chromatin domain of the repressed yeast a cell specific gene *STE6* contains two molecules of the corepressor Tup1p per nucleosome. *EMBO J.*, **19**, 400.
89. Mukai, Y., Matsuo, E., Roth, S. Y., and Harashima, S. (1999) Conservation of histone binding and transcriptional repressor functions in a *Schizosaccharomyces pombe* Tup1p homolog. *Mol. Cell. Biol.*, **19**, 8461.
90. Chen, G., Fernandez, J., Mische, S., and Courey, A. J. (1999) A functional interaction between the histone deacetylase Rpd3 and the corepressor Groucho in *Drosophila* development. *Genes Develop.*, **13**, 2218.
91. Palaparti, A., Baratz, A., and Stifani, S. (1997) The Groucho/transducin-like enhancer of split transcriptional repressors interact with the genetically defined amino terminal silencing domain of histone H3. *J. Biol. Chem.*, **272**, 26604.
92. Wahi, M. and Johnson, A.D. (1995) Identification of genes required for alpha2 repression in *Saccharomyces cerevisiae*. *Genetics*, **140**, 79.
93. Papamichos-Chronakis, M., Conlan, R.S., Gounalaki, N., Copf, T., and Tzamarias, D. (2000) Hrs1/Med3 is a Cyc8–Tup1 corepressor target in the RNA polymerase II holoenzyme. *J. Biol. Chem.*, **275**, 8397.
94. Herschbach, B. M., Arnaud, M. B., and Johnson, A. D. (1994) Transcriptional repression directed by the yeast alpha 2 protein *in vitro*. *Nature*, **370**, 309.
95. Grant, P. A., Duggan, L., Cote, J., Roberts, S., Brownell, J. E., Candau, R., Ohba, R., Owen-Hughes, T., Allis, C. D., Winston, F., et al. (1997) Yeast GCN5 functions in two multi-

subunit complexes to acetylate nucleosomal histones: characterization of an ADA complex and the SAGA (SPT/ADA) complex. *Genes Develop.*, **11**, 1640.

96. Kadosh, D. and Struhl, K. (1997) Repression by UME6 involves recruitment of a complex containing Sin3 corepressor and Rpd3 histone deacetylase to target promoters. *Cell*, **89**, 365.

97. Laherty, C. D., Yang, W.-M., Sun, J.-M., Davie, J. R., Seto, E., and Eisenman, R. N. (1997) Histone deacetylases associated with the mSin3 corepressor mediate Mad transcriptional repression. *Cell*, **89**, 349.

98. Chakravarti, D., Ogryzko, V., Kao, H. Y., Nash, A., Chen, H., Nakatani, Y., and Evans, R. M. (1999) A viral mechanism for inhibition of p300 and PCAF acetyltransferase activity. *Cell*, **96**, 393.

99. Orlando, V. and Paro, R. (1995) Chromatin multiprotein complexes involved in the maintenance of transcription patterns. *Curr. Opin. Genet. Develop.*, **5**, 174.

100. van Holde, K. E. (1989) Chromatin. In *Molecular biology*, (ed. A. Rich). Springer-Verlag, New York.

101. Strahl-Bolsinger, S., Hecht, A., Luo, K., and Grunstein, M. (1997) SIR2 and SIR4 interactions differ in core and extended telomeric heterochromatin in yeast. *Genes Develop.*, **11**, 83.

102. Braunstein, M., Rose, A. B., Holmes, S. G., Allis, C. D., and Broach, J. R. (1993) Transcriptional silencing in yeast is associated with reduced nucleosome acetylation. *Genes Develop.*, **7**, 592.

103. Neely, K. E., Hassan, A. H., Wallberg, A. E., Steger, D. J., Cairns, B. R., Wright, A. P., and Workman, J. L. (1999) Activation domain-mediated targeting of the SWI/SNF complex to promoters stimulates transcription from nucleosome arrays. *Mol. Cell*, **4**, 649.

104. Natarajan, K., Jackson, B. M., Zhou, H., Winston, F., and Hinnebusch, A. G. (1999) Transcriptional activation by Gcn4 involves independent interactions with the SWI/SNF complex and the SRB/mediator. *Mol. Cell*, **4**, 657.

4 | The genetics of chromatin function

FRED WINSTON and M. MITCHELL SMITH

1. Introduction

In the chromatin field today we take for granted the power of both biochemical and genetic approaches for solving problems in transcription. Scientists who consider themselves primarily biochemists or molecular biologists routinely take advantage of histone gene mutations in the budding yeast *Saccharomyces cerevisiae*. Similarly, those who consider themselves primarily geneticists routinely undertake biochemical assays of chromatin structure to test their genetic models. Not too long ago, however, biochemists principally interested in transcription considered chromatin a messy business, and models involving nucleosomes to be the mystery explanation of last resort. After all, *in vitro* transcription systems seemed to be working fine without the cumbersome complications of chromatin. Likewise, geneticists studying transcription found no need to invoke chromatin models since their mutants were identifying a few site-specific DNA-binding transcription factors. However, this picture changed dramatically when unbiased mutational screens in *S. cerevisiae* identified the histone genes themselves as transcription factors (1). The role of the nucleosome in transcription *in vivo* could no longer be ignored.

Genetic analysis continues to provide novel insights into the pathways of chromatin transcription *in vivo*. One of the strengths of mutational studies is the ability to reveal previously unrecognized functional interactions. Thus, this approach provides special direction to transcription research and an important counterpoint to biochemical studies *in vitro*. Our goal in this chapter is, first and foremost, to highlight this principle of discovery. In addition, we hope to provide both a conceptual and factual basis for understanding the emerging genetic picture of chromatin transcription.

Historically, the discovery of large transcription complexes that regulate chromatin structure and the genetic dissection of histone function proceeded hand in hand. In the following sections we begin with a brief summary of histone gene organization in *S. cerevisiae*, since this information is helpful for understanding the genetic screens. Following that background, we will focus first on the genetics of several transcription complexes that act by modulating chromatin structure. Finally, we will discuss the mutational analysis of the targets of these complexes, the histone proteins themselves.

2. Histone gene organization in *S. cerevisiae*

The budding yeast *S. cerevisiae* has several features that make it advantageous for molecular genetic studies of chromatin and transcription (reviewed in 2). Chief among these is the simple organization of its histone genes, which makes it relatively easy to mutate a specific histone gene and then introduce it back into the cell as the only source of that histone.

2.1 The major core histone genes

Most organisms contain many copies of the gene for each of the major histones. In contrast, *S. cerevisiae* contains only two non-allelic copies of each core histone gene organized into four separate genetic loci. Each locus contains two genes, either an H3 (*HHT*) gene paired with an H4 (*HHF*) gene, or an H2A (*HTA*) gene paired with an H2B (*HTB*) gene, and the paired genes are divergently transcribed from a shared promoter region. While at least one copy of each core histone gene is essential, the duplicated gene pairs are functionally redundant and any individual gene or gene set can be deleted without loss of viability. However, the deletion of one of the two H2A–H2B loci, *HTA1–HTB1*, or one of the H3–H4 loci, *HHT2–HHF2*, does cause some mutant phenotypes (1, 3, 4). We'll see shortly that these phenotypes formed the basis of experiments that first linked *HTA1* and *HTB1* with the regulation of transcription.

2.2 Variant histone genes

In addition to genes for the major histones, *S. cerevisiae* also contains two separate unlinked genes that encode variant histones, members of protein families that have been highly conserved during evolution. The first, *CSE4*, encodes a protein related to the CENP-A family of centromere-specific histone H3 proteins. These have now been found in a wide range of eukaryotes from fungi to mammals (5). *CSE4* is an essential single-copy gene required for normal centromere structure and function but with no apparent role in transcription, and it will not be considered further here. The second gene, *HTZ1*, encodes a member of the conserved H2A.F/Z family of variant H2A proteins, which likely represent a separate evolutionary lineage of H2A proteins (6). The *S. cerevisiae* H2A.Z protein shares greater similarity with the mammalian H2A.Z protein than it does with the major H2A protein of *S. cerevisiae* itself. Recent experiments, described in the last section of this chapter, suggest that variant nucleosomes containing H2A.Z play an important role in transcription by modulating chromatin structure.

3. Large transcription complexes that control chromatin structure

The regulation of transcription has been studied extensively in *S. cerevisiae* by the isolation and analysis of regulatory mutations. However, in contrast to what was generally found in the pioneering work in *E. coli*, the *S. cerevisiae* mutants have rarely identified transcription factors that bind directly to DNA sites in promoters. More often, the mutations identified factors that act in a less direct fashion—by participating in large, multiprotein complexes that control transcription of large sets of genes, often by modifying the structure of nucleosomes (7).

Two main classes of nucleosome-modifying activities have been identified—those that remodel nucleosomes in an ATP-dependent fashion and those that affect the acetylation state of histones (8). In *S. cerevisiae*, two of these complexes, the nucleosome-remodelling complex Snf/Swi and the histone acetyltransferase complex SAGA, have been studied extensively (for reviews see 7–9). These studies have yielded significant insights into their functions, both in terms of their mechanisms of action as well as their roles in controlling the transcription of different genes *in vivo*. Such studies have shown that these complexes, although working by different mechanisms, overlap in their roles in ways yet to be determined (10–13). In this section, we will describe genetic studies of these two complexes.

3.1 The Snf/Swi nucleosome remodelling complex

The Snf/Swi complex is the founding member of a conserved family of nucleosome remodelling complexes that exist throughout eukaryotes (reviewed in 9). Snf/Swi has been shown to play important roles in transcription *in vivo*, as described below. The conservation of Snf/Swi complexes have made them the subject of extensive studies in many laboratories. The identification and analysis of the Snf/Swi family has provided new insights into mechanisms of transcriptional regulation and has provided important evidence that chromatin structure plays a critical role in transcriptional control.

3.1.1 The identification and analysis of *snf* and *swi* mutants of *S. cerevisiae*

Several members of the Snf/Swi complex were identified in two genetic screens for regulatory mutants (reviewed in 14). In one screen, mutations were identified that decrease expression of the *SUC2* gene. *SUC2* encodes invertase, an enzyme required for yeast cells to utilize either sucrose or raffinose as a carbon source. *SUC2* transcription is repressed in the presence of high glucose and it is expressed at a high level in low glucose. Mutations that decrease expression of *SUC2* in low glucose were named *snf* (for sucrose *non*-fermenter). The *snf* mutations identified two different classes of genes, based on mutant phenotypes. One of these classes consist of three genes, named *SNF2*, *SNF5*, and *SNF6*, that were subsequently shown to encode components of Snf/Swi. The second class is not directly related to Snf/Swi and is reviewed elsewhere (15). Mutations in *SNF2*, *SNF5*, and *SNF6* were shown to cause a common set of pleiotropic phenotypes, suggesting that they participate in a common function. In the second screen, mutations were identified that impair expression of the *HO* gene, required for mating-type switching. Mutations that decrease *HO* transcription were named *swi* mutants (for *switching*-defective). As for the *snf* screen, this screen also identified multiple classes of genes. One class consisted of *SWI1*, *SWI2*, and *SWI3*.

Mutations in these genes also cause a common set of pleiotropic phenotypes, suggesting that these three genes played a common role in gene expression.

An important breakthrough came when DNA sequence analysis of the *SNF2* and *SWI2* genes showed that they are the same gene. This discovery united the *SNF* and *SWI* sets of genes and suggested that they functioned in a complex. Subsequently, the *S. cerevisiae* complex was purified by standard biochemical steps (16, 17). The biochemical analysis, in combination with additional genetic studies (18), demonstrated that Snf/Swi is a complex of 11 proteins. The biochemical analysis of the Snf/Swi family is described in Chapter 5.

3.1.2 A genetic connection between Snf/Swi and histones

The connection between the function of Snf/Swi and chromatin structure was first suggested by the isolation of mutations that suppress *snf2* and *snf5* mutations. This work identified mutations in two previously identified genes, *SPT5* and *SPT6* (19–22). Although the precise functions of *SPT5* and *SPT6* are not known, it was known that *spt5* and *spt6* mutations cause mutant phenotypes similar to those caused by a deletion of *HTA1–HTB1*, suggesting a functional link between Spt5, Spt6, and histones (23). Thus, since *spt5* and *spt6* mutations suppress *snf2* and *snf5* mutations, it was tested whether the *HTA1–HTB1* deletion would also suppress *snf2* and *snf5* mutations. This analysis demonstrated that, indeed, the histone deletion suppresses the *snf/swi* defect for several phenotypes tested, thereby establishing a genetic relationship between Snf/Swi and histones (24). Subsequently, mutations in genes encoding histones H3 and H4 were also identified as suppressors of *snf/swi* mutations (25; described in Section 4.2).

Motivated by these genetic results, the effect of *snf/swi* mutations on chromatin structure was investigated at the *SUC2* gene, a gene strongly dependent upon Snf/Swi for its transcription (24). In a wild-type strain grown in low glucose (derepressing conditions), the chromatin structure over the *SUC2* promoter region is in an open, 'active' configuation, based on sensitivity to micrococcal nuclease (MNase). However, in *snf/swi* mutants the chromatin is in an inactive, 'closed' conformation, similar to that observed for *SUC2* when grown in high glucose (repressing conditions). Detailed analysis has provided strong evidence for six nucleosomes positioned over the *SUC2* promoter region that are remodelled by Snf/Swi under derepressing conditions (26, 27). Consistent with the genetic results, extensive biochemical analysis has demonstrated that Snf/Swi possesses ATP-dependent nucleosome remodelling activities *in vitro* (Chapter 5).

The demonstration of Snf/Swi-dependent chromatin changes at *SUC2* allowed a test of the cause and effect relationship between the chromatin and transcriptional changes (24). That is, while the chromatin changes at *SUC2* correlated with the level of *SUC2* transcription, two possibilities existed: first, that the changes in chromatin structure cause the changes in transcription and, secondly, that the changes in transcription cause the changes in chromatin structure. To distinguish between these two possibilities, investigators constructed a mutation in the *SUC2* TATA-box, that impaired *SUC2* transcription. Then, they determined whether or not the Snf/Swi-

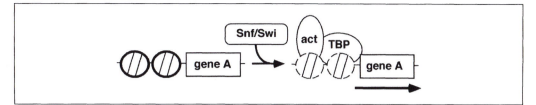

Fig. 1 Model for activation of transcription by the Snf/Swi complex. In the inactive state, nucleosomes, represented by circles, repress transcription of the adjacent gene. Snf/Swi remodels the nucleosomes, allowing binding by a transcriptional activator, TATA-binding protein (TBP), and other factors (not shown), to allow transcription initiation. The nucleosome remodelling could allow the binding of all of these factors. Alternatively, remodelling could allow binding of one factor which, in turn, recruits additional factors.

dependent chromatin changes still occurred in the absence of significant *SUC2* transcription. The results clearly demonstrated the same Snf/Swi-dependent changes in chromatin structure in the *SUC2* TATA mutant, demonstrating that the chromatin changes at *SUC2* occur in the absence of transcription. This result supports very strongly the model that chromatin changes in the *SUC2* promoter are critical for allowing transcripton. These results, in combination with additional genetic and biochemical studies led to a model for activation of transcription by Snf/Swi (Fig. 1).

Although the analysis of chromatin structure at *SUC2* has provided the major evidence that the Snf/Swi complex causes chromatin changes *in vivo*, the exact role of these chromatin changes remains obscure. This is because of the multiple layers by which transcription of *SUC2*, and almost any gene, is regulated *in vivo*. For *SUC2* there is good evidence that chromatin changes are involved at other levels of control. For example, glucose repression of *SUC2* is controlled by the complex Ssn6–Tup1, a global repressor, and there is good evidence that Ssn6–Tup1 also controls *SUC2* chromatin structure (26). In addition, there is evidence for another, uncharacterized mechanism of control at the level of chromatin structure, based on the analysis of a particular class of Snf-like histone H2A mutants described later in this chapter (28). Overall, then, it is likely that there are many levels at which chromatin structure plays a role in the control of transcription of any gene.

3.1.3 Whole-genome expression analysis of Snf/Swi function

The analysis of *snf/swi* mutants raised questions concerning fundamental aspects of Snf/Swi function. First, how many *S. cerevisiae* genes are dependent on Snf/Swi? Secondly, what factors determine whether a gene is Snf/Swi-dependent or Snf/Swi-independent? Thirdly, are all Snf/Swi subunits required similarly *in vivo*? Finally, since Snf/Swi is a nucleosome-remodelling complex, does it remodel chromatin on a gene-specific scale or over a larger domain that might include several genes? Some of these questions were addressed by determining the effect of *snf/swi* mutations on the mRNA levels for all *S. cerevisiae* genes. This approach, whole-genome expression analysis, also revealed unanticipated aspects of Snf/Swi function (29, 30). Some of the results of whole-genome expression analysis, along with related studies are described in this section.

3.1.4 A small percentage of genes require Snf/Swi for normal mRNA levels

The whole-genome expression analysis of *snf/swi* mutants showed that only 1–6% of all *S. cerevisiae* genes require Snf/Swi for normal mRNA levels (29, 30). Interestingly, one study demonstrated that when *S. cerevisiae* cells are grown in two different types of media, the set of genes that require Snf/Swi are overlapping but not identical (29). This result raises the likely possibility that other growth conditions might reveal yet distinct Snf/Swi requirements.

Since only a small percentage of *S. cerevisiae* genes requires Snf/Swi, what are the determinants of Snf/Swi dependence for a gene? Recent studies have shown that Snf/Swi can interact with particular DNA-binding transcriptional activators, providing one obvious mechanism by which Snf/Swi can be recruited to promoters (see Chapter 5). However, Snf/Swi–activator interactions cannot be the sole determinant of Snf/Swi dependence. Recent studies have demonstrated that the *PHO5* gene is Snf/Swi-independent, while the *PHO8* gene is Snf/Swi-dependent, even though both genes are controlled by the Pho4 activator (31). Other studies have shown that both the chromatin structure of a promoter, as well as the affinity of an activator for its site, can determine Snf/Swi dependence (for a review see 32). It seems most likely that a combination of factors will be shown to determine Snf/Swi dependence *in vivo*.

3.1.5 Snf/Swi can repress as well as activate transcription

One of the most startling results to emerge from whole-genome expression analyses was the finding that a significant number of genes appear to be repressed by Snf/Swi (29, 30). This result is in contrast to the models for Snf/Swi function that stipulated that Snf/Swi activates transcription by remodelling nucleosomes, thereby allowing activators and other factors to bind to their sites. There are several possible reasons for the observed repression defect at some genes in *snf/swi* mutants. One possibility, true for any mutant phenotype, is that the repression defect is indirect. By this model, the repression defect is actually caused by an activation defect for a negative regulator. However, a genetic test suggests that this model is not correct (29). This test took advantage of the previous demonstration that mutations in histone genes suppress *snf/swi* activation defects. If the *snf/swi* repression defects occurred as an indirect effect of an activation defect, then the repression defects should also be suppressed by the same histone mutations. However, it was demonstrated that the repression defect is not suppressed by the histone mutations. Therefore, the repression defect observed in *snf/swi* mutants cannot simply be an indirect effect of an activation defect.

There are several possible mechanisms by which Snf/Swi could directly repress transcription. Three of these possibilities will be discussed here (Fig. 2). First, it has been shown that, *in vitro*, nucleosome remodelling by Snf/Swi complexes can occur in either direction (33, 34). If remodelling in both directions occurs *in vivo* as well, then repression could occur by remodelling nucleosomes from an active to an inactive conformation. Secondly, by nucleosome remodelling, Snf/Swi could help a repressor to bind to its site in chromatin. Finally, it is conceivable that Snf/Swi

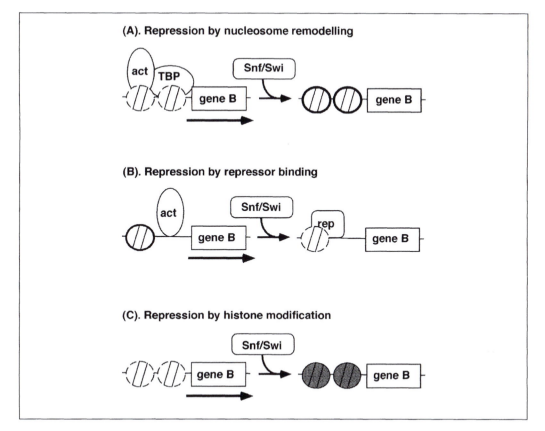

(A). Repression by nucleosome remodelling

(B). Repression by repressor binding

(C). Repression by histone modification

Fig. 2 Models for repression of transcription by the Snf/Swi complex. (A) Repression by nucleosome remodelling. In this model, Snf/Swi remodels nucleosomes from an active state to an inactive state that is not permissive for binding of transcriptional activators. (B) Repression by repressor binding. In this model, Snf/Swi remodels nucleosomes to an active state to facilitate binding by a transcriptional repressor. (C) Repression by histone modification. In this model, Snf/Swi catalyses a histone modification that prohibits binding by transcription factors. Several types of histone modifications have been described that affect transcription (100).

possesses a previously unknown activity that helps to confer repression, such as a particular type of histone modification. This possibility is supported by the discovery of a related class of nucleosome remodelling complexes that contain a Snf2 homologue and also contain a histone deactylase (for a review see 9). Histone deacetylases are believed to be important for transcriptional repression *in vivo* (see Chapter 8).

3.1.6 Summary of genetic studies of Snf/Swi in *S. cerevisiae*

Genetic studies in *S. cerevisiae* have played critical roles in identifying Snf/Swi and in elucidating nucleosome remodelling as an important aspect of transcriptional control in eukaryotes. The identification of *SNF* and *SWI* genes laid the groundwork for all of the *in vivo* analysis and for the biochemical purification of the complex. The isolation and analysis of mutations that suppress *snf/swi* mutations provided the initial connection between Snf/Swi and chromatin structure. Genetic analysis in both *Drosophila* and humans has likewise demonstrated that the Snf/Swi family is important *in vivo* throughout eukaryotes.

One current issue regarding Snf/Swi function is its interactions with other large transcription complexes. How are these complexes coordinated *in vivo* to be used at overlapping sets of promoters? One complex that has been shown to have both overlapping and distinct roles with respect to Snf/Swi is the SAGA complex, the topic of the next section.

3.2 The Spt-Ada-Gcn5 acetyltransferase (SAGA) complex

The SAGA complex of *S. cerevisiae* is a large (1.8 MDa) complex, comprised of approximately 22 different proteins, that plays important roles in transcription *in vivo*. Unlike Snf/Swi, where most of the components are believed to contribute to a common activity (nucleosome remodelling), there is strong evidence that SAGA possesses multiple activities that control transcription by distinct mechanisms. The SAGA activity that has been studied in greatest detail is the histone acetyltransferase (HAT) activity of the SAGA component Gcn5, described in detail in Chapter 7. In this section we will review the history of SAGA and the genetic studies that have led to our current understanding of its roles *in vivo*. Then, we will discuss interactions between SAGA and Snf/Swi.

3.2.1 The identification of the Spt, Ada, and Gcn5 components of SAGA

Although SAGA is now known to contain additional classes of transcription factors, it was initially identified by genetic and biochemical analysis of the three classes for which it was named (SAGA = *Spt-Ada-Gcn5 Acetyltransferase*). These three classes were identified by the isolation of transcription mutants by distinct types of screens and selections. Initially there was little clue that these factors are members of a complex.

3.2.2 Mutations in *SPT* genes cause suppression of promoter insertion mutations

SPT genes were identified by selecting for suppressors of *S. cerevisiae* transposon insertion mutations in promoter regions (for a review see 23). The particular promoter mutations used were insertion mutations in the promoters of the *HIS4* and *LYS2* genes, caused by the *S. cerevisiae* transposable elements Ty1 or Ty2, or by the long terminal repeat (LTR) of these elements. These insertion mutations had been shown to inhibit or otherwise alter transcription of the adjacent gene. Selection for suppressors of such insertion mutations identified mutations in several genes unlinked to the insertion mutation. The genes identified by these suppressor mutations were designated *SPT* (for *suppressor of Ty*).

Analysis of *spt* mutants suggested that the different Spt gene products might be involved in distinct aspects of transcription. Although *spt* mutants were isolated by the same general selection, as suppressors of insertion mutations, they were pleiotropic, with each *spt* mutation causing a set of mutant phenotypes. When the mutant phenotypes caused by each *spt* mutation were compared, it was evident that the mutants fell into distinct phenotypic classes. From this analysis there emerged two main groups that contained the majority of the known *SPT* genes.

The particular aspects of transcription controlled by these two groups of *SPT* genes was suggested by the identification of members of each group as encoding previously identified transcription factors. In one set of *SPT* genes, one gene, *SPT15*, was shown to encode the TATA-binding protein (TBP) (35, 36). This identification provided the initial demonstration that TBP controls transcription *in vivo* and that it is essential for growth. In the other set of *SPT* genes, two genes, *SPT11* and *SPT12*, were shown to encode histones H2A and H2B (1). These genes were already known to be one of the two sets of *S. cerevisiae* genes that encode these histones (37). Based on these analyses, these two sets of *SPT* genes became known as the TBP group and the histone group of *SPT* genes. Each group contains several other members whose functions are unknown.

3.2.3 The histone group of *SPT* genes

The histone group of *SPT* genes contains, in addition to histone genes, three genes named *SPT4*, *SPT5*, and *SPT6*. Since these *SPT* genes do not encode members of SAGA, they will be described only briefly here. These three genes, when mutant, confer phenotypes similar to that caused by a deletion of one of the two pairs of genes that encode histones H2A and H2B. These phenotypes include suppression of insertion mutations and, as described above, suppression of *snf/swi* mutations. The Spt6 protein has been shown to interact with histones, and primarily with histone H3 (38). Biochemical analysis has identified Spt4 and Spt5 as tightly associated in a complex that does not include Spt6, and additional studies have shown that these proteins are required for normal transcription elongation (39, 40). Other studies implicate them in recombination and chromosome segregation (41, 42). Therefore, Spt4, Spt5, and Spt6 may play important roles in several aspects of chromosome biology.

3.2.4 The TBP group of *SPT* genes

The TBP group of *SPT* genes includes *SPT15*, the gene that encodes TBP, as well as four other genes, named *SPT3*, *SPT7*, *SPT8*, and *SPT20* (for a review see 43). While *SPT15* is essential for growth, the other four genes are not. There are two prominent genetic connections among these genes. The first is by mutant phenotypes. Null mutations in *SPT3*, *SPT7*, *SPT8*, and *SPT20* are similar, although as discussed below, not identical (44, 45). These phenotypes include suppression of insertion mutations, mating defects, and sporulation defects. In addition, particular missense mutations in *SPT15* cause phenotypes similar to null mutations in these four *SPT* genes. The mutant phenotypes suggest that the Spt3, Spt7, Spt8, and Spt20 proteins are important for an aspect of TBP function. The second genetic connection is from allele-specific genetic interactions between *spt3* and *spt15* mutations (46). These results strongly suggest a physical interaction between Spt3 and TBP, confirmed by co-immuno-precipitation analysis (46, 47). More recent experiments have suggested a direct role for Spt3 acting both positively and negatively in recruiting TBP to particular pro-moters (48, 49). Thus, several approaches have provided strong evidence for inter-actions among the TBP group of Spt proteins. All of these proteins, except for TBP, were subsequently shown to be members of the SAGA complex, as described later.

3.2.5 The *ADA* and *GCN5* genes

As for *SPT* genes, the *ADA* and *GCN5* set of genes were also identified in various mutant screens for transcriptional defects. These screens were distinct from those done to identify *SPT* genes. *GCN5* was identified as a gene required for maximal activation by the activator Gcn4 (50). The *ADA* genes were identified in a screen for mutations that abolished the toxicity in *S. cerevisiae* caused by high levels of the artificial, powerful activator Gal4-VP16 (reviewed in 51). This screen identified five genes: *ADA1*, *ADA2*, *ADA3*, *ADA4/GCN5*, and *ADA5*. Subsequent analysis suggested that the Ada2, Ada3, and Gcn5 proteins interact and, furthermore, that *ada2*, *ada3*, and *gcn5* mutations cause similar mutant phenotypes. A negative role for *ADA/GCN5* genes was also suggested by the identification of an *ada3* mutation (called *ngg1*) as abolishing repression of a particular gene (52). From genetic characterization, the *ADA1* and *ADA5* genes appeared to be in a distinct class, as *ada1* and *ada5* cause a broader and more severe set of mutant phenotypes (53, 54). Overall, then, genetic and biochemical studies had suggested that the *ADA/GCN5* genes identified two groups of functions, Ada2/Ada3/Gcn5 and Ada1/Ada5.

3.2.6 A surprising connection between *ADA* and *SPT* genes leads to the identification of the SAGA complex

During the analysis of *ADA* and *SPT* genes, a surprising result led to the discovery of the SAGA complex: the sequence determination of *SPT20* and *ADA5* demonstrated that they are the same gene, thus uniting the TBP group of *SPT* genes with the *ADA/GCN5* group of genes (44, 55). Given the existing evidence for a physical interaction among some of the Ada proteins, in combination with the heightened interest in Gcn5 due to its histone acetyltransferase activity, several groups began biochemical purification of Gcn5 from *S. cerevisiae* cell extracts (Chapter 7). In one of these studies, antibodies against several of the Spt proteins were used to demonstrate that they were members of a Gcn5-containing complex which was named SAGA (56). More-over, three different mutations, *spt7*, *spt20*, and *ada1*, were shown to abolish SAGA, while three others, *ada2*, *spt3*, and *spt8*, were shown to alter its mobility (54, 56).

3.2.7 Distinct activities of SAGA

A combination of genetic and biochemical studies have provided strong evidence that SAGA possesses multiple activities. First, analysis of different SAGA mutants has demonstrated that they confer distinct phenotypes. In this way, three classes of mutants have been identified. Mutations in *SPT7*, *SPT20/ADA5*, or *ADA1* cause the broadest and most severe set of phenotypes (12, 53, 54). This result fits well with the observation that these three proteins are required for SAGA integrity (56). In contrast, mutations in two other groups cause distinct but less severe sets of phenotypes. These two groups consist of the *SPT3* and *SPT8* genes in one group and the *ADA2*, *ADA3*, and *GCN5* genes in the other. That these two groups affect a subset of SAGA activities is supported by the observation that *spt3 gcn5* double mutants cause mutant phenotypes very similar to those of *spt7*, *spt20*, or *ada1* mutants (54). Biochemical analysis fits well with the genetic results. When SAGA is analysed from *spt3* mutants, it still possesses Gcn5-dependent HAT activity, demonstrating that the Spt3 and Gcn5 activities are biochemically distinct (54).

3.2.8 Other SAGA proteins

In addition to the SAGA proteins already described, two other classes have been identified. First, there is a subset of TBP-associated factors (TAFs) in SAGA (57). These proteins were initially identified as components of the general transcription factor TFIID. Secondly, the Tra1 protein is a component of SAGA (58, 59). While the function of Tra1 is unknown, it is a homologue of the mammalian TRAAP protein, required for transformation by *myc* oncogene (60).

3.3 Genetic interactions between Snf/Swi and SAGA

Genetic and biochemical analysis has demonstrated the distinct natures of the two transcription complexes described in this chapter, Snf/Swi and SAGA. However, genetic studies have revealed that these two complexes interact *in vivo* to coordinate transcriptional control of at least a subset of genes. These studies have highlighted how eukaryotic promoters are controlled in a combinatorial fashion by distinct transcription complexes.

The first indication of a genetic relationship between Snf/Swi and SAGA came from double mutant analysis (11, 12). These studies showed that, while mutations in genes encoding SAGA components do not cause lethality, they do when in combination with other transcription mutants. Specifically, it was shown that double mutants impairing both SAGA and Snf/Swi or SAGA and Srb/Mediator are inviable. These results suggest that SAGA, Snf/Swi, and Srb/Mediator have partially overlapping functions *in vivo*.

The examination of specific genes provided additional evidence for this partial redundancy between SAGA and Snf/Swi. Analysis of the *SUC2* gene of *S. cerevisiae* showed that its transcription, while reduced in either a *snf/swi* mutant or in a *gcn5* mutant (remember that Gcn5 is the HAT in SAGA), is more severely impaired in a *snf/swi gcn5* double mutant, implying redundant control of *SUC2* by these co-activators (10, 13). Analysis of the *PHO8* gene also showed that both Snf/Swi and Gcn5 are required, and additionally demonstrated that each factor, Snf/Swi and Gcn5, is responsible for a distinct type of change in chromatin structure (31). Finally, a study of expression of the *S. cerevisiae HO* gene revealed that during activation of the gene both Snf/Swi and SAGA are required in a dependent fashion—Snf/Swi is required first, followed by SAGA (61). While it is very likely that the exact nature of the relationship between these complexes will differ at different promoters, it seems clear that *S. cerevisiae* and other eukaryotes have carefully coordinated the use of multiple transcription complexes in overcoming repression by nucleosomes.

3.4 Summary and conclusions

A large body of work in *S. cerevisiae* has shown that multiprotein complexes are required for the transcription of significant subsets of genes in *S. cerevisiae*. The requirement for these complexes was, for the most part, initially elucidated by genetic analysis. Subsequent biochemical analysis has played an essential role in uncovering the mechanisms by which these complexes function. The functional interactions between complexes reveals the intricate network of regulatory processes inside a growing cell. Future studies should help to elucidate these networks in fascinating detail.

4. Genetic analysis of histone function

The ultimate targets of the transcription complexes just described are the nucleosomes and their complement of histone proteins. Mutational studies of the histones have revealed a rich set of functional activities and strengthened the links between nucleosomes and chromatin-modifying complexes. These studies have been based on both site-directed and random mutagenesis, and both strategies have been successful. In the following sections, we begin by examining mutations in the histone tails, then move to the structured domains, and finally consider mutations affecting histone–histone interactions.

4.1 Mutations in the core histone tails

Each core histone is composed of a structured domain, consisting of the histone-fold motif and extra-fold helices, and an unstructured N-terminal domain of 25–40 residues that extends through the DNA gyres and into the space surrounding the nucleosome (see Chapter 1). These terminal domains, or 'tails', of the histone proteins are of considerable genetic interest for a number of reasons. First, since they are largely unordered in both solution and crystal structures, genetic approaches are particularly important for understanding their functions. Secondly, the tails contain sites for a variety of post-translational modifications, including acetylation, phosphorylation, and methylation, which are implicated in chromatin-mediated regulation of transcription. Mutational studies provide a major approach for understanding these pathways *in vivo*. Finally, there is mounting evidence that the histone terminal domains may function through complex interactions of multiple tail modifications. Thus, unravelling the combinatorial codes governing these interactions poses a fascinating genetic challenge. Site-directed mutagenesis of the N-terminal histone tails has provided a wealth of new information on the functional roles of these domains in transcription.

4.1.1 Silencing and gene repression

The best-characterized functions of the histone tails are in transcriptional repression, gene silencing, and heterochromatin (see Chapters 3, 8, and 10). The first clue to the importance of the tails in gene repression came from the remarkable discovery that a mutant H4 allele, encoding an N-terminal truncation of the protein, resulted in the loss of permanent gene repression at the silent mating type loci of *S. cerevisiae* (62). Subsequent site-directed point mutations succeeded in defining the specificity of this repression and in mapping a specific repression domain within the H4 N-terminal tail extending from residues 16–29 (63–66). This repression domain was found to be essential both for silencing and telomere position effect repression (67). Subsequent chromatin immunoprecipitation experiments in *S. cerevisiae* and indirect immuno-fluorescence microscopy in mammalian cells found a correlation between the acetylation state of lysines in the H4 tails and silencing and heterochromatin (68, 69). The N-terminal domain of histone H3 also participates in silencing and, although less well characterized, a specific repression domain has been roughly mapped by deletion analysis to residues 1–20 (70). The activity of the H3 tail is partially redundant with the H4 tail and it appears to be less important than H4 in silencing, but more important in Tup1–Ssn6 repression (70–72).

How do the H3 and H4 tails contribute to gene repression and silencing? The results of a variety of genetic and biochemical studies suggest that repression is mediated through direct protein–protein interactions between silencing and repressing complexes and the tails themselves. A major clue to this functional pathway came from the identification of the *SIR3* gene in a genetic screen for mutations that suppress the silencing defect in an H4 tail mutant (65). The Sir3 protein was known to be required for silencing and this discovery led to the hypothesis that the Sir complex might interact directly with the histone tails. Subsequent biochemical studies have confirmed that both silencing and repression involve direct protein–protein interactions between the histone tails and specific protein complexes; Tup1–Ssn6 for repression, and the Sir3–Sir4 complex for silencing. Importantly, the specificity of these binding interactions was confirmed using the well-characterized H3 and H4 tail mutants available from the earlier studies (72, 73).

4.1.2 Transcription activation

In addition to their roles in repression and silencing, mutational studies have shown that the core histone tail domains also have important roles in regulating gene transcription. These interactions appear to be more complex than those mediating silencing. At present they are only just beginning to be understood at the level of molecular pathways. Early studies with deletion derivatives of histones H3 and H4 highlighted this complexity. Mutants with a deletion of the H4 N-terminal domain were found to be severely impaired for transcriptional activation of both *GAL1* and *PHO5* (74). This activation function was mapped to the first 23 residues of the tail and is distinct from the repression domain required for silencing. In contrast, mutants deleted for the H3 tail have exactly the opposite effect on induction of the *GAL1* gene, which was inhibited approximately 20-fold, but the deletion had little effect on *PHO5* induction

(70). Point mutations directed to alter the lysines in both the H3 and H4 tails confirmed the results of these deletion experiments. However, these mutants were not easily interpreted in terms of lysine modifications and the molecular pathways for H3 and H4 tails in transcription remained speculative (70, 74).

The reversible acetylation of N-terminal lysines was first correlated with transcriptional activation in pioneering experiments by Allfrey (75). The landmark discovery that *GCN5* encodes a transcriptional adapter protein with histone acetyltransferase (HAT) activity immediately suggested a simple model in which HAT enzymes would acetylate lysines in the N-terminal domains and thus activate transcription (76). Indeed, extensive biochemical experiments, summarized elsewhere in this volume, leave no doubt that HAT and histone deacetylase (HDAC) complexes target the histone lysines and have central roles in transcriptional regulation (see Chapters 7 and 8). However, the results of genetic tests of this hypothesis have been anything but simple. If the function of a transcriptional HAT is to acetylate specific lysines in the N-terminal tails, then a mutation that changes those lysines to arginines should produce the same phenotype as loss of function mutations in the HAT enzyme itself, since arginine is not acetylated. In fact, it has been remarkably difficult to demonstrate this simple relationship. The most extensive experiments to date have been carried out by Zhang *et al.* (77). Point mutations that change lysine residues within the H3 and H4 tails did not generally mimic the phenotypes produced by *gcn5* mutations. For example, the major substrate site for recombinant Gcn5 *in vitro* is lysine 14 in the histone H3 tail (78). Thus, the H3 K14R mutations might be expected to mimic the loss of function of Gcn5. However, using a Gal4-VP16 reporter that is dependent on Gcn5, it was found that transcription was decreased roughly ninefold by a *gcn5* deletion, but was barely changed in the H4 K14R mutant. In fact, a key discovery in these experiments was that combining the *gcn5* deletion with point mutations in the tail lysines revealed complex interactions, often giving a synthetic phenotype, that is, one strikingly different from either single mutation on it own. One interpretation of these results is that Gcn5-containing HAT complexes must target other substrates in addition to the histone tails, a prediction consistent with biochemical results (79). In addition, there are multiple HAT complexes in the cell and the extent to which these interact *in vivo* in acetylating multiple lysines in the histone tails, redundantly or antagonistically, is currently an open question (80, 81). It is also likely that an additional difficulty stems from the combinatorial complexity of the tails themselves. In fact, one specific combination of H3 and H4 point mutations behaved exactly as predicted by the model. Mutations that mimic acetylated lysine, such as glutamine, might be expected to bypass the requirement for Gcn5 acetylation. Neither the single K14Q mutation in histone H3 alone, nor the double K8Q K16Q mutations in histone H4 alone, eliminated the requirement of Gcn5 for transcription by Gal4-VP16. However, when these two mutant H3 and H4 alleles were combined in the same cell they completely bypassed the requirement for Gcn5. Other combinations of H3 and H4 point mutations failed to effect this suppression. While it is not certain that these point mutations are actually modelling lysine acetylation, these results suggest that decoding the rules of lysine signalling in the nucleosome is likely an achievable goal.

Given the importance of histone acetylation in transcription, the molecular mechanisms through which this modification actually modulates transcription is currently not well understood. Until recently, lysine acetylation was viewed as a mechanism for altering the charge density of the tail domains. Since acetylation neutralizes the positive charge of lysine residues, this might weaken the interactions between the histone tails and the negatively charged DNA backbone. This was imagined to create a 'looser' or more 'open' chromatin configuration for the entry of transcription factors. By this rationale, mutations that replaced lysines with neutral polar residues, such as glutamine, were thought to mimic the acetylated state, while those that replaced lysine with arginine were thought to mimic the unacetylated state. In some cases this interpretation fits well with the data. For example, K16R and K16Q substitutions in histone H4 produce quite different silencing phenotypes. However, in many instances the phenotypes produced by arginine and glutamine substitutions are identical and can not be easily explained by a charge model (70, 77). The recent discovery that a specific protein motif, the bromo domain, can bind to acetyl-lysine (82) suggests a second mechanism for histone-tail function in which lysine acetylation sites may serve to signal the recruitment of transcription factors and their binding partners through the bromo domain. In this model, mutations that replace lysine with either arginine or glutamine might produce very similar phenotypes since both would block the binding of bromo-domain signalling molecules.

A third mechanism for histone-tail function is suggested by the surprising results of a multicopy suppressor screen. Starting with a *gcn5* deletion strain, Pollard *et al.* (83) selected for genes that, at high copy, could restore transcription of an *HO–LacZ* reporter. One rationale for such a screen would be to identify genes encoding other HAT enzymes able to substitute for Gcn5 when overexpressed. The screen was successful but it did not identify a HAT. Instead, it identified *ARG3*, the gene encoding ornithine transcarbamoylase, a mitochondrial enzyme that converts ornithine to citrulline on the arginine biosynthetic pathway. How could a mitochondrial metabolic enzyme bypass the requirement for histone modification? Ornithine is also an intermediate in the biosynthesis of polyamines, which previously had been implicated in chromatin higher-order structure. Thus, Pollard *et al.* proposed that the overexpression of Arg3 depletes the cellular pool of ornithine, decreasing the synthesis of polyamines, which in turn suppresses *gcn5*. Two genetic predictions of this model have been satisfied: first, the addition of ornithine to the media eliminates suppression by high-copy *ARG3*; and secondly, other mutations that decrease polyamine biosynthesis also suppress the *gcn5* deletion. Thus, polyamines act as negative regulators of *HO–LacZ* transcription and one function of Gcn5 is to antagonize this repression, presumably through acetylation of the histone tails.

4.2 Mutations in the structured domains

Mutational studies have also revealed the importance of the structured domain of the histones for normal transcription. In this case, unbiased screens were particularly important for linking nucleosome structure to transcription. With hindsight, we now know that these experiments were probing genetic interactions between nucleosomes and multiprotein complexes directly involved in modulating chromatin structure.

4.2.1 The Sin class of mutations

Deletion of the *SUC2* UAS renders the gene uninducible and thus *suc2ΔUAS* mutants are unable to grow on medium containing sucrose as the sole carbon source. This system provides a strong selection for mutations that bypass the requirement for up-stream activation factors. These genes were named *BUR* (for *bypasses UAS* requirement) (83a). One of these genes, *BUR5*, was found to be allelic to the histone H3 gene *HHT2*. The *bur5–1* mutation encoded a T119I substitution within the organized core domain of H3.

The screen for suppressors of *swi* mutations, described earlier in this chapter, found a similar class of mutants. These genes were called *SIN* (for *switch independent*). When one of these, *SIN2*, was cloned by complementation, it was also found to be allelic with *HHT2* (25). A strong prediction of these results was that mutations in other histones, particularly H4, would also confer a Sin phenotype. Indeed, direct mutagenesis of the H3 and H4 genes identified multiple additional alleles of both histone genes capable of bypassing the requirement for Swi1 at the HO promoter. One of these carried the same T119I change that was found in *bur5–1*.

How do histone Sin mutations bypass the normal requirements for the upstream transcriptional machinery? The answer to this question is important because it promises to shed light on the mechanisms of normal chromatin remodelling. Many histone gene mutations have now been found to generate alleles in the Sin class. However, understanding how they suppress *snf* and *swi* mutations is complicated by the diversity of their structural roles in the histone octamer. A striking feature of the original Sin H3 and H4 mutations is that they cluster together on the surface of the nucleosome, and three of the five mutations are predicted to alter histone–DNA contacts (25). On the other hand, the Spt and Sin H2A and H2B deletion mutants described earlier, and specific Sin mutations in both H4 and H2B, compromise the integrity of the interactions between the histone H3–H4 tetramer and the H2A–H2B dimers (24, 84, 85). Furthermore, one of the Sin H2B alleles is predicted to affect the association of H2A and H2B in the dimer subunits (85). Finally, an allele of H4 originally isolated in a screen for mitotic segregation mutants is also Sin and encodes two amino-acid substitutions within the fold domain but outside the original cluster of Sin mutations (86).

One possibility is that all of the Sin histone mutations share a common mechanism of suppression through the disruption of histone–DNA interactions. In all four core histones, the deletion of a short peptide segment at the junctions of the N-terminal domain with the central ordered domain results in derepression of *GAL1* and *PHO5* under repressing and non-inducing conditions. These segments are at the histone–DNA interface of the nucleosome (87). Furthermore, the original *bur5–1* Sin mutation was found to derepress *GAL1* and *PHO5* transcription in this same system, suggesting that these two different assays may be monitoring the same transcription defect. This model of a single Sin mechanism requires that the dimer–tetramer interface mutants act by indirectly disrupting histone–DNA interactions through a perturbation of octamer structure. Alternatively, there may be two or more Sin bypass pathways for the histones, at least one involving histone–DNA contacts and another involving dimer–tetramer interactions.

4.2.2 The Snf class of mutations

Not all mutations in the ordered histone domains produce a Sin phenotype. An intriguing class of H2A mutations, alluded to earlier, actually have the opposite effect and produce a Snf-like phenotype at a subset of Snf/Swi-dependent genes (28). These mutations were identified by mutagenesis of *HTA1*, which encodes histone H2A, to produce a library of randomly mutated H2A alleles. Cells carrying a member of this library as the only source of histone H2A were then screened for mutants unable to grow on raffinose, reflecting an inability to transcribe *SUC2*. Sixteen *hta1* alleles, and a total of 14 different amino-acid substitutions, were recovered in this screen. Nine of these 14 mutations occur towards the N- and C-termini of the protein, between residues 20–30 and 114–121. Some of these are semi-dominant, suggesting that they are gain of function alleles and actively interfere with transcriptional activation. Although they produce a Snf phenotype, these H2A mutations are functionally distinct from mutations in the *SNF* and *SWI* genes. First, they do not affect all the genes dependent on Snf/Swi function. Secondly, extragenic suppressors in the Sin class, such as *spt6* mutations, fail to bypass the *hta1* transcriptional defect. Finally, the chromatin structure of the *SUC2* promoter is in the active conformation in the *hta1* mutants, while it remains in the inactive conformation in *snf* and *swi* mutants. These novel properties suggest that H2A plays a role in transcriptional activation downstream or independent of Snf/Swi nucleosome remodelling at *SUC2*. An attractive hypothesis is that these mutations block a second nucleosome reorganization step.

4.3 Mutations in the dimer–tetramer interfaces

At a higher level of organization, the histone octamer is composed of three thermodynamic units, the central H3–H4 tetramer flanked by two H2A–H2B dimers (see Chapter 1). Several observations suggested that the interactions between these subunits within the nucleosome might play important roles in transcription. First, as we have already seen, alterations in the relative stoichiometry of the histone gene pairs cause Spt and Sin transcription phenotypes (1, 24). Furthermore, some of the founding H3 and H4 Sin mutations are located at the dimer–tetramer interface (25). Finally, biochemical experiments suggested that H2A–H2B dimers might be in dynamic exchange in chromatin, at least a part of which might depend on transcription (88). Thus, it became of interest to test the function of the dimer–tetramer interactions by site-directed mutations.

Each histone H4 in the central tetramer interacts with both of the H2A–H2B dimer sets through two protein–protein interfaces (see Chapter 1). Three tyrosine residues in H4, at positions 72, 88, and 98, were predicted to participate in these binding surfaces, based on both the crystal structure and chemical modification studies of the octamer. Two of these tyrosines, Y72 and Y88, contact one H2A–H2B dimer, while the other, Y98, contacts the other H2A–H2B dimer. By testing a number of site-directed mutations, conditional temperature-sensitive (Ts) alleles were ultimately obtained at each position, yielding the three single substitution mutations Y72G, Y88G, and Y98H (84). Several lines of genetic evidence argue that these mutations actually perturb the dimer–tetramer interfaces *in vivo*. Most importantly, all three mutants display Spt and Sin phenotypes, as expected for alterations in dimer–tetramer interactions. Secondly, overexpression of the wild-type H2A and H2B genes is specifically lethal in combination with the Y72G and Y98H mutations, demonstrating a genetic defect in the interactions between the dimer–tetramer sets.

What are the consequences of these mutations on transcription? Remarkably, mutants expressing the Y72G and Y88G mutations arrested growth uniformly at the G_1/S restriction point in the *S. cerevisiae* cell cycle. This cell division cycle arrest was shown to result from a failure to transcribe the G_1 cyclin genes *CLN1*, *CLN2*, and *CLN3*, and their associated transcription factor genes *SWI4* and *SWI6*. Overexpression of wild-type *CLN2* from the *GAL1* promoter, or expression of a stable dominant active allele of *CLN2*, was sufficient to drive the *hhf1* mutants out of G_1 arrest, showing that the limitation of G_1 cyclin protein is sufficient to explain the arrest phenotype. However, expression of *CLN2* was not sufficient to restore viability, and these cells subsequently died throughout the cell cycle. The precise molecular defect blocking the expression of the G_1 cyclin genes in the Y72 mutant is currently unknown. However, the phenotype is strikingly similar to *spt16* (*cdc68*) mutants, which are also Ts, have Sin and Spt phenotypes, and fail to transcribe the genes for the G_1 cyclin machinery at restrictive temperature (89–91). Recently, mammalian Spt16 was discovered as part of a protein complex called FACT. This complex is capable of facilitating transcriptional elongation on chromatin templates *in vitro* (92). Recently, the *S. cerevisiae* homologues of this mammalian complex, the Spt16 and Pob3 proteins, have been discovered to be associate with the NuA3 HAT complex, suggesting a direct connection between components of FACT and histone modification *in vivo* (92a). Thus, an attractive model for the Y72G and Y88G H4 interface mutants is that FACT or NuA3 functions to facilitate elongation of the transcription machinery through nucleosomes, and that this activity depends on the wild-type dynamics of dimer–tetramer histone interactions.

What about the interface defined by the Y98H mutation? This mutant was also Spt and Sin at the *INO1* gene, but at restrictive temperature cells did not arrest at any particular place in the cell division cycle. While these results were consistent with a general defect in gene expression, they did not provide an obvious candidate for the transcriptional pathway. The answer to this problem came from a genetic screen designed to find genes that would suppress the Ts phenotype of the Y98H mutants when present at increased gene dosage. The rationale for this screen was that at elevated temperature the Y98H mutation might weaken the interaction of some component of the transcription machinery with chromatin. In such a case, overexpression of that component, because of its higher gene dosage, might suppress the Ts phenotype by driving interactions through mass action. The surprising result of that screen was the identification of the gene encoding H2A.Z, a conserved variant of the H2A protein family introduced at the beginning of this chapter (Fig. 3).

4.4 Histone-variant nucleosomes and transcription

In vertebrates, H2A.Z accounts for roughly 5–10% of the total H2A in chromatin and is known to be associated with H2B in the nucleosome (93, 94). Thus, H2A.Z must be present in a fraction, but not an inconsequential fraction, of total nucleosomes. The first clues to the importance of the H2A.Z variant came from the discoveries that the gene was essential in *Drosophila* (95) and *Tetrahymena* (96). More recent experiments demonstrate that it is also essential in mouse (97). Thus, nucleosomes containing H2A.Z perform one or more essential functions, but the identity of those functions remained unclear.

The *S. cerevisiae* H2A.Z protein is encoded by the *HTZ1* (*HTA3*) gene (98, 99). *HTZ1* is not essential under normal growth conditions; however, the null mutant is severely impaired for growth at either elevated or reduced temperatures. Thus, the Ts defect of the H4 Y98H mutant can be explained by a disruption of its interaction with dimers containing H2A.Z, a phenocopy of the H2A.Z null mutant. Interestingly, while overexpression of H2A.Z suppressed the Ts phenotype of the H4 Y98H mutant, it did not suppress the Spt or Sin phenotypes. In fact, overexpression of H2A.Z actually enhanced some of these defects. One interpretation of these genetic interactions is that the Spt and Sin phenotypes derive from defects in tetramer binding to the major H2A–H2B dimers, while the Ts phenotype derives from defects in binding to the variant H2A.Z–H2B dimers (Fig. 3). These results suggested that H2A.Z might define a novel transcriptional pathway.

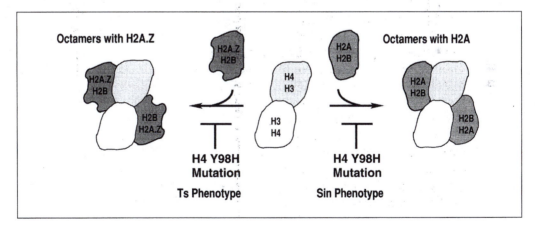

Fig. 3 Genetic interactions for major and variant histone octamers. In this model an individual H3–H4 tetramer is proposed to associate with either H2A–H2B dimers or H2A.Z–H2B dimers. The Y98H mutation in histone H4 partially blocks the association with both major and variant dimer sets. However, in the case of the major H2A–H2B dimers this causes a constitutive Sin phenotype, while in the case of the variant H2A.Z–H2B dimers it produces a Ts phenotype. The Ts phenotype, but not the Sin phenotype, is suppressed by overexpression of histone H2A.Z.

The *htz1Δ* single mutant did not exhibit a Snf phenotype at either *SUC2* or *INO1* in a wild-type background. Thus, the next logical step was to test the *htz1Δ* mutant for a Sin phenotype; that is, could elimination of H2A.Z nucleosomes bypass the requirement for Snf/Swi in transcription? The results were surprising. Unlike Sin mutations in the major histone genes, which partially bypass the requirement for Snf/Swi function, the *htz1Δ* deletion actually increases the requirement for Snf/Swi. Genetic crosses designed to combine the *htz1Δ* allele with a *snf2Δ* allele demonstrated that the *htz1Δ snf2Δ* double mutant is extremely sick, or even inviable in some strain backgrounds. Thus, in the absence of nucleosome remodelling by the Snf/Swi complex the elimination of nucleosomes containing H2A.Z is lethal, and vice versa. How about other chromatin modification complexes? A number of genes encoding components of the Gcn5-containing family of HAT complexes were tested, and these also exhibit synthetic defects in combination with *htz1Δ*. The *htz1* deletion is lethal in a strain carrying a deletion of the *SPT20 (ADA5)* gene, encoding a subunit of the SAGA complex. The *htz1Δ gcn5Δ* double mutant is viable but is extremely slow growing, as are *htz1Δ ada2Δ* and *htz1Δ ada3Δ* double mutants.

These results provide compelling evidence that nucleosomes containing the H2A.Z variant play an important role in transcription. The synthetic lethality between *htz1Δ* and components of nucleosome remodelling and histone modification complexes suggests that the function of H2A.Z is partially redundant with these complexes. Thus, H2A.Z may define in a third interacting pathway of transcription regulation by modulating chromatin structure through histone variants.

5. Final comments

Genetic approaches in *S. Cerevisiae* have made fundamental contributions to our understanding of the roles of histones and large, multiprotein complexes in transcriptional control. The types of studies described in this chapter point towards the directions that genetic studies are likely to go in the near future. With the increased number of genome sequences available, along with advances in bioinformatics, additional genes that play important roles in transcription will be identified based on their DNA sequence. Moreover, as whole-genome expression studies become a commonly available method, the cataloguing of transcription effects and our understanding of the relationships among different transcription factors will increase hugely. However, these modern approaches will need to be combined with classical genetic approaches, including mutant isolation and analysis, in order to learn about the control of transcription *in vivo*. With a combination of these methods, then, we can be hopeful of understanding the complex network of factors that have evolved to control transcription.

6. Discussion

In e-mails with other authors it was pointed out that yeast have been the organism of choice for genetic studies of histones for two main reasons. First, many types of genetic manipulations in yeast are much easier to perform than in larger eukaryotes. Secondly, yeast have only two copies of each core histone gene, rather than the large number of copies found in other eukaryotes. Therefore, the phenotypes caused by histone mutants can be detected after removing just a second copy.

One issue of fundamental interest is the requirement for so many members in each complex. In SAGA it is likely due to multiple functions. In Snf/Swi it remains to be determined. A second issue of interest is how cells have evolved and have coordinated multiple activities to modify and control nucleosome structure and function, including remodelling, histone modification, and use of histone variants.

Acknowledgements

We are grateful to Sharon Roth and Wolfram Hörz for their comments on this chapter. Work from our labs is supported by the NIH.

References

1. Clark-Adams, C. D., Norris, D., Osley, M. A., Fassler, J. S., and Winston, F. (1988) Changes in histone gene dosage alter transcription in yeast. *Genes Develop., 2*, 150.
2. Smith, M. M. and Santisteban, M. S. (1998) Genetic dissection of histone function. *Methods, 15*, 269.
3. Norris, D. and Osley, M. A. (1987) The two gene pairs encoding H2A and H2B play different roles in the *Saccharomyces cerevisiae* life cycle. *Mol. Cell. Biol., 7*, 3473.
4. Smith, M. M. and Stirling, V. B. (1988) Histone H3 and H4 gene deletions in *Saccharomyces cerevisiae. J. Cell Biol., 106*, 557.
5. Henikoff, S., Ahmad, K., Platero, J. S., and Steensel, B. V. (2000) Heterochromatin deposition of centromeric histone H3-like protein. *Proc. Natl. Acad. Sci., USA, 97*, 716.
6. van Daal, A., White, E. M., Elgin, S. C., and Gorovsky, M. A. (1990) Conservation of intron position indicates separation of major and variant H2As is an early event in the evolution of eukaryotes. *J. Mol. Evol., 30*, 449.
7. Hampsey, M. (1998) Molecular genetics of the RNA polymerase II general transcriptional machinery [in process citation]. *Microbiol. Mol. Biol. Rev., 62*, 465.
8. Workman, J. L. and Kingston, R. E. (1998) Alteration of nucleosome structure as a mechanism of transcriptional regulation. *Annu. Rev. Biochem., 67*, 545.
9. Kingston, R. E. and Narlikar, G. J. (1999) ATP-dependent remodeling and acetylation as regulators of chromatin fluidity. *Genes Develop., 13*, 2339.
10. Biggar, S. R. and Crabtree, G. R. (1999) Continuous and widespread roles for the Swi–Snf complex in transcription. *EMBO J., 18*, 2254.
11. Pollard, K. J. and Peterson, C. L. (1997) Role for ADA/GCN5 products in antagonizing chromatin-mediated transcriptional repression. *Mol. Cell. Biol., 17*, 6212.
12. Roberts, S. M. and Winston, F. (1997) Essential functional interactions of SAGA, a *Saccharomyces cerevisiae* complex of Spt, Ada, and Gcn5 proteins, with the Snf/Swi and Srb/mediator complexes. *Genetics, 147*, 451.
13. Sudarsanam, P., Cao, Y., Wu, L., Laurent, B. C., and Winston, F. (1999) The nucleosome remodeling complex, Snf/Swi, is required for the maintenance of transcription *in vivo* and is partially redundant with the histone acetyltransferase, Gcn5. *EMBO J., 18*, 3101.

14. Winston, F. and Carlson, M. (1992) Yeast SNF/SWI transcriptional activators and the SPT/SIN chromatin connection. *Trends Genet.*, **8**, 387.

15. Johnston, M. and Carlson, M. (1992) In *The molecular and cellular biology of the yeast* Saccharomyces. Jones, E. W., Pringle, J. R., and Broach, J. R. (ed.). Cold Spring Harbor Laboratory Press, Cold Spring Harbor, New York, Vol. 2, p. 193.

16. Peterson, C. L., Dingwall, A., and Scott, M. P. (1994) Five *SWI/SNF* gene products are components of a large multisubunit complex required for transcriptional enhancement *Proc. Natl Acad. Sci., USA*, **91**, 2905.

17. Cairns, B. R., Kim, Y. J., Sayre, M. H., Laurent, B. C., and Kornberg, R. D. (1994) A multi-subunit complex containing the SWI1/ADR6, SWI2/SNF2, SWI3, SNF5, and SNF6 gene products isolated from yeast. *Proc. Natl Acad. Sci., USA*, **91**, 1950.

18. Treich, I., Cairns, B. R., de los Santos, T., Brewster, E., and Carlson, M. (1995) SNF11, a new component of the yeast SNF–SWI complex that interacts with a conserved region of SNF2. *Mol. Cell. Biol.*, **15**, 4240.

19. Neigeborn, L., Rubin, K., and Carlson, M. (1986) Suppressors of snf2 mutations restore in-vertase derepression and cause temperature-sensitive lethality in yeast. *Genetics*, **112**, 741.

20. Neigeborn, L., Celenza, J. L., and Carlson, M. (1987) *SSN20* is an essential gene with mutant alleles that suppress defects in *SUC2* transcription in *Saccharomyces cerevisiae*. *Mol. Cell. Biol.*, **7**, 672.

21. Clark-Adams, C. D. and Winston, F. (1987) The *SPT6* gene is essential for growth and is required for delta- mediated transcription in *Saccharomyces cerevisiae*. *Mol. Cell. Biol.*, **7**, 679. [Published erratum appears in *Mol. Cell. Biol.*, **7**, (5), 2035.]

22. Swanson, M. S., Malone, E. A., and Winston, F. (1991) *SPT5*, an essential gene important for normal transcription in *Saccharomyces cerevisiae*, encodes an acidic nuclear protein with a carboxy-terminal repeat. *Mol. Cell. Biol.*, **11**, 3009. [Published erratum appears in *Mol. Cell. Biol.*, **11**, (8), 4286.]

23. Winston, F. (1992) In *Transcriptional regulation*. McKnight, S. L. and Yamamoto, K. R. (ed.). Cold Spring Harbor Laboratory Press, Cold Spring Harbor, New York, p. 1271.

24. Hirschhorn, J. N., Brown, S. A., Clark, C. D., and Winston, F. (1992) Evidence that SNF2/SWI2 and SNF5 activate transcription in yeast by altering chromatin structure. *Genes Develop.*, **6**, 2288.

25. Kruger, W., Peterson, C. L., Sil, A., Coburn, C., Arents, G., Moudrianakis, E. N., and Herskowitz, I. (1995) Amino acid substitutions in the structured domains of histones H3 and H4 partially relieve the requirement of the yeast SWI/SNF complex for transcription *Genes Develop.*, **9**, 2770.

26. Gavin, I. M. and Simpson, R. T. (1997) Interplay of yeast global transcriptional regulators Ssn6p–Tup1p and Swi–Snf and their effect on chromatin structure. *EMBO J.*, **16**, 6263.

27. Wu, L. and Winston, F. (1997) Evidence that Snf–Swi controls chromatin structure over both the TATA and UAS regions of the SUC2 promoter in *Saccharomyces cerevisiae*. *Nucleic Acids Res.*, **25**, 4230.

28. Hirschhorn, J. N., Bortvin, A. L., Ricupero-Hovasse, S. L., and Winston, F. (1995) A new class of histone H2A mutations in *Saccharomyces cerevisiae* causes specific transcriptional defects *in vivo*. *Mol. Cell. Biol.*, **15**, 1999.

29. Sudarsanam, P., Iyer, V. R., Brown, P. O., and Winston, F. (2000) Whole-genome ex-pression analysis of *snf/swi* mutants of *S. cerevisiae*. *Proc. Natl Acad. Sci., USA*, **97**, 3364.

30. Holstege, F. C., Jennings, E. G., Wyrick, J. J., Lee, T. I., Hengartner, C. J., Green, M. R., Golub, T. R., Lander, E. S., and Young, R. A. (1998) Dissecting the regulatory circuitry of a eukaryotic genome. *Cell*, **95**, 717.

31. Gregory, P. D., Schmid, A., Zavari, M., Munsterkotter, M., and Horz, W. (1999) Chromatin remodelling at the PHO8 promoter requires SWI–SNF and SAGA at a step subsequent to activator binding. *EMBO J.*, **18**, 6407.

32. Vignali, M., Hassan, A. H., Neely, K. E., and Workman, J. L. (2000) ATP-dependent chromatin-remodeling complexes *Mol. Cell. Biol.*, **20**, 1899.

33. Lorch, Y., Cairns, B. R., Zhang, M., and Kornberg, R. D. (1998) Activated RSC–nucleosome complex and persistently altered form of the nucleosome. *Cell*, **94**, 29.

34. Schnitzler, G., Sif, S., and Kingston, R. E. (1998) Human SWI/SNF interconverts a nucleosome between its base state and a stable remodeled state. *Cell*, **94**, 17.

35. Hahn, S., Buratowski, S., Sharp, P. A., and Guarente, L. (1989) Isolation of the gene encoding the yeast TATA binding protein TFIID: a gene identical to the *SPT15* suppressor of Ty element insertions. *Cell*, **58**, 1173.

36. Eisenmann, D. M., Dollard, C., and Winston, F. (1989) *SPT15*, the gene encoding the yeast TATA binding factor TFIID, is required for normal transcription initiation *in vivo*. *Cell*, **58**, 1183.

37. Hereford, L., Fahrner, K., Woolford, J. Jr, and Rosbash, M. (1979) Isolation of yeast histone genes H2A and H2B. *Cell*, **18**, 1261.

38. Bortvin, A. L. and Wintston, F. (1996) Evidence that Spt6p controls chromatin structure by a direct interaction with histones. *Science*, **272**, 1473.

39. Hartzog, G. A., Wada, T., Handa, H., and Winston, F. (1998) Evidence that Spt4, Spt5, and Spt6 control transcription elongation by RNA polymerase II in *Saccharomyces cerevisiae*. *Genes Develop.*, **12**, 357.

40. Wada, T., Takagi, T., Yamaguchi, Y., Ferdous, A., Imai, T., Hirose, S., Sugimoto, S., Yano, K., Hartzog, G. A., Winston, F., *et al.* (1998) DSIF, a novel transcription elongation factor that regulates RNA polymerase II processivity, is composed of human Spt4 and Spt5 homologs. *Genes Develop.*, **12**, 343.

41. Malagon, F. and Andres, A. (1996) Differential intrachromosomal hyper-recombination phenotype of *spt4* and *spt6* mutants of *S. cerevisiae*. *Curr. Genet.*, **30**, 101.

42. Basrai, M. A., Kingsbury, J., Koshland, D., Spencer, F., and Hieter, P. (1996) Faithful chromosome transmission requires Spt4p, a putative regulator of chromatin structure in *Saccharomyces cereivsiae*. *Mol. Cell. Biol.*, **16**, 2838.

43. Winston, F. and Sudarsanam, P. (1998) The SAGA of Spt proteins and transcriptional analysis in yeast: past, present, and future. *Cold Spring Harbor Symp. Quant. Biol.*, **63**, 553.

44. Roberts, S. M. and Winston, F. (1996) SPT20/ADA5 encodes a novel protein functionally related to the TATA-binding protein and important for transcription in *Saccharomyces cerevisiae*. *Mol. Cell. Biol.*, **16**, 3206.

45. Gansheroff, L. J. (1995) The *Saccharomyces cerevisiae SPT7* gene encodes a very acidic protein important for transcription *in vivo*. *Genetics*, **139**, 523.

46. Eisenmann, D. M., Arndt, K. M., Ricupero, S. L., Rooney, J. W., and Winston, F. (1992) SPT3 interacts with TFIID to allow normal transcription in *Saccharomyces cerevisiae*. *Genes Develop.*, **6**, 1319.

47. Lee, T. I. and Young, R. A. (1998) Regulation of gene expression by TBP-associated proteins. *Genes Develop.*, **12**, 1398.

48. Dudley, A. M., Rougeulle, C., and Winston, F. (1999) The Spt components of SAGA facilitate TBP binding to a promoter at a post-activator-binding step *in vivo*. *Genes Develop.*, **13**, 2940.

49. Belotserkovskaya, R., Sterner, D. E., Deng, M., Sayre, M. H., Lieberman, P. M., and Berger, S. L. (2000) Inhibition of TATA-binding protein function by SAGA subunits Spt3 and Spt8 at Gcn4-activated promoters *Mol. Cell. Biol.*, **20**, 634.

50. Georgakopoulos, T. and Thireos, G. (1992) Two distinct yeast transcriptional activators require the function of the GCN5 protein to promote normal levels of transcription. *EMBO J.*, **11**, 4145.

51. Grant, P. A., Sterner, D. E., Duggan, L. J., Workman, J. L., and Berger, S. L. (1998) The SAGA unfolds: convergence of transcription regulators in chromatin-modifying complexes. *Trends Cell Biol.*, **8**, 193.

52. Brandl, C. J., Furlanetto, A. M., Martens, J. A., and Hamilton, K. S. (1993) Characterization of NGG1, a novel yeast gene required for glucose repression of GAL4p-regulated transcription. *EMBO J.*, **12**, 5255.

53. Horiuchi, J., Silverman, N., Pina, B., Marcus, G. A., and Guarente, L. (1997) ADA1, a novel component of the ADA/GCN5 complex, has broader effects than GCN5, ADA2, or ADA3. *Mol. Cell. Biol.*, **17**, 3220.

54. Sterner, D. E., Grant, P. A., Roberts, S. M., Duggan, L. J., Belotserkovskaya, R., Pacella, L. A., Winston, F., Workman, J. L., and Berger, S. L. (1999) Functional organization of the yeast SAGA complex: Distinct components involved in structural integrity, nucleosome acetylation, and TBP binding. *Mol. Cell. Biol.*, **19**, 86.

55. Marcus, G. A., Horiuchi, J., Silverman, N., and Guarente, L. (1996) ADA5/SPT20 links the ADA and SPT genes, which are involved in yeast transcription. *Mol. Cell. Biol.*, **16**, 3197.

56. Grant, P. A., Duggan, L., Cote, J., Roberts, S. M., Brownell, J. E., Candau, R., Ohba, R., Owen-Hughes, T., Allis, C. D., Winston, F., *et al.* (1997) Yeast Gcn5 functions in two multi-subunit complexes to acetylate nucleosomal histones: characterization of an Ada complex and the SAGA (Spt/Ada) complex. *Genes Develop.*, **11**, 1640.

57. Grant, P. A., Schieltz, D., Pray-Grant, M. G., Steger, D. J., Reese, J. C., Yates, J. R., and Workman, J. L. (1998) A subset of TAFIIs are integral components of the SAGA complex required for nucleosome acetylation and transcriptional stimulation. *Cell*, **94**, 45.

58. Saleh, A., Schieltz, D., Ting, N., McMahon, S. B., Litchfield, D. W., Yates, J. R. 3rd, Lees-Miller, S. P., Cole, M. D., and Brandl, C. J. (1998) Tra1p is a component of the yeast Ada.Spt transcriptional regulatory complexes. *J. Biol. Chem.*, **273**, 26559.

59. Grant, P. A., Schieltz, D., Pray-Grant, M. G., Yates, J. R. 3rd, and Workman, J. L. (1998) The ATM-related cofactor Tra1 is a component of the purified SAGA complex. *Mol. Cell*, **2**, 863.

60. McMahon, S. B., Van Buskirk, H. A., Dugan, K. A., Copeland, T. D., and Cole, M. D. (1998) The novel ATM-related protein TRRAP is an essential cofactor for the c-Myc and E2F oncoproteins. *Cell*, **94**, 363.

61. Cosma, M. P., Tanaka, T., and Nasmyth, K. (1999) Ordered recruitment of transcription and chromatin remodeling factors to a cell cycle- and developmentally regulated promoter. *Cell*, **97**, 299.

62. Kayne, P. S., Kim, U.-J., Han, M., Mullen, J. R., Yoshizaki, F., and Grunstein, M. (1988) Extremely conserved histone H4 N terminus is dispensable for growth but essential for repressing the silent mating loci in yeast. *Cell*, **55**, 27.

63. Megee, P. C., Morgan, B. A., Mittman, B. A., and Smith, M. M. (1990) Genetic analysis of histone H4: essential role of lysines subject to reversible acetylation. *Science*, **247**, 841.

64. Park, E.-C. and Szostak, J. W. (1990) Point mutations in the yeast histone H4 gene prevent silencing of the silent mating type locus *HML*. *Mol. Cell. Biol.*, **10**, 4932.

65. Johnson, L. M., Kayne, P. S., Kahn, E. S., and Grunstein, M. (1990) Genetic evidence for an interaction between SIR3 and histone H4 in the repression of the silent mating loci in *Saccharomyces cerevisiae*. *Proc. Natl Acad. Sci., USA*, **87**, 6286.

66. Johnson, L. M., Fisher-Adams, G., and Grunstein, M. (1992) Identification of a non-basic domain in the histone H4 N-terminus required for repression of the yeast silent mating loci. *EMBO J.*, **11**, 2201.

67. Gottschling, D. E., Aparicio, O. M., Billington, B. L., and Zakian, V. A. (1990) Position effect at *S. cerevisiae* telomeres: reversible repression of Pol II transcription. *Cell*, **63**, 751.

68. Braunstein, M., Rose, A. B., Holmes, S. G., Allis, C. D., and Broach, J. R. (1993) Transcriptional silencing in yeast is associated with reduced nucleosome acetylation. *Genes Develop.*, **7**, 592.

69. Jeppesen, P. and Turner, B. M. (1993) The inactive X chromosome in female mammals is distinguished by a lack of histone H4 acetylation, a cytogenetic marker for gene expression. *Cell*, **74**, 281.

70. Mann, R. K. and Grunstein, M. (1992) Histone H3 N-terminal mutations allow hyperactivation of the yeast *GAL1* gene *in vivo*. *EMBO J.*, **11**, 3297.

71. Morgan, B. A., Mittman, B. A., and Smith, M. M. (1991) The highly conserved N-terminal domains of histones H3 and H4 are required for normal cell cycle progression. *Mol. Cell. Biol.*, **11**, 4111.

72. Edmondson, D. G., Smith, M. M., and Roth, S. Y. (1996) Repression domain of the yeast global repressor Tup1 interacts directly with histones H3 and H4. *Genes Develop.*, **10**, 1247.

73. Hecht, A., Larouche, T., Strahl-Bolsinger, S., Gasser, S. M., and Grunstein, M. (1995) Histone H3 and H4 N-termini interact with SIR3 and SIR4 proteins: a molecular model for the formation of heterochromatin in yeast. *Cell*, **80**, 583.

74. Durrin, L. K., Mann, R. K., Kayne, P. S., and Grunstein, M. (1991) Yeast histone H4 N-terminal sequence is required for promoter activation *in vivo*. *Cell*, **65**, 1023.

75. Allfrey, V., Falukner, R. M. and Mirsky, A. E. (1964) Acetylation and methylation of histones and their possible role in the regulation of RNA synthesis. *Proc. Natl Acad. Sci., USA*, **51**, 786.

76. Brownell, J. E., Zhou, J., Ranalli, T., Kobayashi, R., Edmondson, D. G., Roth, S. Y., and Allis, C. D. (1996) Tetrahymena histone acetyltransferase A: a homolog to yeast Gcn5p linking histone acetylation to gene activation. *Cell*, **84**, 843.

77. Zhang, W., Bone, J. R., Edmondson, D. G., Turner, B. M., and Roth, S. Y. (1998) Essential and redundant functions of histone acetylation revealed by mutation of target lysines and loss of the Gcn5 acetyltransferase. *EMBO J.*, **17**, 3155.

78. Kuo, M.-H., Brownwell, J. E., Sobel, R. E., Ranalli, T. A., Cook, R., G, Edmondson, D. G., Roth, S. Y., and Allis, C. D. (1996) Transcription-linked acetylation by Gcn5p of histones H3 and H4 at specific lysines. *Nature*, **383**, 269.

79. Imhof, A., Yang, X. J., Ogryzko, V. V., Nakatani, Y., Wolffe, A. P., and Ge, H. (1997) Acetylation of general transcription factors by histone acetyltransferases. *Curr. Biol.*, **7**, 689.

80. Eberharter, A., John, S., Grant, P. A., Utley, R. T., and Workman, J. L. (1998) Identification and analysis of yeast nucleosomal histone acetyltransferase complexes. *Methods*, **15**, 315.

81. Grant, P. A., Eberharter, A., John, S., Cook, R. G., Turner, B. M., and Workman, J. L. (1999) Expanded lysine acetylation specificity of Gcn5 in native complexes. *J. Biol. Chem.*, **274**, 5895.

82. Dhalluin, C., Carlson, J. E., Zeng, L., He, C., Aggarwal, A. K., and Zhou, M. M. (1999) Structure and ligand of a histone acetyltransferase bromodomain. *Nature*, **399**, 491.

83. Pollard, K. J., Samuels, M. L., Crowley, K. A., Hansen, J. C., and Peterson, C. L. (1999) Functional interaction between GCN5 and polyamines: a new role for core histone acetylation. *EMBO J.*, **18**, 5622.

83a. Prelich G. and Winston, F. (1993) Mutations that suppress the deletion of an upstream activating sequence in yeast: involvement of a protein kinase and histone H3 in repressing transcription *in vivo. Genetics,* **135**, 665.

84. Santisteban, M. S., Arents, G., Moudrianakis, E. N., and Smith, M. M. (1997) Histone octamer function *in vivo*: mutations in the dimer–tetramer interfaces disrupt both gene activation and repression. *EMBO J.,* **16**, 2493.

85. Recht, J. and Osley, M. A. (1999) Mutations in both the structured domain and N-terminus of histone H2B bypass the requirement for Swi–Snf in yeast. *EMBO J.,* **18**, 229.

86. Smith, M. M., Yang, P., Santisteban, M. S., Boone, P. W., Goldstein, A. T., and Megee, P. C. (1996) A novel histone H4 mutant defective for nuclear division and mitotic chromosome transmission. *Mol. Cell. Biol.,* **16**, 1017.

87. Lenfant, F., Mann, R. K., Thomsen, B., Ling, X., and Grunstein, M. (1996) All four core histone N-termini contain sequences required for the repression of basal transcription in yeast. *EMBO J.,* **15**, 3974.

88. Jackson, V. (1990) *In vivo* studies on the dynamics of histone–DNA interaction: evidence for nucleosome dissolution during replication and transcription and a low level of dissolution independent of both. *Biochemistry,* **29**, 719.

89. Malone, E. A., Clark, C. D., Chiang, A., and Winston, F. (1991) Mutations in *SPT16/CDC68* suppress *cis*- and *trans*-acting mutations that affect promoter function in *Saccharomyces cerevisiae. Mol. Cell. Biol.,* **11**, 5710.

90. Rowley, A., Singer, R. A., and Johnston, G. C. (1991) CDC68, a yeast gene that affects regulation of cell proliferation and transcription, encodes a protein with a highly acidic carboxyl terminus. *Mol. Cell. Biol.,* **11**, 5718.

91. Lycan, D., Mikesell, G., Bunger, M., and Breeden, L. (1994) Differential effects of Cdc68 on cell cycle-regulated promoters in *Saccharomyces cerevisiae. Mol. Cell. Biol.,* **14**, 7455.

92. Orphanides, G., Wu, W. H., Lane, W. S., Hampsey, M., and Reinberg, D. (1999) The chromatin-specific transcription elongation factor FACT comprises human SPT16 and SSRP1 proteins. *Nature,* **400**, 284.

92a. John, S., Howe, I., Tafrov, S. T., Grant, P. A., Sternglanz, R., and Workman, J. L. (2000) The something about silencing protein, Sas3, is the catalytic subunit of NuA3, a yTAF(II)30-containing HAT complex that interacts with the SpH6 subunit of the yeast CP (Cdc68/Pob3)-FACT complex. *Genes Develop.* **14**, 1196.

93. West, M. H. and Bonner, W. M. (1980) Histone 2A, a heteromorphous family of eight protein species *Biochemistry,* **19**, 3238.

94. Hatch, C. L., Bonner, W. M., and Moudrianakis, E. N. (1983) Minor histone 2A variants and ubiquinated forms in the native H2A:H2B dimer. *Science,* **221**, 468.

95. van Daal, A. and Elgin, S. C. (1992) A histone variant, H2AvD, is essential in *Drosophila melanogaster. Mol. Biol. Cell,* **3**, 593.

96. Liu, X., Li, B., and Gorovsky, M. A. (1996) Essential and nonessential histone H2A variants in *Tetrahymena thermophila. Mol. Cell. Biol.,* **16**, 4305.

97. Clarkson, M. J., Wells, J. R., Gibson, F., Saint, R., and Tremethick, D. J. (1999) Regions of variant histone His2AvD required for *Drosophila* development. *Nature,* **399**, 694.

98. Jackson, J. D., Falciano, V. T., and Gorovsky, M. A. (1996) A likely histone H2A.F/Z variant in *Saccharomyces cerevisiae. Trends Biochem. Sci.,* **21**, 466.

99. Santisteban, M. S., Kalashnikova, T., and Smith, M. M. (2000) Histone H2A.Z regulates transcription and is redundant with nucleosome remodeling complexes, *Cell,* in press.

100. Strahl, B. D. and Allis, C. D. (2000) The language of covalent histone modifications. *Nature,* **403**, 41.

5 | The SWI/SNF family of remodelling complexes

BRADLEY R. CAIRNS and ROBERT E. KINGSTON

1. Introduction

Chromatin is a dynamic material with important roles in gene regulation. Chromatin structures formed to repress transcription are remodelled in response to environmental cues or cell cycle progression. Importantly, these structural transitions are rapid, reversible, and not merely a consequence of transcription, as promoter restructuring occurs in the absence of RNA synthesis (see Chapter 3). Within the past several years, factors specialized for chromatin interconversion have been revealed through genetic and biochemical experiments in yeast, *Drosophila*, and human cells, and emerging evidence suggests their involvement in both establishing and removing repressive chromatin. Although most is known regarding their roles in transcription, they may also be utilized for controlling other important aspects of chromosomal biology, such as recombination, replication, DNA repair, chromosome condensation/decondensation, and chromatin assembly.

Eukaryotic cells contain two general classes of chromatin-modifying factors; those that covalently modify the amino-terminal 'tails' of histone proteins, and those that utilize ATP hydrolysis to remodel or reposition nucleosomes. The first class includes protein complexes that acetylate or deacetylate lysine residues present in the amino termini of histone proteins (termed histone acetyltransferases (HATs) and histone deacetylases (HDACs), respectively). Histone acetylation is correlated with active transcription, whereas deacetylation is correlated with transcriptional silencing (see Chapters 7 and 8, respectively). This class may also expand to include proteins responsible for other modifications of histones, such as phosphorylation and methylation. The second class of factors is composed of ATP-dependent chromatin remodelling complexes (remodellers), which alter nucleosome structure or positioning. These factors have been separated into two families that are defined by the different features of their ATPase subunits. The 'ISWI' family of complexes (covered in Chapter 6) all contain an ATPase related to the *Drosophila* ISWI protein, whereas the 'SWI/SNF' family of complexes, the subject of this chapter, all contain a protein related to the yeast protein Swi2/Snf2.

An emerging theme in chromatin regulation is that chromatin-modifying proteins

(HATs, HDACs, and remodellers) all function in large protein complexes. This enables many important aspects of their function to be coordinately regulated, with certain subunits playing distinct roles in complex regulation, targeting, localization, or substrate specificity. Here, we will discuss these important aspects of SWI/SNF complex function, as well as their remarkable nucleosome remodelling activities. In addition, we will address how SWI/SNF complexes may function together with HAT complexes to antagonize chromatin repression and help establish or maintain active gene expression. We will also discuss emerging connections of SWI/SNF to cell-cycle regulation and cancer.

2. Structure/function of the SWI/SNF family

2.1 Isolation of the SWI/SNF family of remodellers

The detection and isolation of SWI/SNF family complexes came from a combination of genetic and biochemical approaches. Genetic experiments in yeast revealed sucrose non-fermentation (*snf*) mutants which fail to transcribe the gene encoding invertase (*SUC2*). A separate screen revealed mutants defective in mating-type switching (*swi* mutants), which fail to transcribe the endonuclease *HO* important for the switching process. Genetic experiments in *Drosophila* revealed a set of mutants termed trithorax (*trx*) that are defective in maintaining patterns of expression of homeotic genes. The action of the yeast Swi/Snf was established through biochemical purification, which revealed a stable 11-protein complex (ySWI/SNF) with potent DNA-dependent ATPase activity and ATP-dependent chromatin-remodelling activity (Table 1) (1–3). Yeast also contain the more abundant RSC complex, which includes many subunits that are similar or identical to members of ySWI/SNF (4). Human cells contain several SWI/SNF-related complexes (termed hSWI/SNF) that are similar to each other and to their yeast counterparts in both composition and *in vitro* activities (5–7). Interestingly, *Drosophila* cells appear to utilize a single complex, termed dSWI/SNF (8, 9). Human and *Drosophila* SWI/SNF complexes are present at approximately 100 000 per cell, suggesting a major role for these factors in chromatin regulation.

Table 1 Related components of SWI/SNF complexes

ySWI/SNF	RSC	hSWI/SNF	dSWI/SNF	Core member	Domains/ function
Snf2/Swi2	Sth1	Brg1, hBrm	Brm	Yes	ATPase
Swi3	Rsc8	Baf155, Baf170	Bap155	Yes	Coiled coils, SANT
Snf5	Sfh1	hSnf5/Ini1	Bap45/Snr1	Yes	Tumour suppressor
Swp73/Snf12	Rsc6	Baf60a,b,c	Bap60	Yes	Activator interaction?
Swi1	none	Baf250	?	?	ARID/dead-ringer
Arp7, Arp9	Arp7, Arp9	Baf53	Bap55	?	Actin-related proteins
None	None	Actin	Actin/Bap47	?	Nuclear matrix assoc.?
None	Rsc1, Rsc2	Polybromo	?	No	Bromodomains, BAH
None	None	Baf57	Bap111	No	HMG domain
+4	+7	0	+1		Unique proteins

2.2 Subunits of SWI/SNF-related complexes

2.2.1 ATPase subunits

All SWI/SNF-related complexes contain a highly conserved ATPase subunit that is essential for all aspects of complex function. The yeast remodellers are built around one of two different ATPases, Snf2/Swi2 for SWI/SNF itself (10), and Sth1 for RSC (4, 11). Likewise, human cells contain two classes of SWI/SNF-related complexes, built around the ATPases Brg1 (12) and hBrm (13, 14), whereas *Drosophila* dSWI/SNF utilizes the ATPase Brahma (15). Although their conserved ATPase domains are also similar to domains in known helicases, helicase activity has not been detected. ATPase activity of SWI/SNF complexes is stimulated 20–50-fold by single-stranded, double-stranded, or structured DNA (four-way junction) (1, 2, 6, 16). Although nucleosomes are not a preferred effector of the ATPase, the presence of ATP increases the affinity of the RSC complex for nucleosomes but not naked DNA, suggesting some recognition of nucleosome structure (17).

ATP hydrolysis is essential for chromatin remodelling by ySWI/SNF, as mutations in the ATPase eliminate both ATPase and remodelling activity, but do not affect complex assembly (2). How ATP hydrolysis is coupled to chromatin remodelling has not been determined and is the subject of active study (described below). The SWI/SNF ATPases all contain a bromodomain, which binds to acetylated lysine residues present in histone amino-terminal 'tails', and may help link these complexes to chromatin (discussed below). In addition, Brg1, hBrm, and Snf2 all contain an AT hook within 30 residues of the bromodomain. The AT hook is an HMG-I(Y)-related domain that binds to the minor groove of AT-rich DNA. AT hooks are found in several transcription factors, and the domain may help connect these complexes to DNA (18).

Genetic experiments have established an essential role for the ATPase in complex function. In yeast, deletion of *SWI2/SWI2* (ySWI/SNF) prevents the activation of several genes (19), whereas deletion of *STH1* (RSC) causes lethality (11, 20). Likewise, mice lacking Brm are viable but show defects in cell proliferation (21), whereas deletion of Brg in murine F9 carcinoma cells causes inviability (22). For *Drosophila*, Brahma is essential for proper development and survival (23). Taken together, these results show that SWI/SNF-related remodellers are essential and abundant factors.

2.2.2 Core subunits

SWI/SNF complexes appear to consist of:

(1) a central core that includes an ATPase and three other polypeptides (homologues of the yeast Snf5, Swp73, and Swi3 proteins) important for remodelling activity, and

(2) several other tightly associated proteins important for complex targeting or regulation.

Yeast SWI/SNF complex does not assemble properly if yeast lack one of these conserved core subunits, and their absence also confers a complete (or nearly complete) *swi/snf* null phenotype (3). Together, these observations suggest the presence of a conserved 'core' of four subunits crucial for complex assembly and function. In human cells, certain core subunits (i.e. the Swp73/BAF60 family) are encoded by a family of related genes, which may enable certain cells to tailor the composition of their remodeller (7, 24).

Presently, little is known about how each member of the conserved core contributes to the function of SWI/SNF-related subunits. For human SWI/SNF, the ATPase subunits alone display modest amounts of chromatin remodelling activity *in vitro*, and these activities can be stimulated by the human SWI3 homologues (BAF155, BAF170) and the Snf5 homologue (Ini1) (25). It is not known how these non-ATPase subunits stimulate remodelling activity; they could either contribute directly to function, could stimulate the activity of the ATPase subunit, or could perform both functions. The Swi3 family members all contain a leucine zipper that may mediate their homodimerization, and a SANT domain of unknown function (7). The Snf5/ Ini1 component has been shown to interact with the SET domain (26), a motif present in several chromatin/transcriptional regulators, including *Drosophila* trithorax (*trx*) and its mammalian counterpart HRX/MLL/ALL, which is linked to leukaemogenesis. Whether these domains contribute directly to remodelling activity, or instead play a regulatory role, is not known. The mechanistic role of the other extensively conserved subunit (Swp73 in yeast, BAF60 in humans) has not been studied biochemically. However, the yeast Swp73 subunit is not required for the response of SWI/SNF to certain activators, and SWI/SNF is at least partially assembled in *swp73* mutants, suggesting a more specialized function for this core member (24).

2.2.3 Other conserved subunits: potential roles in targeting or regulation

SWI/SNF core proteins are tightly associated with a diverse set of polypeptides. For example, the BAF57 subunit of human SWI/SNF contains a non-specific DNA-binding (high-mobility group/HMG) domain and domain related to the motor protein kinesin (27). The yeast complexes lack a similar subunit, suggesting that human SWI/SNF complexes may perform certain unique functions. ySWI/SNF and RSC also contain unique members; only RSC contains subunits (Rsc1 and Rsc2) with multiple chromatin-recognition domains (bromodomains, discussed below) that may be important for RSC targeting or regulation. Interestingly, the mammalian homologue of Rsc1/Rsc2, termed polybromo, is only associated with the essential Brg complex and not the non-essential Brm complex, suggesting that the human Brg complex is more related to yeast RSC than to ySWI/SNF (W. Wang, in preparation). Taken together, SWI/SNF complexes consist of a central core that includes an ATPase and three other polypeptides important for remodelling activity, as well as several other tightly associated proteins that contribute to complex targeting or regulation.

2.2.4 Actin-related protein (ARP) subunits

Remarkably, all SWI/SNF-related complexes contain either actin, an actin-related protein (ARP), or both (9, 28–30). Though neither is required for the basic remodelling reactions, they may provide essential regulatory functions. Yeast SWI/SNF and

RSC complexes share two essential ARPs, Arp7 and Arp9, but lack actin entirely (28). Arp7 and Arp9 are both present in these complexes, and the two proteins may interact both physically and functionally. Human and *Drosophila* SWI/SNF complexes contain the ARPs BAF53 and BAP55, respectively (9, 29). All ARPs show only limited identity to actin, which is localized to regions important for the overall fold. The actin fold is utilized in a large group of diverse ATP-binding and hydrolysing proteins, including heat-shock proteins and sugar kinases. However, the human ARP BAF53 lacks detectable ATP-binding activity (29). Furthermore, purified yeast Arp7 and Arp9 lack ATPase activity *in vitro*, and extensive mutagenesis of their putative ATPase domains did not impair their function *in vivo*, strongly suggesting that these ARPs bear a structural rather than enzymatic similarity to actin (28).

Surprisingly, *Drosophila* dSWI/SNF and certain preparations of human SWI/SNF also contain actin. Although the role of actin in motility and cytokinesis is well known, a role for actin in the nucleus had not been defined. Brg1 appears to interact directly with both BAF53 and actin, and both proteins were capable of stimulating Brg1 ATPase activity, suggesting that actin and actin-related proteins may affect activity of SWI/SNF. In addition, Brg1 is necessary for actin and BAF53 association with the nuclear matrix, suggesting that their association is required for proper SWI/SNF localization (29).

2.3 ATP-dependent remodelling activities of SWI/SNF family complexes

SWI/SNF family complexes are able to perform several different types of reactions, described below, that alter the structure of either mononucleosomes or nucleosomal arrays (Fig. 1). All of these reactions require ATP hydrolysis. How these reactions relate to each other is an open question, as is the relevance of each of these reactions to the function of these complexes *in vivo*.

2.3.1 Remodelling reactions

SWI/SNF complexes alter the path of DNA on a mononucleosome in an ATP-dependent manner (2, 31). This is displayed most dramatically when rotationally positioned mononucleosomes are assembled using end-labelled DNA and digested with limited amounts of DNAase. In mononucleosomes, the side of DNA that faces the octamer is protected from cleavage. This protection restricts DNAase cleavage of the minor groove to one cleavage per helical turn, resulting in a characteristic 10 bp repeat cleavage pattern. However, this characteristic cleavage pattern is significantly altered by the remodelling activity of SWI/SNF complexes, indicating a change in histone–DNA contacts (Fig. 1A). Restriction enzyme cleavage and the binding of transcription factors such as the yeast transcription factor GAL4 and the TATA-binding protein TBP are also generally increased following remodelling by SWI/SNF (2, 5, 32, 33). Thus, ATP-dependent remodelling by SWI/SNF makes mononucleosomal DNA more accessible to a variety of proteins. This might play a role in increasing the efficiency of steps required for transcriptional initiation.

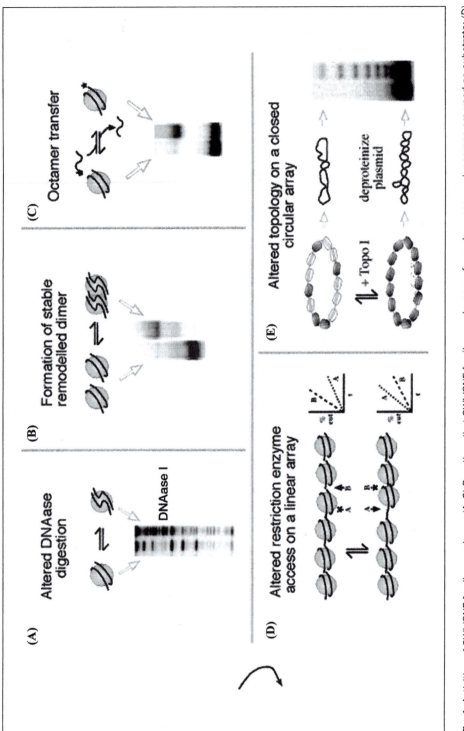

Fig. 1 Activities of SWI/SNF family complexes. (A–C) Reactions that SWI/SNF family complexes perform when mononucleosomes are used as substrate; (D) and (E) reactions performed when arrays are used as substrate. In all panels except (D), the left lane shows the starting substrate, the right lane shows the result after remodelling. In (B) the reactions always produce a mixture of the two species shown; the relative ratio of the two species will vary with solution condition. In (D) the top structure and graph show the substrate before remodelling, the bottom structure and graph show the substrate after remodelling.

Remodelling of closed circular nucleosomal arrays by SWI/SNF complexes causes significant changes in topology, indicating that remodelling also alters the path of DNA on arrays (6, 34). Deproteinization of a standard closed circular nucleosomal fragment results in the formation of one negative supercoil for each nucleosome. Remodelling of a closed circular plasmid by SWI/SNF decreases the resulting number of negative supercoils (Fig. 1E). Whether this is caused by a decrease in the amount of DNA that is associated with histones or a change in the topological path of the DNA on the histones (or a combination of the two effects) is under active investigation.

Changed access to restriction enzyme cleavage on nucleosomal arrays is another hallmark of SWI/SNF activity (33). Nucleosomes normally inhibit the ability of restriction enzymes to cleave DNA. Arrays of nucleosomes in defined positions (this is usually accomplished by using templates with repeats of the sea urchin 5S nucleosome positioning sequence) have restriction enzyme sites within the arrays that are inaccessible to cleavage (Fig. 1D). SWI/SNF complexes can increase cleavage at these sites in an ATP-dependent manner. Intriguingly, other restriction enzyme sites that are normally accessible (because they are in linker DNA between nucleosomes) can become less accessible following SWI/SNF action. These results are most simply interpreted as indicating that SWI/SNF redistributes nucleosome position on arrays (see also 35), although direct visualization of nucleosome position on remodelled arrays is needed to test this hypothesis.

2.3.2 Octamer sliding, octamer transfer, and conformational changes

The results described above demonstrate that SWI/SNF complexes alter histone–DNA contacts. This could be accomplished solely by changes in histone–DNA contacts, or could also involve changes in histone–histone contacts that result in altered association of the histones with DNA. These possibilities are not mutually exclusive and there are data that support each possibility.

SWI/SNF complexes are able to move histone octamers from one piece of DNA to another, a process called 'octamer transfer' (36) (Fig. 1C). This process is inefficient, however, resulting in transfer of considerably less than 5% of nucleosomes. The low efficiency of the reaction might reflect a lack of *in vitro* optimization, or it might be that this reaction is not a major mechanism of remodelling. SWI/SNF complexes can also move nucleosomes along DNA in *cis*, a process that occurs more efficiently than octamer transfer (35). This process, termed sliding, is also seen with other ATP-dependent remodelling complexes (37, 38), and may reflect a common mechanistic capability of ATP-dependent remodellers. Whether or not different mechanisms are used by SWI/SNF to transfer octamers in *trans* and in *cis* is not clear.

There are also data that are consistent with SWI/SNF causing conformational changes in the nucleosome (17, 39). Remodelling of mononucleosomes by SWI/SNF complexes creates a stably remodelled species that contains all of the components of two nucleosomes (Fig. 1B). One possibility is that the conformation of the histone octamer is changed to promote association of stably remodelled species. A second possibility is that the remodelled dimer contains histone octamers that are in a standard conformation, but that are tethered together by DNA that has shifted position. Cross-linking studies show that histone H2A contacts change upon remodelling, which might reflect either a shift in position of H2A or a movement of DNA relative to H2A (40). Importantly, remodelling can occur on one portion of a nucleosome even after cross-linking has locked the DNA in place in a different portion of the same nucleosome. This indicates that significant movement of the entire strand of nucleosomal DNA is not required to observe remodelling.

2.3.3 SWI/SNF as an enzyme

Several studies imply that SWI/SNF functions catalytically to remodel chromatin structure. Most of the reactions described above have been shown to occur at molar ratios of less than one SWI/SNF complex per nucleosome. Perhaps the most convincing argument for catalytic function comes from studies examining restriction enzyme access to nucleosomal arrays that show that addition of fresh, non-remodelled template to a reaction will result in a new burst of remodelling activity (33).

As with all enzymes, the simplest hypothesis for SWI/SNF action is that SWI/SNF binds nucleosomes and forms a single high-energy intermediate state. A complex containing the yeast RSC complex and nucleosome(s) has been identified on native gels that might reflect such an intermediate (17). The reaction characteristics for octamer transfer and for the creation of stably remodelled dimers are consistent with the existence of a single intermediate that can resolve to either a stably remodelled state, transfer of octamer position, or back to the starting structure (Fig. 2A). All of the data concerning altered accessibility of nucleases and binding factors, as well as the topological effects of remodelling, are also consistent with a single intermediate. On nucleosomal arrays, and *in vivo*, this intermediate species might resolve to nucleosomes with either altered position or altered conformation (Fig. 2B).

2.3.4 Reversibility of the SWI/SNF remodelling reaction

A key aspect of remodelling by SWI/SNF is that this family of complexes appears to catalyse both formation of remodelled nucleosome structures and the 'reverse' reaction, where remodelled structures are reconverted to standard nucleosomes. This is most easily shown in experiments that use the stably remodelled dimer that is created by SWI/SNF action on mononucleosomes. This species can be isolated, and SWI/SNF will reconvert this species back to a standard nucleosome in a process that requires ATP (17, 39). The ratio of 'standard' and 'remodelled' structures that are formed in this reaction appears similar no matter whether the starting substrate is, so this appears to reflect an equilibrium process. While it is difficult to determine directly whether SWI/SNF interconverts remodelled and standard states on nucleosomal arrays, all current data are at least consistent with this possibility. These data suggest the possibility that SWI/SNF uses the energy of ATP hydrolysis to create a high-energy intermediate state that has altered histone–DNA contacts, and this high-energy state can resolve to either a remodelled state or to a standard nucleosomal state (Fig. 2).

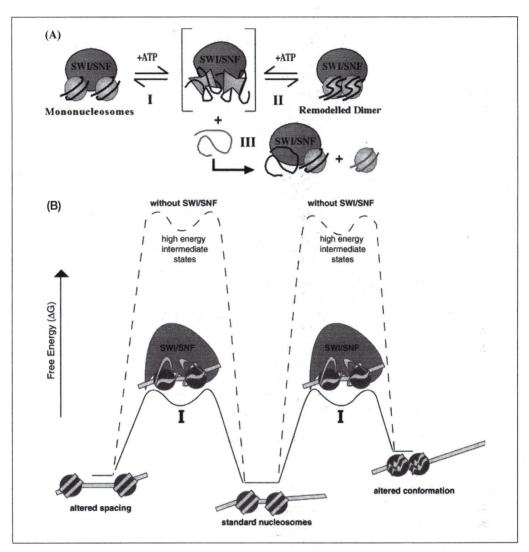

Fig. 2 SWI/SNF as an enzyme. (A) Reactions that can be catalysed by SWI/SNF when mononucleosomes are used as substrate, depicting a single hypothetical activated intermediate state. Two binding sites for nucleosomes are depicted because that is the simplest hypothesis to account for the reactions shown. Bare DNA might or might not occupy the same pocket that is normally bound by a nucleosome. (B) A hypothetical scheme for how a single activated intermediate state is used in alteration of nucleosome position and alteration of nucleosome conformation.

2.3.5 Stability of the remodelled state

The stability of the remodelled state that is created by SWI/SNF action has been examined by depleting ATP from the reaction, by titrating SWI/SNF away from the reacted template using excess DNA competitor, and by a combination of these approaches (34, 39, 41, 42). These data imply that the remodelled state is stable on nucleosomal arrays for at least 30 minutes following ATP depletion or SWI/SNF removal. Several experiments suggest that the remodelled state gradually decays to a standard nucleosomal state, and that the rate of reversion will differ with template and with reaction condition; however, this issue has not yet been explored in detail.

3. Targeting of SWI/SNF activity

Two fundamental issues regarding SWI/SNF function are:

(1) understanding how SWI/SNF is targeted to particular locations in the genome, and

(2) identifying the function(s) SWI/SNF performs during the transcription cycle.

Although there are putative DNA-binding domains present on several SWI/SNF family members, all those tested have shown only non-specific affinity for DNA, suggesting that these complexes lack the ability to recognize specific DNA sequences. However, these complexes can be targeted to particular DNA sequences through interaction with site-specific DNA-binding proteins. For example, ySWI/SNF interacts with the Swi5 activator protein for recruitment to the HO promoter (discussed in greater detail below) (43). Interestingly, ySWI/SNF is also recruited by HIR proteins, known repressors of yeast histone genes (44). In this instance, ySWI/SNF is not required to establish repression, but rather to reverse HIR-mediated silencing. However, the temporal relationship between HIR recruitment of SWI/SNF and promoter activity remains unresolved.

Several studies show that there is direct contact between SWI/SNF and transcriptional activators. The acidic activators Swi5, Gcn4, Gal4-VP16, and Hap4 interact with yeast SWI/SNF, and these activators are capable of targeting SWI/SNF activity to chromatin templates (43, 45, 46). In *Drosophila*, genetic and biochemical experiments support an interaction of dSWI/SNF with the homeotic transcriptional regulator trx (covered in Chapter 11). In human cells, the activator EKLF recruits hSWI/SNF to chromatin templates bearing the β-globin promoter, and promotes both chromatin structural changes and transcriptional activation *in vitro* (47). The mammalian activator C/EBPB contacts hSWI/SNF, perhaps directing its association with myeloid lineage genes such as *mim-1* (48). SWI/SNF is also required for activation by the glucocorticoid receptor, which can direct remodelling activity to a mononucleosome *in vitro* (49, 50).

SWI/SNF complexes all contain proteins with one or more bromodomains (BD), which bind acetyl lysine and may help target these complexes to genes with acetylated histone tails (51, 52). For example, RSC complex contains proteins with two BDs (Rsc1, Rsc2) as well as Sth1, which bears a single BD (53, 54). An important unresolved question is whether bromodomains are utilized for the initial targeting of SWI/SNF complexes to certain promoters bound by acetylated histones, or for their persistent interaction with such promoters following recruitment by activators.

Somewhat surprisingly, DNA microarray experiments with yeast *swi/snf* mutants show increased transcription of many genes, raising the possibility that SWI/SNF may be involved in establishing the repressed state (55, 56). In support of this model, certain *rsc* mutants are defective in repression of the *CHA1* gene (57), and hSWI/SNF function is required for repression of the c-*fos* gene (58). Although it has not been determined whether these effects are a direct result of SWI/SNF action at these promoters, it raises the possibility that SWI/SNF complexes remodel promoters without regard to the transcriptional consequence, with the outcome depending on the action of other factors. The ability of SWI/SNF to catalyse both 'forward' and 'reverse' remodelling reactions might allow SWI/SNF to be involved in different types of gene regulatory interactions.

4. Timing of SWI/SNF function

Genes require diverse factors for their transcriptional activation, including DNA-binding activators, HATs, ATP-dependent remodellers, initiation machinery, and elongation factors. As nucleosomes are present on the promoter and body of the gene during both the active and repressed state, remodellers could perform roles at one or more times during the transcription cycle. Therefore, understanding their temporal recruitment and action is of central importance. One model for SWI/SNF function is to assist activator binding in the context of chromatin (2, 59). Support is provided by the *in vitro* experiments described above, where SWI/SNF assists either Gal4 derivatives or TBP in binding to nucleosomal DNA. However, other experiments support a role for SWI/SNF at a step subsequent to activator binding, possibly through promoter restructuring to facilitate the binding of additional transcription factors or through effects on transcriptional elongation (60, 61).

Studies of the yeast HO promoter have advanced our understanding of timing significantly. Activation of HO occurs in a window of the G_1 phase of the cell cycle and requires the activators Swi5 and Swi4/Swi6, the remodeller ySWI/SNF, and the HAT complex SAGA (SAGA is covered in detail in Chapter 7). To establish the order in which these factors occupy the HO promoter, chromatin immunoprecipitation experiments were performed with synchronized yeast cells (62, 63). Swi5 binds HO first, and this event is required for ySWI/SNF occupancy, suggesting that Swi5 recruits ySWI/SNF to HO. Recruitment of ySWI/SNF is required for the subsequent association of SAGA with HO; however, it is not clear whether it is the physical presence of, or remodelling by, SWI/SNF that enables SAGA association. In addition, the presence of SAGA is correlated with promoter acetylation. Interestingly, all these events occur before RNA synthesis, suggesting that these chromatin remodelling events poise the promoter for subsequent activation. Action by ySWI/SNF and SAGA is required for the association of the activators Swi4/Swi6 (62). Presently, it is not known which factors recruit the transcription machinery, or when this occurs in relation to other factors. These studies suggest that SWI/SNF acts prior to HATs, and that SWI/SNF action is required for establishing the active state of the gene. However, it is not yet clear whether this order of action is applicable to all, or even most, promoters, as different promoters may utilize SWI/SNF differently.

5. Maintaining active transcription

Genetic studies indicate that SWI/SNF family complexes play an important role in maintaining active transcriptional states. This role becomes increasingly important in complex organisms such as *Drosophila* and humans, as appropriate development requires not only the establishment of tissue-specific transcription states, but also the maintenance if those states as the organism develops. Genetic studies imply a role for the genes *brahma* and *moira* (both part of the core of the dSWI/SNF complex) in maintaining gene expression patterns in *Drosophila* (64). Recent studies that use a combination of genetic techniques and direct analysis of gene expression demonstrate that ySWI/SNF is also required to maintain active transcriptional states of several genes (65, 66).

The mechanisms that are used to maintain active states are not understood, and will constitute an important research area in years to come. One possibility is that SWI/SNF complexes are continually associated with appropriate active genes, and their association allows the efficient re-activation of genes that are temporarily inactivated by stochastic events. Many SWI/SNF family complexes are extremely abundant, and thus there are sufficient complexes to maintain association with a large number of genes. Alternatively, SWI/SNF could be transiently recruited to maintain active states by sequence-specific DNA binding factors (e.g. activators) that maintain a stable association. Finally, the increased association of SWI/SNF with acetylated histones might be crucial to maintenance. Expression patterns are maintained across DNA replication, so whatever mechanism is used for maintenance must account for this property.

6. Regulation of SWI/SNF across the cell cycle

SWI/SNF family complexes play a role in regulating the ability of cells to traverse the cell cycle (see below), and it appears that cell-cycle-dependent changes in phosphorylation in turn regulate SWI/SNF function. The Brg1 subunit of human SWI/SNF is phosphorylated by cyclin E/cdk2 in G_1 (67). Co-transfection studies have led to the suggestion that the ability of SWI/SNF to function in repression might be altered by this phosphorylation event (68); however, the nature and the activity of phosphorylated complexes in G_1 has not yet been addressed directly. Biochemical studies suggest that SWI/SNF is able to remodel chromatin at this stage of the cell cycle, so it is unlikely that phosphorylation inhibits this function, but phosphorylation might alter the ability of SWI/SNF to play specific regulatory roles by altering targeting or by altering its ability to associate with other proteins such as deacetylases (68).

As cells enter mitosis, human SWI/SNF complexes are phosphorylated by apparently different kinases than phosphorylate in G_1, and in this case phosphorylation inactivates remodelling activity (69, 70). Both human ATPase subunits (BRG1 and hBrm) become phosphorylated, as does at least one of the SWI3 homologues. The MAP kinase, ERK1, is able to perform this phosphorylation *in vitro*, and is one candidate for the kinase that functions *in vivo* at this stage of cell cycle (70). As cells leave mitosis, SWI/SNF is dephosphorylated and re-activated. One potential reason for this regulation of SWI/SNF activity is that chromosomes condense in mitosis and decondense following mitosis, and regulation of SWI/SNF chromatin remodelling activity may be required for appropriate condensation/decondensation.

7. Connections of SWI/SNF complex to cancer

hSWI/SNF displays properties characteristic of tumour suppressors. Many tumour suppressors, such as the retinoblastoma (Rb) gene product, inhibit proliferation by controlling cell-cycle transitions. As described above, SWI/SNF is regulated as cells traverse the cell cycle (69, 70), and a role for SWI/SNF in controlling proliferation is emerging. For example, full cell-cycle (G_1) arrest by Rb requires the presence of, and interaction with, hSWI/SNF (71, 72). In addition, Rb cannot induce growth arrest in the presence of a dominant-negative form of hSWI/SNF, further suggesting cooperativity between these factors Their cooperation and association requires the presence of a 'pocket' on Rb which also mediates binding of the viral protein E7. Furthermore, embryonic fibroblasts derived from Brm–/– mice lack the ability to arrest in G_1 in response to DNA damage, or in G_0 in response to confluency (21). Finally, transformation by *ras* may depend, at least in part, on decreased abundance of the hBrm form of hSWI/SNF, as *ras* expression lowers hBrm levels, whereas ectopic expression of hBrm leads to a partial reversion of the transformed phenotype (73).

Active tumour suppressors inhibit oncogenesis, and lose this property when mutated. Likewise, truncation or deletion mutations in the conserved hSnf5/Ini1 component of hSWI/SNF are associated with the formation of malignant rhabdoid tumours (74). In all tumours tested, both alleles were either truncated or deleted, whereas constitutionally matched DNA lacked such alterations. These data strongly suggest that rhabdoid tumours arise from bi-allelic somatic mutations in the hSNF5/INI1 gene, and support a 'two-hit' recessive model for tumour formation. Also, deletions of hSNF5/INI1 are frequent (36%) in patients with chronic myeloid leukaemia. Taken together, hSWI/SNF appears to have an important role in cell-cycle control, cell proliferation, and tumour suppression.

8. Summary and discussion

Chromatin is dynamic—its form and composition change during the cell cycle and also in response to the cell's environment. Two types of factors play central roles in the remodelling process: chromatin modifiers such as histone acetyltransferases and deacetylases, and ATP-dependent chromatin remodelling machines such as the highly conserved SWI/SNF family of complexes. These factors appear to work in concert to establish or maintain chromatin in either an active or silent state. Although significant progress has been made in understanding the function of the SWI/SNF family, many fundamental questions remain and are the subject of active study. Among them are:

(1) the precise mechanism of nucleosome remodelling;

(2) the unique functions provided by each subunit;

(3) how interactions of activators and SWI/SNF are regulated;

(4) how the action of chromatin modifiers such as HATs and HDACs are co-ordinated with the action of SWI/SNF;

(5) how SWI/SNF helps control the cell cycle, and how its misregulation can contribute to cancer.

Answers will continue to emerge through the combined efforts of researchers using both model organisms and human cells.

References

1. Cairns, B. R., *et al.* (1994) A multisubunit complex containing the SWI1/ADR6, SWI2/SNF2, SWI3, SNF5, and SNF6 gene products isolated from yeast. *Proc. Natl Acad. Sci., USA*, **91**, 1950.
2. Cote, J., *et al.* (1994) Stimulation of GAL4 derivative binding to nucleosomal DNA by the yeast SWI/SNF complex. *Science*, **265**, 53.
3. Peterson, C. L., Dingwall, A., and Scott, M. P. (1994) Five SWI/SNF gene products are components of a large multiprotein complex required for transcriptional enhancement. *Proc. Natl Acad. Sci., USA*, **91**, 2905.
4. Cairns, B. R., *et al.* (1996) RSC, an essential, abundant chromatin-remodeling complex. *Cell*, **87**, 1249.
5. Imbalzano, A. N., *et al.* (1994) Facilitated binding of TATA-binding protein to nucleosomal DNA. *Nature*, **370**, 481.
6. Kwon, H., *et al.* (1994) Nucleosome disruption and enhancement of activator binding by a human SWI/SNF complex. *Nature*, **370**, 477.
7. Wang, W., *et al.* (1996) Diversity and specialization of mammalian SWI/SNF complexes. *Genes Develop.*, **10**, 2117.
8. Dingwall, A. K., *et al.* (1995) The *Drosophila* snr1 and brm proteins are related to yeast SWI/SNF proteins and are components of a large protein complex. *Mol. Biol. Cell*, **6**, 777.
9. Papoulas, O., *et al.* (1998) The *Drosophila* trithorax group proteins BRM, ASH1 and ASH2 are subunits of distinct protein complexes. *Development*, **125**, 3955.
10. Laurent, B. C., Treich, I., and Carlson, M. (1993) The yeast SNF2/SWI2 protein has DNA-stimulated ATPase activity required for transcriptional activation. *Genes Develop.*, **7**, 583.
11. Laurent, B. C., Yang, X., and Carlson, M. (1992) An essential *Saccharomyces cerevisiae* gene homologous to SNF2 encodes a helicase-related protein in a new family. *Mol. Cell. Biol.*, **12**, 1893.
12. Khavari, P. A., *et al.* (1993) BRG1 contains a conserved domain of the SWI2/SNF2 family necessary for normal mitotic growth and transcription. *Nature*, **366**, 170.
13. Muchardt, C. and Yaniv, M. (1993) A human homologue of *Saccharomyces cerevisiae* SNF2/SWI2 and *Drosophila brm* genes potentiates transcriptional activation by the glucocorticoid receptor. *EMBO J.*, **12**, 4279.

14. Wang, W., *et al.* (1996) Purification and biochemical heterogeneity of the mammalian SWI–SNF complex. *EMBO J.,* **15**, 5370.

15. Tamkun, J. W., *et al.* (1992) brahma: a regulator of *Drosophila* homeotic genes structurally related to the yeast transcription activator SNF2/SWI2. *Cell,* **68**, 561.

16. Quinn, J., *et al.* (1996) The yeast SWI/SNF complex has DNA binding properties similar to HMG-box proteins. *Nature,* **379**, 844.

17. Lorch, Y., *et al.* (1998) Activated RSC–nucleosome complex and persistently altered form of the nucleosome. *Cell,* **94**, 29.

18. Aravind, L. and Landsman, D. (1998) AT-hook motifs in a wide variety of DNA-binding proteins. *Nucleic Acids Res.,* **26**, 4413.

19. Winston, F. and Carlson, M. (1992) Yeast SNF/SWI transcriptional activators and the SPT/SIN chromatin connection. *Trends Genet.,* **8**, 387.

20. Tsuchiya, E., *et al.* (1992) The *Saccharomyces cerevisiae NPS1* gene, a novel CDC gene which encodes a 160 kDa nuclear protein involved in G2 phase control. *EMBO J.,* **11**, 4017.

21. Reyes, J. C., *et al.* (1998) Altered control of cellular proliferation in the absence of mammalian brahma (SNF2alpha). *EMBO J.,* **17**, 6979.

22. Sumi, I. C., *et al.* (1997) SNF2beta-BRG1 is essential for the viability of F9 murine embryonal carcinoma cells. *Mol. Cell. Biol.,* **17**, 5976.

23. Elfring, L. K., *et al.* (1998) Genetic analysis of brahma: the *Drosophila* homolog of the yeast chromatin remodeling factor SWI2/SNF2. *Genetics,* **148**, 251.

24. Cairns, B. R., *et al.* (1996) Essential role of Swp73p in the function of yeast Swi/Snf complex. *Genes Develop.,* **10**, 2131.

25. Phelan, M. L., *et al.* (1999) Reconstitution of a core chromatin remodeling complex from SWI/SNF subunits. *Mol. Cell,* **3**, 247.

26. Rozenblatt-Rosen, O., *et al.* (1998) The C-terminal SET domains of ALL-1 and TRITHORAX interact with the INI1 and SNR1 proteins, components of the SWI/SNF complex. *Proc. Natl Acad. Sci., USA,* **95**, 4152.

27. Wang, W., *et al.* (1998) Architectural DNA binding by a high-mobility-group/kinesin-like subunit in mammalian SWI/SNF-related complexes. *Proc. Natl Acad. Sci., USA,* **95**, 492.

28. Cairns, B. R., *et al.* (1998) Two actin-related proteins are shared functional components of the chromatin-remodeling complexes RSC and SWI/SNF. *Mol. Cell,* **2**, 639.

29. Zhao, K., *et al.* (1998) Rapid and phosphoinositol-dependent binding of the SWI/SNF-like BAF complex to chromatin after T lymphocyte receptor signaling. *Cell,* **95**, 625.

30. Peterson, C. L., Zhao, Y., and Chait, B. T. (1998) Subunits of the yeast SWI/SNF complex are members of the actin-related protein (ARP) family. *J. Biol. Chem.,* **273**, 23641.

31. Kwon, H., *et al.* (1994) Nucleosome disruption and enhancement of activator binding by a human SW1/SNF complex. *Nature,* **370**, 477.

32. Owen-Hughes, T. and Workman, J. L. (1996) Remodeling the chromatin structure of a nucleosome array by transcription factor-targeted trans-displacement of histones. *EMBO J.,* **15**, 4702.

33. Logie, C. and Peterson, C. L. (1997) Catalytic activity of the yeast SWI/SNF complex on reconstituted nucleosome arrays. *EMBO J.,* **16**, 6772.

34. Imbalzano, A. N., Schnitzler, G. R., and Kingston, R. E. (1996) Nucleosome disruption by human SWI/SNF is maintained in the absence of continued ATP hydrolysis. *J. Biol. Chem.,* **271**, 20726.

35. Whitehouse, I., *et al.* (1999) Nucleosome mobilization catalyzed by the yeast SWI/SNF complex. *Nature,* **400**, 784.

36. Lorch, Y., Zhang, M., and Kornberg, R. D. (1999) Histone octamer transfer by a chromatin-remodeling complex. *Cell,* **96**, 389.

37. Hamiche, A., *et al.* (1999) ATP-dependent histone octamer sliding mediated by the chromatin remodeling complex NURF [in process citation]. *Cell*, **97**, 833.

38. Langst, G., *et al.* (1999) Nucleosome movement by CHRAC and ISWI without disruption or trans- displacement of the histone octamer [in process citation]. *Cell*, **97**, 843.

39. Schnitzler, G., Sif, S., and Kingston, R. E. (1998) Human SWI/SNF interconverts a nucleosome between its base state and a stable remodeled state. *Cell*, **94**, 17.

40. Lee, K.-M., *et al.* (1999) hSWI/SNF disrupts interactions between the H2A N-terminal tail and nucleosomal DNA. *Biochemistry*, **38**, 8423.

41. Cote, J., Peterson, C. L., and Workman, J. L. (1998) Perturbation of nucleosome core structure by the SWI/SNF complex persists after its detachment, enhancing subsequent transcription factor binding. *Proc. Natl Acad. Sci., USA*, **95**, 4947.

42. Owen-Hughes, T., *et al.* (1996) Persistent site-specific remodeling of a nucleosome array by transient action of the SWI/SNF complex. *Science*, **273**, 513.

43. Neely, K., *et al.* (1999) Activation domain-mediated targeting of the SWI/SNF complex to promoters stimulates transcription from nucleosome arrays. *Mol. Cell*, **4**, 649.

44. Dimova, D., *et al.* (1999) A role for transcriptional repressors in targeting the yeast SWI/SNF complex. *Mol. Cell*, **4**, 75.

45. Natarajan, K., *et al.* (1999) Transcriptional activation by gcn4p involves independent interactions with the SWI/SNF complex and the SRB/mediator. *Mol. Cell*, **4**, 657.

46. Yudkovsky, N., *et al.* (1999) Recruitment of the SWI/SNF chromatin remodeling complex by transcriptional activators. *Genes Develop.*, **13**, 2369.

47. Armstrong, J. A., Bieker, J. J., and Emerson, B. M. (1998) A SWI/SNF-related chromatin remodeling complex, E-RC1, is required for tissue-specific transcriptional regulation by EKLF *in vitro*. *Cell*, **95**, 93.

48. Kowenz-Leutz, E. and Leutz, A. (1999) A C/EBP isoform recruits the SWI/SNF complex to activate myeloid genes. *Mol. Cell*, **4**, 735.

49. Ostlund, F. A., *et al.* (1997) Glucocorticoid receptor–glucocorticoid response element binding stimulates nucleosome disruption by the SWI/SNF complex. *Mol. Cell. Biol.*, **17**, 895.

50. Fryer, C. J. and Archer, T. K. (1998) Chromatin remodelling by the glucocorticoid receptor requires the BRG1 complex. *Nature*, **393**, 88.

51. Jeanmougin, F., *et al.* (1997) The bromodomain revisited. *Trends Biochem. Sci.*, **22**, 151.

52. Dhalluin, C., *et al.* (1999) Structure and ligand of a histone acetyltransferase bromodomain. *Nature*, **399**, 491.

53. Cairns, B., *et al.* (1999) Two functionally distinct forms of the RSC nucleosome-remodeling complex, containing essential AT-hook, BAH, and bromodomains. *Mol. Cell*, **4**, 715.

54. Du, J., *et al.* (1998) Sth1p, a *Saccharomyces cerevisiae* Snf2p/Swi2p homolog, is an essential ATPase in RSC and differs from Snf/Swi in its interactions with histones and chromatin-associated proteins. *Genetics*, **150**, 987.

55. Holstege, F. C., *et al.* (1998) Dissecting the regulatory circuitry of a eukaryotic genome. *Cell*, **95**, 717.

56. Sudarsanam, P., *et al.* (2000) Whole-genome expression analysis of *snf/swi* mutants of *Saccharomyces cerevisiae*. *Proc. Natl Acad. Sci., USA*, **97**, 3364.

57. Moreira, J. M. and Holmberg, S. (1999) Transcriptional repression of the yeast CHA1 gene requires the chromatin-remodeling complex RSC. *EMBO J.*, **18**, 2836.

58. Murphy, D. J., Hardy, S., and Engel, D. A. (1999) Human SWI–SNF component BRG1 represses transcription of the c-*fos* gene. *Mol. Cell. Biol.*, **19**, 2724.

59. Burns, L. G. and Peterson, C. L. (1997) The yeast SWI–SNF complex facilitates binding of a transcriptional activator to nucleosomal sites *in vivo*. *Mol. Cell. Biol.*, **17**, 4811.

60. Ryan, M. P., Jones, R., and Morse, R. H. (1998) SWI–SNF complex participation in transcriptional activation at a step subsequent to activator binding. *Mol. Cell. Biol.*, **18**, 1774.

61. Brown, S. A., Imbalzano, A. N., and Kingston, R. E. (1996) Activator-dependent regulation of transcriptional pausing on nucleosomal templates. *Genes Develop.*, **10**, 1479.

62. Cosma, M. P., Tanaka, T., and Nasmyth, K. (1999) Ordered recruitment of transcription and chromatin remodeling factors to a cell cycle- and developmentally regulated promoter [in process citation]. *Cell*, **97**, 299.

63. Krebs, J., *et al.* (1999) Cell cycle-regulated histone acetylation required for expression of the yeast HO gene. *Genes Develop.*, **13**, 1412.

64. Kennison, J. A. (1995) The Polycomb and trithorax group proteins of *Drosophila*: trans-regulators of homeotic gene function. *Annu. Rev. Genet.*, **29**, 289.

65. Sudarsanam, P., *et al.* (1999) The nucleosome remodeling complex, Snf/Swi, is required for the maintenance of transcription *in vivo* and is partially redundant with the histone acetyltransferase, Gcn5. *EMBO J.*, **18**, 3101.

66. Biggar, S. R. and Crabtree, G. R. (1999) Continuous and widespread roles for the Swi–Snf complex in transcription. *EMBO J.*, **18**, 2254.

67. Shanahan, F., *et al.* (1999) Cyclin E associates with BAF155 and BRG-1, components of the mammalian SWI–SNF complex, and alters the ability of BRG1 to induce growth arrest. *Mol. Cell. Biol.*, **19**, 1460.

68. Zhang, H. S., *et al.* (2000) Exit from G1 and S phase of the cell cycle is regulated by repressor complexes containing HDAC–Rb–hSWI/SNF and Rb–hSWI/SNF. *Cell*, **101**, 78.

69. Muchardt, C., *et al.* (1996) The hbrm and BRG-1 proteins, components of the human SNF/SWI complex, are phosphorylated and excluded from the condensed chromosomes during mitosis. *EMBO J.*, **15**, 3394.

70. Sif, S., *et al.* (1998) Mitotic inactivation of a human SWI/SNF chromatin remodeling complex. *Genes Develop.*, **12**, 2842.

71. Dunaief, J. L., *et al.* (1994) The retinoblastoma protein and BRG1 form a complex and cooperate to induce cell cycle arrest. *Cell*, **79**, 119.

72. Strober, B. E., *et al.* (1996) Functional interactions between the hBRM/hBRG1 transcriptional activators and the pRB family of proteins. *Mol. Cell. Biol.*, **16**, 1576.

73. Muchardt, C., *et al.* (1998) *ras* transformation is associated with decreased expression of the brm/SNF2 ATPase from the mammalian SWI–SNF complex. *EMBO J.*, **17**, 223.

74. Versteege, I., *et al.* (1998) Truncating mutations of hSNF5/INI1 in aggressive paediatric cancer. *Nature*, **394**, 203.

6 | ATP-dependent chromatin remodelling by the ISWI complexes

CARL WU, PETER B. BECKER, and TOSHIO TSUKIYAMA

1. Introduction

Recent advances in the isolation and characterization of ATP-dependent chromatin remodelling and assembly factors provide compelling evidence for the view that the structural organization of nucleosomes in chromatin can be highly dynamic. The concept of an active chromatin architecture, fuelled by the free energy of ATP hydrolysis, was first proposed by Abraham Worcel. In the 1980s, Worcel's laboratory at the University of Rochester extended the *Xenopus* cell-free chromatin assembly system introduced by Ronald Laskey (1), by defining parameters for the reconstitution of chromatin in a high-speed supernatant prepared from *Xenopus* oocyte extracts. Using the cell-free system, Worcel and co-workers made the key observation that ATP was required for the generation of regular spacing between nucleosomes when plasmid DNA was assembled into chromatin (2). This finding provided the seed for the biochemical analysis of chromatin dynamics, now a vigorous sub-field of chromatin research centred on the ISWI (imitation *switch*), SWI2/SNF2 (*switch*/*sucrose nonfermenting*), and related members of the SWI2/SNF2 superfamily, and the protein complexes they form. Given the growing appreciation of the roles of such complexes in a variety of chromosomal processes, it is instructive to recall the summary of the first report showing ATP-dependent changes in chromatin architecture:

We describe and characterize a complex reaction that catalyzes DNA supercoiling and chromatin assembly *in vitro*. A *Xenopus* oocyte extract supplemented with ATP and Mg^{++} converts DNA circles into minichromosomes that display a native, 200 bp periodicity. When supercoiled DNA is added to this extract it undergoes a time-dependent series of topological changes, which precisely mimic those found when the DNA is microinjected into oocytes. As judged by the conformation of the subsequently deproteinized DNA, the supercoiled DNA is first relaxed, in a reaction that takes 4 min, and then it is resupercoiled in a slower process that takes 4 hr. The relaxation is partially inhibited by EDTA, to an extent that suggests that it is catalyzed by a type I DNA topoisomerase. The resupercoiling, on the other hand, requires

ATP and Mg^{++}, is completely inhibited by EDTA, and is inhibited by novobiocin in a manner that suggests it is catalyzed by a type II DNA topoisomerase. These findings, and the ones reported in the preceding paper (3), lead us to propose that chromatin assembly is an active, ATP-driven process.

(Glikin, G. C., Ruberti, I., and Worcel, A. (1984) Chromatin assembly in *Xenopus* oocytes: *in vitro* studies. *Cell*, **37**, 33–41, copyright, Cell Press)

The requirement for ATP to generate periodically spaced nucleosomes during assembly in a cell-free extract was subsequently confirmed, and assembly conditions improved in Worcel's laboratory (4, 5) and also demonstrated by other groups (6, 7). However, identification of the source of the ATP requirement, at the time thought to be the type II DNA topoisomerase, was greatly hampered by Worcel's death. Although there was progress towards this goal from studies continued by his colleagues (e.g. ref. 8), the pace did not accelerate until similar chromatin assembly systems were derived from *Drosophila* embryos (9, 10). The *Drosophila* embryo extract, presently utilized to prepare chromatin substrates for *in vitro* reactions (11, 12), proved to be a rich and more tractable source of multiple biochemical activities that could dynamically restructure chromatin. Thus, not only could the *Drosophila* extract produce regularly spaced arrays of nucleosomes, but it also allowed the disruption of a pre-assembled nucleosome array in a process requiring ATP and a promoter-specific DNA-binding protein. (13). Such a local perturbation, which creates a DNAase I hypersensitive chromatin structure (14, 15), is loosely termed chromatin remodelling.

As described below, three biochemical assays utilizing the *Drosophila* chromatin assembly system were established independently to recapitulate different aspects of chromatin dynamics: promoter-specific disruption by a DNA-binding factor, global restriction enzyme accessibility, and periodic nucleosome spacing. These assays allowed the identification and purification by chromatography of distinct chromatin remodelling and assembly factors, protein complexes comprising 2–5 subunits—NURF (*nucleosome remodelling factor*) (16), ACF (*ATP-utilizing chromatin assembly and remodelling factor*) (17), and CHRAC (*chromatin accessibility complex*) (18). NURF, ACF, and CHRAC each were found to contain a common component, the ISWI ATPase protein, originally identified by sequence similarity to Brahma, the *Drosophila* counterpart of yeast SWI2/SNF2 (19, 20) (see Chapter 5). The catalysis of chromatin remodelling by several protein complexes harbouring the identical ISWI ATPase suggests that the underlying mechanism is fundamentally the same.

2. Biochemical assays for remodelling factors using nucleosome arrays

'Nucleosome remodelling' is a fuzzy term that describes an alteration of nucleosome structure in the absence of further knowledge about the nature of the structural transition or the underlying mechanism. Nucleosome remodelling can be observed for single nucleosomes or at the level of an oligonucleosomal array. The *Drosophila* embryo extract, which allowed the reconstitution of long, regular arrays of nucleosomes under physiological conditions, seemed ideally suited as a substrate to monitor the 'remodelling' of nucleosomes within an array. Consequently, assays were developed that focused on this particular aspect of chromatin dynamics, and led to the identification of remodelling factors that apparently leave the individual nucleosome intact, but alter the position of nucleosomes within an extended array.

2.1 Promoter-specific remodelling in concert with DNA-binding proteins

The interaction of transcription factors with their binding sites in promoters and enhancers is frequently inhibited by the wrapping of the DNA around histone octamers. Consequently, active promoters in nuclei are not organized as canonical nucleosomes, but display more accessible structures. *Drosophila* embryo extracts not only contain the machinery to assemble chromatin with physiological properties, but also enzymes that facilitate the interaction of proteins with nucleosomal DNA. A key experiment involved the reconstitution of nucleosomal arrays on a model promoter such that regulatory elements were covered by nucleosomes. Next, the binding of a transcription factor with this DNA sequence was monitored and the stability of the underlying nucleosome was evaluated by probing accessibility of the DNA to micrococcal nuclease (MNase). MNase is unable to cleave nucleosome-associated DNA but digests the more accessible linker DNA between nucleosomes. Partial digestion of a nucleosomal array yields a 'ladder' of DNA fragments of defined size ranges. 150–180 bp fragments represent DNA that is protected due to its association with a single nucleosome (Fig. 1A, 'mono'), ~360 bp fragments correspond to protected DNA from di-nucleosomes (Fig. 1A, 'di'), and larger multiples of ~180 bp correspond to nucleosome trimers, tetramers, etc. (Fig. 1A, 'tri, tetra', etc). In order to test whether a DNA sequence, such as a promoter, is wrapped around a histone octamer a 'Southern blot' of such a gel is hybridized with a radiolabelled oligonucleotide probe corresponding to the region of interest (probe 'a' in Fig. 1B). Autoradiography reveals the nucleosomal fragment ladder if the target DNA is organized in a periodically spaced nucleosome array. Any change in nucleosome structure towards increased accessibility will disrupt the regular nucleosomal array and is detected by a smearing of the DNA ladder.

This assay led to the identification of NURF (13, 16) as a biochemical entity that facilitates disruption of nucleosome order in an ATP-dependent manner. Nucleosome remodelling as detected by this assay is local, since re-hybridizing the blot with an oligonucleotide probe far away from the promoter (probe 'b' in Fig. 1B) revealed no perturbance of the array at the remote site. In remodelling nucleosomes, NURF can cooperate with a wide range of sequence-specific DNA-binding proteins in a rather promiscuous manner, including those that are not native to *Drosophila*. Hence, under assay conditions, NURF remodels nucleosomes at sites determined by the sequence-specific factors.

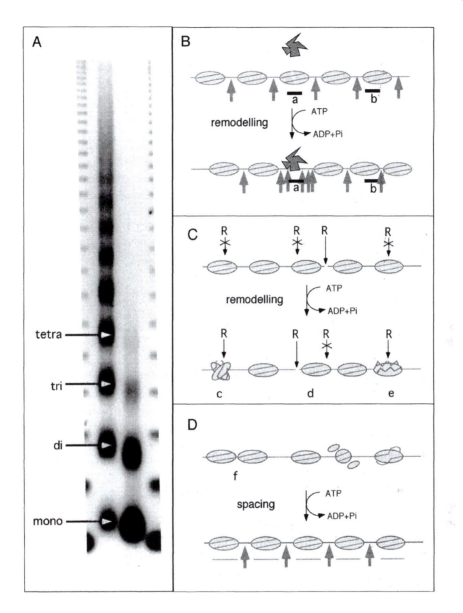

Fig. 1 Three assays to monitor ATP-dependent remodelling of nucleosomes within an array. (A) Pattern generated by digestion of chromatin with micrococcal nuclease (MNase) and separation of the resulting DNA fragments on an agarose gel. A regular fragment 'ladder' is only obtained if the nucleosomes are evenly spaced. For a detailed description, see text. (B) Nucleosome remodelling associated with transcription factor binding. Vertical arrows symbolize cleavage of the nucleosomal array with MNase. Horizontal bars (a, b) indicate the location of hybridization probes. The irregular object represents a DNA-binding protein. (C) Restriction enzyme (R) accessibility assay. In principle, energy-dependent nucleosome unfolding (c), nucleosome movement (d), and disruption of interactions between a histone octamer and DNA (e) may lead to increased accessibility to endonucleases. (D) Nucleosome spacing assay. The assembly of spaced chromatin requires the proper wrapping of DNA around histone octamers and the establishment of even distances between neighbouring nucleosomes.

2.2 Increased global accessibility to restriction enzymes

How does remodelling by NURF lead to the alteration of only those nucleosomes that cover the transcription factor binding sites? Two different mechanisms could be envisaged. Remodelling could be highly site-specific due to the targeting of NURF by the sequence-specific DNA-binding protein. Alternatively, the remodelling factor might not be targeted, but 'free running' in chromatin, generating transient access at many sites. A DNA-binding protein in the vicinity at the time of the stochastic re-modelling of the nucleosome could take advantage of this 'window of opportunity' to interact stably with DNA, thus preventing the reformation of the inaccessible structure. The latter scenario predicts that chromatin should be transiently accessible at many sites, as long as ATP is available, independent of the presence of a sequence-specific transcription factor.

The possibility of this scenario provided the impetus for the development of a restriction enzyme accessibility assay (Fig. 1C). Chromatin reconstituted on plasmid DNA was tested for transient accessibility at many sites employing restriction endo-nucleases. Like MNase, such site-specific nucleases are often unable to cleave DNA on the surface of a nucleosome, but can cut the accessible linker DNA. Cleavage of DNA in chromatin is very inefficient in the absence of ATP. By contrast, a profound increase in global accessibility is observed if nucleosome remodelling activities were activated by the addition of ATP. By tracking ATP-dependent remodelling activities in chromatographic fractions with this assay, the chromatin accessibility complex (CHRAC) was identified (18). Like the MNase digestion assay, the restriction enzyme assay does not illuminate the basis for the global increase in accessibility. In prin-ciple, access could be generated via the transient unfolding of nucleosomal structures ('c' in Fig. 1C), the relocation of intact nucleosomes ('d' in Fig. 1C), or the transient disruption of histone–DNA contacts ('e' in Fig. 1C).

2.3 Imposition of regular spacing on nucleosome arrays

If nucleosomes are assembled in embryo extracts in the absence of ATP, no regular pattern is detectable by MNase digestion analysis, indicating that the establishment of a nucleosomal array with regular spacing, i.e. with equal distances between neigh-bouring nucleosomes, is an active process requiring energy input (Fig. 1D). The search for the biochemical activities in embryo extracts that could catalyse the establishment of nucleosomal arrays with regular spacing identified ACF (17) and CHRAC (18). ACF also possesses promoter-specific nucleosome remodelling activity, as displayed by NURF. The observation that factors involved with nucleosome assembly and spacing may also function to some extent in nucleosome remodelling is not without precedent. For example, nucleoplasmin, a histone chaperone, may catalyse the assembly of nucleosomes or their disassembly (21). Thus ACF and CHRAC might catalyse forward and backward reactions, depending on reaction conditions. In the context of generating regularity in the spacing between adjacent nucleosomes, ACF and CHRAC may render intact nucleosomes mobile to facilitate their relocation on DNA. They may also help the assembly of histone octamers, or assist the proper winding of DNA around the octamer.

3. ISWI remodelling complexes purified from *Drosophila*, budding yeast, and mammal

3.1 Structure of ISWI proteins

Analysis of purified ACF, CHRAC, and NURF revealed that they were each composed of several subunits, with one common component, the ISWI ATPase. ISWI is a member of the SWI2/SNF2 superfamily, which consists of a large, conserved group of putative ATPase/helicases (22). Members of this superfamily can be subdivided into several families based on protein sequence similarity, e.g. the SNF2, ISWI (SNF2L), CHD1 (chromodomain–helicase–DNA-binding protein), Mot1 (*modulator of transcription*), and RAD54 (*radiation-sensitive*) families. The founding member of the ISWI family, *Drosophila ISWI* (*dISWI*) was identified by cross-hybridization of cDNA clones to the ATPase domain of *brahma* (*brm*) (20), a fly homologue of *SWI2/SNF2* (19). dISWI protein is highly homologous to Brahma and SWI2/SNF2p only within the ATPase domain (50% amino-acid identity) (20). ISWI-family members do not carry the bromodomain motif (23) characteristically found in the SNF2 family, but they possess a SANT domain (24), which is contained in several transcription regulators or proteins implicated in chromatin functions (SWI3, ADA2, N-CoR, TFIIIB and c-Myb). *ISWI* genes have been found in budding yeast and humans; the corresponding proteins show strong sequence similarity to dISWI over the entire open reading frame, including the ATPase and SANT domains.

3.2 ISWI complexes from *Drosophila*

At least three ISWI complexes have been purified from *Drosophila* extracts (Fig. 2). NURF has a native molecular mass of *c.* 500 kDa, and is composed of four subunits: p215, dISWI (25), p55, and p38. The p55 subunit is a *Drosophila* homologue of RbAp46/48 (retinoblastoma-*associated protein*) (26). RbAp46/48 family proteins are highly conserved from budding yeast to human, and are found in multiple chromatin regulators, including human and yeast chromatin assembly factor-I (CAF-I, see Chapter 2) (27, 28), and yeast histone acetyltransferase 1 (29). It has been demonstrated that human RbAp46/48 proteins interact directly with histone H4 (30), suggesting a corresponding interaction for p55 in NURF. The p38 subunit of NURF is an inorganic pyrophosphatase (PPase) (31), of which only a small percentage of the total cellular enzyme is contained in NURF. The function of p38 in remodelling is unclear; the PPase activity appears to be dispensable for chromatin remodelling *in vitro* (31).

CHRAC was purified from *Drosophila* embryo extracts by tracking its ability to facilitate access of restriction enzymes to chromatin (see Section 2.2) (18). The most purified fraction of CHRAC has a native molecular mass of *c.* 670 kDa, and is com-

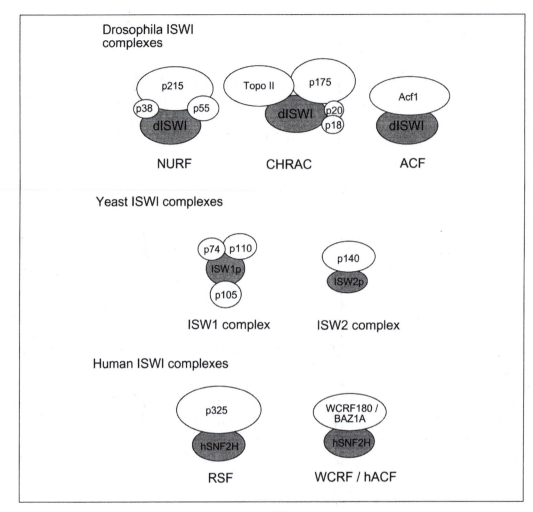

Fig. 2 Diagram showing *Drosophila*, yeast, and human ISWI complexes.

posed of 175, 160 (topoisomerase II (topo II)), 130 (dISWI), 20, and 18 kDa proteins. The topoisomerase activity is dispensable for chromatin remodelling by CHRAC (18). Topo II may recruit the CHRAC complex to specific sites on chromosomes, or chromatin remodelling by CHRAC may facilitate access of topo II to DNA in chromatin.

ACF was purified from *Drosophila* embryos by following its ATP-dependent nucleosome spacing and deposition activities (32). ACF is comprised of two subunits, dISWI (140 kDa) and Acf1 (185 and/or 170 kDa) (33). The 185 kDa and 170 kDa forms of Acf1 are encoded by the same gene. Multiple sequence motifs have been found within Acf1, including two PHD (*plant homeodomain*) fingers and a bromodomain. PHD finger motifs have been identified in several transcription factor and chromatin regulators; their functions remain to be determined (34). Bromodomain motifs have also been identified in multiple chromatin or transcription regulators (23), and the bromodomain of the histone acetyltransferase Gcn5p (general control non-derepressible) has been show to interact physically with the N-terminal tail of histone H4 (35, 36). In addition, two novel motifs named WAC (WSTF/Acf1/cbp146) and WAKZ (WSTF/Acf1/KIAA0314/ZK783.4) are found in Acf1. It is of interest that every conserved motif in Acf1 is also found in human WSTF (Williams syndrome transcription factor) (37), which is encoded by one of many genes deleted in individuals with Williams syndrome, a complex developmental disorder with multi-systemic defects.

3.3 Yeast ISW1 and ISW2 complexes

Two *ISWI* genes in budding yeast, *ISW1* and *ISW2*, were identified by their homology to *dISWI* (38), and the corresponding proteins also form complexes (Fig. 2). However, despite extensive sequence similarity between ISW1p and ISW2p, the two complexes do not share common subunits. ISW1p co-purifies with three proteins, p110, p105, and p74, and ISW2p co-purifies with a single protein, p140 (38). None of these subunits show extensive sequence similarity with known subunits of the *Drosophila* ISWI complexes, except for a small domain shared between p140 of the ISW2 complex, Acf1, and WCRF/hACF (Toshi Tsukiyama, unpublished observations; see below). The yeast ISW1 and ISW2 complexes have distinct chromatin remodelling activities —only the ISW1 complex exhibits promoter-specific remodelling activity, while both complexes can generate evenly spaced nucleosomes with different periodicities (175 bp for the ISW1 complex, and 210 bp for the ISW2 complex). The ISW1 and ISW2 complexes may have partially overlapping functions *in vivo*, since *isw1* and *isw2* mutations cause synthetic stress-sensitive phenotypes (see Section 6.1) (38).

3.4 Human ISWI complexes

As in yeast, two *ISWI* genes have been identified in human. *hSNF2L* (39) was isolated in a genomic walk, while *hSNF2H* was identified from a cDNA library based on its homology to *dISWI* and *hSNF2L* (40). Both hSNF2L and hSNF2H proteins have extensive homology to dISWI (more than 70% amino-acid identity), making them the closest known relatives of dISWI. hSNF2H is contained in at least two complexes (Fig. 2). RSF (remodelling and spacing factor) (41) consists of hSNF2H and a 325 kDa polypeptide, exhibits promoter-specific remodelling and nucleosome spacing activities, and stimulates transcription of chromatin *in vitro*. hACF (42)/WCRF (Williams syndrome transcription factor-related chromatin remodelling factor) (43) consists of hSNF2H associated with WCRF180 (37)/BAZ1A (42). WCRF180/BAZ1A is highly similar to WSTF, and shares all the conserved motifs found in Acf1. hACF/WCRF displays chromatin remodelling and assembly activities similar to those of *Drosophila* ACF. Complexes containing the other human ISWI protein, hSNF2L, have not yet been identified.

3.5 Related remodelling factors

A gene required for maintenance of genomic methylation, *DDM1* (*decreased DNA methylation*) was recently identified in *Arabidopsis* (44). The DDM1 protein has relatively high similarity (*c.* 45% identity) to hSNF2H in the ATPase domain, and is more related to the ISWI family than to any other group in the genome databases. Although the DDM1 protein has not yet been characterized biochemically, genetic evidence suggests that *DDM1* facilitates maintenance DNA methylation by regulating chromatin structure. One possible model is that the DDM1 protein or protein complex remodels chromatin to facilitate the access of maintenance DNA methyltransferases.

By two independent phylogenetic analyses, the family of CHD1 proteins is most closely related to the ISWI family (22). The vertebrate Mi-2 proteins, members of the CHD1 family, form complexes that have ATP-dependent chromatin remodelling activity as well as histone deacetylase activity (45–48) (see Chapter 8). The yeast *CHD1* gene shows strong genetic interactions with *ISW1* and *ISW2* genes (38), suggesting that it may also function in modulating chromatin structure. There are many additional SWI2/SNF2 superfamily members in the yeast genome database that have not been characterized, raising the possibility that some of those proteins may also function as ATP-dependent chromatin remodelling machines.

4. Mechanism of nucleosome remodelling—catalysed mobility of histone octamer

4.1 General considerations

How can ISWI complexes increase access to nucleosomal DNA? ISWI complexes are able to trigger the repositioning of nucleosomes if a boundary, such as a high-affinity DNA-binding protein, is newly integrated into chromatin (16, 17, 49–51). In these reactions, the positions of nucleosomes change, presumably to avoid steric clashes with DNA-binding motifs. Thus, nucleosomes remodelled by ISWI complexes are apparently not terminally disassembled or destroyed, but remain intact after the re-modelling process. Moreover, ACF and CHRAC (but not NURF), yeast ISW1 and ISW2 complexes, and hACF are able to improve the quality of chromatin by catalysing the ordering of nucleosomes from an irregular arrangement into a regularly spaced array (see Section 2.3). Increased access to nucleosomal DNA can be achieved con-comitantly with an improvement of the regularity of nucleosome spacing if one assumes that these factors catalyse movement of the histone octamer on DNA. Such movements would assure that any given DNA sequence would be stochastically exposed at least transiently in the more accessible linker DNA between nucleosomes. How can nucleosome mobility be brought about? Two scenarios can be envisaged:

(1) nucleosomes could be moved without disruption of the histone octamer in a process that could best be described as 'sliding'; and

(2) they could be disassembled into subnucleosomal parts and rapidly reassembled at a vicinal site on DNA.

It turns out that the former possibility is correct.

4.2 Histone octamer sliding

An analysis of nucleosome mobility requires that nucleosomes are positioned to begin with and that the change in position can be monitored. Although most DNA sequences can be wrapped around a histone octamer, some DNA sequences can bend easier than others, and preferentially position nucleosomes (for a discussion of the principles that govern nucleosome positioning, see Chapter 1). Therefore, if mono-nucleosomes are reconstituted by salt gradient dialysis on short DNA fragments they frequently adopt preferred positions, which depend on the underlying sequence. These positions can be determined by 'footprinting' techniques that employ nucleases, such as MNase and restriction endonucleases. Because nucleosomal DNA is effectively bent when associated with the histone octamer, the nucleosomes positioned close to a DNA fragment end are conformationally distinct from those that occupy a more central position (Fig. 3A). The different nucleosome conformations are electrophoretically separable on a native gel. Since nucleosomes are very stable entities, each conformationally distinct species can be purified by gel elution without a positional change. Any alternation in position caused by remodelling factors can then be monitored easily by repeating the gel electrophoresis, and further verified by footprinting techniques. This sensitive assay allows one to test the effect of nucleosome remodelling factors on nucleosome distributions.

When mononucleosomes are reconstituted on a 360 bp DNA fragment derived from the *Drosophila* hsp70 promoter, a number of preferred positions are adopted. In addition to nucleosomes that abut the fragment ends (a favoured location with little physiological relevance) nucleosomes also occupy several internal positions, covering most of the promoter region (52). In the presence of hydrolysable ATP, purified NURF is able to move nucleosomes from specific positions in the promoter to a more upstream position (Fig. 3B). The repositioning of nucleosomes renders accessible critical regulatory elements, including the binding sites for the regulators GAGA and heat-shock factor. In an analogous set of experiments, CHRAC catalysed the repositioning of nucleosomes from the ends of an rDNA promoter fragment to a central, preferred position (51) (Fig. 3C). During the remodelling process, nucleosomes are not disassembled and reassembled by NURF or CHRAC, as measured by various criteria. Rather, nucleosomes move along DNA as intact entities in a process that can be best described as 'sliding'. Competition experiments with free DNA *in trans* show that nucleosomes never lose contact with the DNA while sliding (51, 52). It is likely that ACF and the yeast and human ISWI complexes remodel nucleosomes by a similar mechanism.

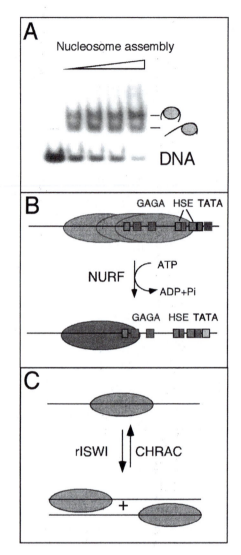

Fig. 3 ISWI complexes catalyse ATP-dependent nucleosome sliding. (A) Nucleosomes reconstituted at the end or in the centre of a DNA fragment can be separated on a native polyacrylamide gel. (B) NURF catalyses the sliding of nucleosomes, exposing some of the regulatory elements of the *Drosophila* hsp70 promoter. (C) CHRAC catalyses the movement of nucleosomes from a fragment end to central positions. By contrast, recombinant ISWI mobilizes nucleosomes from a central position to fragment ends.

4.3 ISWI is the engine of nucleosome mobility

Although the presence of ISWI as a common component of ACF, CHRAC, and NURF suggests a critical role for this ATPase, direct proof of this notion came from an analysis of recombinant ISWI in isolation. ISWI itself is able to recognize nucleosome structure—the basal ATPase activity is strongly stimulated by addition of nucleosomes to the reaction, whereas inclusion of free DNA or free histones have little effect (53). Furthermore, recombinant ISWI alone, albeit at concentrations stoichiometric to nucleosomes, is able to carry out the basics of nucleosome remodelling—generating access to DNA, repositioning nucleosomes, establishing regular spacing between nucleosomes after assembly, and moving nucleosomes on short fragments (51–53).

However, there is little free ISWI in extracts, and the subunits that associate with ISWI in the respective complexes clearly affect nucleosome remodelling quantitatively and qualitatively. Native remodelling complexes are more efficient than the recombinant ISWI alone (51, 52). Moreover, the addition of Acf1 to ISWI to form the ACF complex reveals that the two components function synergistically in chromatin assembly (33). A comparison of nucleosome mobility catalysed by CHRAC, NURF, and recombinant ISWI also pointed to qualitative differences, as seen by the directionality of nucleosome movement. For example, while CHRAC caused nucleosomes located at fragment ends to slide to central positions on the DNA, ISWI was unable to mobilize peripheral nucleosomes, but triggered the relocation of centrally positioned particles to end positions (Fig. 3C). Although fragment ends are unlikely to have physiological relevance, the changes in directionality underscore the modulation of ISWI activity by associated subunits.

4.4 Similarities and differences in remodelling by ISWI and SWI/SNF-like complexes

How does nucleosome remodelling by ISWI complexes compare to analogous reactions carried out by the other major class of remodelling ATPases, represented by SWI/SNF, RSC, and their counterparts in other species? The SWI/SNF complex is able to catalyse a repositioning of nucleosomes on DNA (54). Nevertheless, a number of phenomena that are characteristic of the SWI/SNF type of nucleosome remodelling reactions (see Chapter 5) have not been observed in the nucleosome sliding reactions catalysed by ISWI:

(1) Remodelling by SWI/SNF factors leads to an obvious disruption of histone–DNA interactions throughout the nucleosome.

(2) SWN/SNF factors cause a significant loss of constrained superhelicity from a topologically closed minichromosome in the presence of topoisomerase, again pointing to a dramatic loss of histone–DNA interactions.

(3) The 'remodelled' state of the nucleosome, as determined by these altered histone–DNA interactions, is stable once the energy source is removed from the reaction.

(4) The remodelled state has been correlated to the formation of an atypical 'dinucleosomal' particle that may represent a kinetic intermediate of the remodelling reaction.

(5) Finally, SWI/SNF-type factors, particularly RSC, have been shown to cause the 'eviction' of nucleosomes from a DNA fragment and their transfer to competing DNA.

Although these differences have to be interpreted with caution until the various factors have been tested side-by-side in standardized reactions, it appears that nucleosome remodelling by SWI/SNF-like factors involves drastic disruption of histone–DNA interactions that may be due, in part, to high-affinity binding of the complexes.

By contrast, a stable interaction between nucleosomes and ISWI complexes has not been detected, and remodelling by ISWI complexes appears to happen in many small steps, each involving the disruption of only few bonds at a time, but preserving overall nucleosome structure (55, 56) (see Chapter 1). DNA may be twisted at the sites where it enters its winding path around the histone octamer. Since small distortions in nucleosome structure can be accommodated (57), ISWI complexes may catalyse the propagation of a change in the twist of the DNA helix over the nucleosomal surface, effectively helping the DNA to 'screw' over the surface of the histone octamer. ISWI complexes also show a preferential recognition of the nucleosome as a substrate, as indicated from the stimulation of ATPase activity by nucleosomes but not by free DNA. An element of recognition includes the histone tails, removal of which from nucleosomes, or presentation as glutathione S-transferase (GST)–peptide competitors, abrogates the stimulation.

5. Nucleosome mobility can potentially affect nuclear functions

What might be the functional consequences of nucleosome mobilization? A priori, any process involving the readout of DNA information by a macromolecular enzyme could be facilitated by an increase in nucleosome dynamics. The ISWI complexes ACF, NURF, CHRAC, RSF, and hACF/WCRF, have each been shown to affect a variety of DNA transactions when analysed by biochemical assays. However, it should be noted that the *in vitro* reactions described below, while potentially revealing, may not necessarily represent the *in vivo* functions of ISWI complexes.

5.1 Transcription of chromatin templates *in vitro*

ACF, hACF, NURF, and RSF can facilitate formation of a preinitiation complex on chromatin templates and allow transcription up to *c*.100 nucleotides downstream from the transcription start site (17, 41, 58). These studies utilized a plasmid carrying a simple model promoter consisting of five tandemly repeated GAL4-binding sites upstream of a minimal adenovirus E4 core promoter (59). Typically, plasmid DNA is assembled into chromatin and purified by gel filtration. Alternatively, chromatin reconstitution can be conducted using a purified system consisting of core histones, NAP-1 (*n*ucleosome *a*ssembly *p*rotein), and ACF (see below). Chromatin remodelling then proceeds in the presence of ATP, ISWI complex, and GAL4-activator protein, before introduction into a crude or purified transcription system (Fig. 4A).

In such experiments, reconstitution of the model template into chromatin dramatically represses its ability to be transcribed when compared to transcription of free

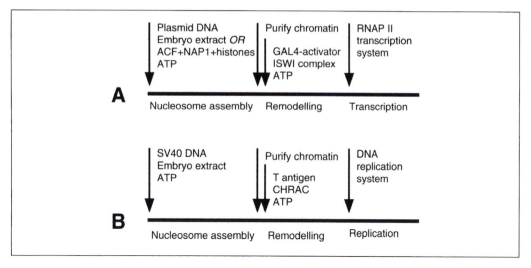

Fig. 4 Effects of chromatin remodelling on DNA metabolism. (A) Protocol for ATP-dependent chromatin assembly, ATP-dependent chromatin remodelling by ISWI complexes, and *in vitro* transcription. (B) Protocol for ATP-dependent chromatin assembly, ATP-dependent chromatin remodelling by CHRAC, and DNA replication *in vitro*.

DNA. Activation, or alleviation of repression, is dependent on the action of the ISWI complex ACF (17). Similar results are obtained with NURF (58), RSF (41), and hACF (42). Interestingly, activation is not observed when the GAL4 DNA-binding domain without activator is employed in the assay, despite clearly effective remodelling over promoter sequences. This finding indicates that ATP-dependent remodelling is necessary, but not sufficient for activation. Activation regions apparently play an important role in helping alleviate chromatin-mediated repression, perhaps by recruitment of basal transcription factors. In addition, while transcriptional activation of chromatin can be observed for ACF, hACF, NURF, and RSF, it is not detectable for CHRAC, or the yeast SWI/SNF complex (58). While differences may be related to assay conditions, they may also signify functional specialization between remodelling machines.

5.2 T-antigen-dependent DNA replication

The initiation of DNA replication *in vitro* from a nucleosomal origin is another transaction that requires an ISWI complex. When SV40 DNA is reconstituted into chromatin, the presence of CHRAC and ATP is necessary to allow efficient SV40 T-antigen-dependent replication to proceed from the origin (60). For these experiments, the fraction of reconstituted viral chromatin containing an accessible origin is eliminated by digestion with a restriction enzyme that cuts SV40 DNA once within origin sequences. The viral chromatin is then incubated with CHRAC and ATP, followed by further incubation with SV40 T-antigen, and introduction into a DNA replication system (61) (Fig. 4B). It can be shown that chromatin remodelling at origin sequences mediated by the ATP-dependent action of CHRAC facilitates binding of T antigen, and leads to stimulation of DNA replication. Moreover, the stimulatory effects of CHRAC are entirely due to initiation of replication rather than elongation, as indicated by pulse–chase experiments.

5.3 Assembly and spacing of nucleosomes

Purified ACF is essential for the assembly of extended, periodic nucleosome arrays in a reaction containing supercoiled plasmid DNA, DNA topoisomerase I (to relax unconstrained supercoils), ATP, purified core histones, and a histone chaperone dNAP-1 (62) (Fig. 4A). Importantly, the level of ACF relative to nucleosomes required for the assembly of periodically spaced arrays is of the order of 1:90 (assuming 1 ISWI polypeptide/ACF complex), indicating that ACF acts in a catalytic fashion, in contrast to the stoichiometric requirements for dNAP-1. In addition, ACF appears to act synergistically with dNAP-1 in the assembly reaction, suggesting that ACF facilitates both the deposition of histones and their subsequent mobilization on DNA to generate periodic spacing. Interestingly, dNAP-1 can be interchanged with another histone chaperone, CAF-1 (12, 61, 63), in this reaction. The human counterpart of *Drosophila* ACF, hACF, displays a similar ability to assemble periodically spaced nucleosomes (42).

The cloning and expression of the gene for the large subunit of ACF, Acf1 (33), has allowed reconstitution of full ACF activity (33) from recombinant Acf1 and recombinant ISWI (52, 53). The two subunits of ACF act cooperatively in catalysing the assembly of periodic nucleosome arrays, with Acf1 providing additional functionality to the ISWI engine of remodelling. The reconstitution of recombinant ACF is a key advance that establishes a purified system for the assembly of native chromatin *in vitro*. The purified system should provide defined chromatin substrates for a broad range of DNA transactions, and facilitate further dissection of the chromatin assembly process *in vitro*.

6. *In vivo* functions of ISWI

6.1 Physiological roles of yeast ISW1 and ISW2

In budding yeast, an *isw1* null mutation causes a weak temperature-sensitive (Ts) phenotype (38). An *isw2* null mutation does not exhibit detectable growth defects, but an *isw1 isw2* double mutant has a Ts phenotype that is stronger than that of the single *isw1* mutant. Furthermore, an *isw1 isw2 chd1* triple mutation causes an even stronger Ts phenotype (10^{-4}–10^{-5} viability compared to wild type at 38.5 °C). The triple mutant shows sensitivity to other environmental stresses, such as the presence of formamide and ethanol in culture media. The mechanisms underlying this stress sensitivity, which is not caused by an inability to induce the heat-shock response, are unknown (Toshi Tsukiyama, unpublished observations). It is possible that the absence of yeast ISWIp and ISW2p alters transcription of certain yeast genes required for cell viability under environmentally stressful conditions. The expression patterns of the *ISW1* and *ISW2* genes respond differently to environmental changes—*ISW1* mRNA decreases upon starvation and sporulation, while *ISW2* mRNA increases upon heat shock and sporulation (65). A recent report reveals that the *isw2* homozygous mutant exhibits sporulation defects at early stages of meiosis (64). Furthermore, during meiosis, the intracellular localization of ISW2p changes, suggesting that the localization and activity of the ISW2 complex may be subject to regulation.

6.2 *Drosophila* ISWI is required for transcription and chromosome structure

The *Drosophila ISWI* gene is essential for both normal development and cell viability (66). Homozygous *ISWI* null mutants die at late larval or early pupal stages. A large amount of maternally deposited ISWI protein (~100 000 molecules/nucleus (25)) is probably responsible for survival until relatively late in development, a period when gross morphological patterns of the fly are likely to have been established. Thus, homozygous *ISWI* null mutants die without any obvious morphological abnormalities. The requirement of ISWI protein for cell viability is also shown by the expression of a dominant negative form of ISWI protein, resulting in diminished eyes in adult flies due to cell lethality. *ISWI* mutants also show a strikingly altered structure of the male X chromosome, which becomes much shorter and broader than the normal X (66). Transcription from the male X chromosome is about twofold higher than that from the female X, due to dosage compensation, and the male X is also known to be hyperacetylated at lysine 16 of histone H4. It is possible that the mechanisms for dosage compensation somehow make the male X more sensitive to the loss of ISWI. In addition, the ISWI protein is localized at discrete heterochromatic regions (66). Taken together, the data suggest that ISWI is involved in the maintenance of chromosome architecture.

There is initial evidence supporting a physiological role for the ISWI protein in gene expression, since the level of engrailed and ultrabithorax proteins in imaginal discs are dramatically reduced in *ISWI* mutants (66). However, immunostaining of polytene chromosomes shows that ISWI protein and RNA polymerase II are not generally co-localized, raising the prospect of a more complex role for ISWI (at least the fraction of ISWI that is stably associated with polytene chromosomes). It is possible that ISWI is involved in both positive and negative regulation of transcription. Because ISWI is contained in multiple protein complexes, it is not yet possible to assign the different phenotypes of *ISWI* mutants to a specific ISWI-containing complex. For this purpose, it will be important to study the genes coding for subunits that are exclusively present in each complex.

7. Summary and discussion

Biochemical studies have led to the identification and purification of distinct ISWI protein complexes. ISWI complexes contain the ISWI ATPase, which is conserved from yeast to human, and one or more additional components. ISWI complexes participate in chromatin dynamics by catalysing nucleosome remodelling in an ATP-dependent fashion. Remodelling involves histone octamer sliding, which may be caused by propagation of a twist or bulge along nucleosomal DNA. Nucleosome movement by the ISWI complexes can affect several DNA processing activities, including transcription, replication, and chromatin assembly *in vitro*. Initial genetic studies of *Drosophila* and yeast *ISWI* mutants reveal phenotypes that are generally consistent with the biochemical activities.

Although our understanding of chromatin dynamics has been greatly increased by the studies described in this chapter, there are important issues to be resolved, and many new and interesting questions. Perhaps most important is the need to elucidate the *in vivo* function(s) of each ISWI complex, by conducting genetic studies for sub-units that are exclusively present in, for example, ACF, NURF, and CHRAC. Understanding the mechanism of nucleosome movement in molecular detail also presents a formidable challenge. In addition, biochemical and genetic studies are required to assess whether ISWI complexes operate in a global manner throughout the genome, or are targeted to specific promoters by transcription factors, as has been demonstrated for the SWI2/SNF2 complexes (see Chapter 5). It is conceivable that both local and global mechanisms are utilized to some extent in the life of an organism, depending on the endogenous concentration of complex. Also, little is known about how the nucleosome assembly and nucleosome spacing activities possessed by an ISWI complex are related, and how usage is coordinated between different families of ATP-dependent remodelling complexes. Connections between the ISWI complexes, which alter chromatin non-covalently, and the covalent modification of histones (e.g. by histone acetyltransferases and histone deacetylases) have not been explored. Given that acetylated histones have affinity for a bromodomain (see Chapter 7), the presence of such a motif in Acf1 and WCRF180/BAZ1A provides an intriguing intersection between two avenues of chromatin modification. From a biological perspective, it is also important to learn how the expression, localization, and activity of ISWI complexes are controlled during growth, differentiation, and development. These and other issues provide outstanding opportunities for future study in chromatin dynamics.

Acknowledgements

We wish to thank members of our laboratories for their contributions, and acknowledge the support of the Intramural Research Program of the National Cancer Institute, USA (CW), the European Molecular Biology Laboratory and the Deutsche Forschungsgemeinschaft (PBB), and the Pew Charitable Trust and the National Institutes of Health, USA—GM58465–01 (TT).

References

1. Laskey, R. A., Mills, A. D. and Morris, N. R. (1977) Assembly of SV40 chromatin in a cell-free system from *Xenopus* eggs. *Cell*, **10**, 237.

2. Glikin, G. C., Ruberti, I., and Worcel, A. (1984) Chromatin assembly in *Xenopus* oocytes: *in vitro* studies. *Cell*, **37**, 33.

3. Ryoji, M. and Worcel, A. (1984) Chromatin assembly in *Xenopus* oocytes: *in vivo* studies. *Cell*, **37**, 21.

4. Ruberti, I. and Worcel, A. (1986) Mechanism of chromatin assembly in *Xenopus* oocytes. *J. Mol. Biol.*, **189**, 457.

5. Shimamura, A., Tremethick, D., and Worcel, A. (1988) Characterization of the repressed 5S DNA minichromosomes assembled *in vitro* with a high-speed supernatant of *Xenopus laevis* oocytes. *Mol. Cell. Biol.*, **8**, 4257.

6. Almouzni, G. and Mechali, M. (1988) Assembly of spaced chromatin promoted by DNA synthesis in extracts from *Xenopus* eggs. *EMBO J.*, **7**, 665.

7. Knezetic, J. A. and Luse, D. S. (1986) The presence of nucleosomes on a DNA template prevents initiation by RNA polymerase II *in vitro*. *Cell*, **45**, 95.

8. Tremethick, D. J. and Frommer, M. (1992) Partial purification, from *Xenopus laevis* oocytes, of an ATP-dependent activity required for nucleosome spacing *in vitro*. *J. Biol. Chem.*, **267**, 15041.

9. Kamakaka, R. T., Bulger, M., and Kadonaga, J. T. (1993) Potentiation of RNA polymerase II transcription by Gal4-VP16 during but not after DNA replication and chromatin assembly. *Genes Develop.*, **7**, 1779.

10. Becker, P. B. and Wu, C. (1992) Cell-free system for assembly of transcriptionally repressed chromatin from *Drosophila* embryos. *Mol. Cell. Biol.*, **12**, 2241.

11. Becker, P. B., Tsukiyama, T., and Wu, C. (1994) Chromatin assembly extracts from *Drosophila* embryos. *Methods Cell Biol.*, **44**, 207.

12. Bulger, M., Ito, T., Kamakaka, R. T., and Kadonaga, J. T. (1995) Assembly of regularly spaced nucleosome arrays by *Drosophila* chromatin assembly factor 1 and a 56-kDa histone-binding protein. *Proc. Natl Acad. Sci., USA*, **92**, 11726.

13. Tsukiyama, T., Becker, P. B., and Wu, C. (1994) ATP-dependent nucleosome disruption at a heat-shock promoter mediated by binding of GAGA transcription factor. *Nature*, **367**, 525.

14. Wu, C. (1984) Two protein-binding sites in chromatin implicated in the activation of heat-shock genes. *Nature*, **309**, 229.

15. Wu, C. (1980) The 5' ends of *Drosophila* heat shock genes in chromatin are hypersensitive to DNase I. *Nature*, **286**, 854.

16. Tsukiyama, T. and Wu, C. (1995) Purification and properties of an ATP-dependent nucleosome remodeling factor. *Cell*, **83**, 1011.

17. Ito, T., Bulger, M., Pazin, M. J., Kobayashi, R., and Kadonaga, J. T. (1997) ACF, an ISWI-containing and ATP-utilizing chromatin assembly and remodeling factor. *Cell*, **90**, 145.

18. Varga-Weisz, P. D., Wilm, M., Bonte, E., Dumas, K., Mann, M., and Becker, P. B. (1997) Chromatin-remodelling factor CHRAC contains the ATPases ISWI and topoisomerase II. *Nature*, **388**, 598.

19. Tamkun, J. W., Deuring, R., Scott, M. P., Kissinger, M., Pattatucci, A. M., Kaufman, T. C., and Kennison, J. A. (1992) brahma: a regulator of *Drosophila* homeotic genes structurally related to the yeast transcriptional activator SNF2/SWI2. *Cell*, **68**, 561.

20. Elfring, L. K., Deuring, R., McCallum, C. M., Peterson, C. L., and Tamkun, J. W. (1994) Identification and characterization of *Drosophila* relatives of the yeast transcriptional activator SNF2/SWI2. *Mol. Cell. Biol.*, **14**, 2225.

21. Chen, H., Li, B., and Workman, J. L. (1994) A histone-binding protein, nucleoplasmin, stimulates transcription factor binding to nucleosomes and factor-induced nucleosome disassembly. *EMBO J.*, **13**, 380.

22. Eisen, J. A., Sweder, K. S., and Hanawalt, P. C. (1995) Evolution of the SNF2 family of proteins: subfamilies with distinct sequences and functions. *Nucleic Acids Res.*, **23**, 2715.

23. Haynes, S. R., Dollard, C., Winston, F., Beck, S., Trowsdale, J., and Dawid, I. B. (1992) The bromodomain: a conserved sequence found in human, *Drosophila* and yeast proteins. *Nucleic Acids Res.*, **20**, 2603.

24. Aasland, R., Stewart, A. F., and Gibson, T. (1996) The SANT domain: a putative DNA-binding domain in the SWI–SNF and ADA complexes, the transcriptional co-repressor N-CoR and TFIIIB. *Trends Biochem. Sci.*, **21**, 87.

25. Tsukiyama, T., Daniel, C., Tamkun, J., and Wu, C. (1995) ISWI, a member of the SWI2/SNF2 ATPase family, encodes the 140 kDa subunit of the nucleosome remodeling factor. *Cell*, **83**, 1021.

26. Martinez-Balbas, M. A., Tsukiyama, T., Gdula, D., and Wu, C. (1998) *Drosophila* NURF-55, a WD repeat protein involved in histone metabolism. *Proc. Natl Acad. Sci., USA*, **95**, 132.

27. Verreault, A., Kaufman, P. D., Kobayashi, R., and Stillman, B. (1996) Nucleosome assembly by a complex of CAF-1 and acetylated histones H3/H4. *Cell*, **87**, 95.

28. Kaufman, P. D., Kobayashi, R., and Stillman, B. (1997) Ultraviolet radiation sensitivity and reduction of telomeric silencing in *Saccharomyces cerevisiae* cells lacking chromatin assembly factor-I. *Genes Develop.*, **11**, 345.

29. Parthun, M. R., Widom, J., and Gottschling, D. E. (1996) The major cytoplasmic histone acetyltransferase in yeast: links to chromatin replication and histone metabolism. *Cell*, **87**, 85.

30. Verreault, A., Kaufman, P. D., Kobayashi, R., and Stillman, B. (1998) Nucleosomal DNA regulates the core-histone-binding subunit of the human Hat1 acetyltransferase. *Curr. Biol.*, **8**, 96.

31. Gdula, D. A., Sandaltzopoulos, R., Tsukiyama, T., Ossipow, V., and Wu, C. (1998) Inorganic pyrophosphatase is a component of the *Drosophila* nucleosome remodeling factor complex. *Genes Develop.*, **12**, 3206.

32. Ito, T., Tyler, J. K., Bulger, M., Kobayashi, R., and Kadonaga, J. T. (1996) ATP-facilitated chromatin assembly with a nucleoplasmin-like protein from *Drosophila melanogaster*. *J. Biol. Chem.*, **271**, 25041.

33. Ito, T., Levenstein, M. E., Fyodorov, D. V., Kutach, A. K., Kobayashi, R., and Kadonaga, J. T. (1999) ACF consists of two subunits, Acf1 and ISWI, that function cooperatively in the ATP-dependent catalysis of chromatin assembly. *Genes Develop.*, **13**, 1529.

34. Aasland, R., Gibson, T. J., and Stewart, A. F. (1995) The PHD finger: implications for chromatin-mediated transcriptional regulation. *Trends Biochem. Sci.*, **20**, 56.

35. Ornaghi, P., Ballario, P., Lena, A. M., Gonzalez, A., and Filetici, P. (1999) The bromo-domain of Gcn5p interacts *in vitro* with specific residues in the N terminus of histone H4. *J. Mol. Biol.*, **287**, 1.

36. Dhalluin, C., Carlson, J. E., Zeng, L., He, C., Aggarwal, A. K., and Zhou, M. M. (1999) Structure and ligand of a histone acetyltransferase bromodomain. *Nature*, **399**, 491.

37. Lu, X., Meng, X., Morris, C. A., and Keating, M. T. (1998) A novel human gene, WSTF, is deleted in Williams syndrome. *Genomics*, **54**, 241.

38. Tsukiyama, T., Palmer, J., Landel, C. C., Shiloach, J., and Wu, C. (1999) Characterization of the imitation switch subfamily of ATP-dependent chromatin-remodeling factors in *Saccharomyces cerevisiae*. *Genes Develop.*, **13**, 686.

39. Okabe, I., Bailey, L. C., Attree, O., Srinivasan, S., Perkel, J. M., Laurent, B. C., Carlson, M., Nelson, D. L., and Nussbaum, R. L. (1992) Cloning of human and bovine homologs of SNF2/SWI2: a global activator of transcription in yeast *S. cerevisiae*. *Nucleic Acids Res.*, **20**, 4649.

40. Aihara, T., Miyoshi, Y., Koyama, K., Suzuki, M., Takahashi, E., Monden, M., and Nakamura, Y. (1998) Cloning and mapping of SMARCA5 encoding hSNF2H, a novel human homologue of *Drosophila* ISWI. *Cytogenet. Cell Genet.*, **81**, 191.

41. LeRoy, G., Orphanides, G., Lane, W. S., and Reinberg, D. (1998) Requirement of RSF and FACT for transcription of chromatin templates *in vitro*. *Science*, **282**, 1900.

42. LeRoy, G., Loyola, A., Lane, W. S., and Reinberg, D. (2000) Purification and characterization of a human factor that assembles and remodels chromatin. *J. Biol. Chem.*, **275**, 14787.

43. Bochar, D. A., Savard, J., Wang, W., Lafleur, D. W., Moore, P., Cote, J., and Shiekhattar, R. (2000) A family of chromatin remodeling factors related to Williams syndrome transcription factor. *Proc. Natl Acad. Sci., USA*, **97**, 1038.

44. Jeddeloh, J. A., Stokes, T. L., and Richards, E. J. (1999) Maintenance of genomic methylation requires a SWI2/SNF2-like protein. *Nature Genet.*, **22**, 94.

45. Wade, P. A., Jones, P. L., Vermaak, D., and Wolffe, A. P. (1998) A multiple subunit Mi-2 histone deacetylase from *Xenopus laevis* cofractionates with an associated Snf2 superfamily ATPase. *Curr. Biol.*, **8**, 843.

46. Zhang, Y., LeRoy, G., Seelig, H. P., Lane, W. S., and Reinberg, D. (1998) The dermatomyositis-specific autoantigen Mi2 is a component of a complex containing histone deacetylase and nucleosome remodeling activities. *Cell*, **95**, 279.

47. Tong, J. K., Hassig, C. A., Schnitzler, G. R., Kingston, R. E., and Schreiber, S. L. (1998) Chromatin deacetylation by an ATP-dependent nucleosome remodelling complex. *Nature*, **395**, 917.

48. Xue, Y., Wong, J., Moreno, G. T., Young, M. K., Cote, J., and Wang, W. (1998) NURD, a novel complex with both ATP-dependent chromatin-remodeling and histone deacetylase activities. *Mol. Cell*, **2**, 851.

49. Wall, G., Varga-Weisz, P. D., Sandaltzopoulos, R., and Becker, P. B. (1995) Chromatin remodeling by GAGA factor and heat shock factor at the hypersensitive *Drosophila* hsp26 promoter *in vitro*. *EMBO J.*, **14**, 1727.

50. Varga-Weisz, P. D., Blank, T. A., and Becker, P. B. (1995) Energy-dependent chromatin accessibility and nucleosome mobility in a cell-free system. *EMBO J.*, **14**, 2209.

51. Langst, G., Bonte, E. J., Corona, D. F., and Becker, P. B. (1999) Nucleosome movement by CHRAC and ISWI without disruption or trans-displacement of the histone octamer. *Cell*, **97**, 843.

52. Hamiche, A., Sandaltzopoulos, R., Gdula, D. A., and Wu, C. (1999) ATP-dependent histone octamer sliding mediated by the chromatin remodeling complex NURF. *Cell*, **97**, 833.

53. Corona, D. F., Langst, G., Clapier, C. R., Bonte, E. J., Ferrari, S., Tamkun, J. W., and Becker, P. B. (1999) ISWI is an ATP-dependent nucleosome remodeling factor. *Mol. Cell*, **3**, 239.

54. Whitehouse, I., Flaus, A., Cairns, B. R., White, M. F., Workman, J. L., and Owen-Hughes, T. (1999) Nucleosome mobilization catalysed by the yeast SWI/SNF complex. *Nature*, **400**, 784.

55. Varga-Weisz, P. D. and Becker, P. B. (1998) Chromatin-remodeling factors: machines that regulate? *Curr. Opin. Cell Biol.*, **10**, 346.

56. Yager, T. D. and van Holde, K. E. (1984) Dynamics and equilibria of nucleosomes at elevated ionic strength. *J. Biol. Chem.*, **259**, 4212.

57. Luger, K., Mader, A. W., Richmond, R. K., Sargent, D. F., and Richmond, T. J. (1997) Crystal structure of the nucleosome core particle at 2.8 Å resolution. *Nature*, **389**, 251.

58. Mizuguchi, G., Tsukiyama, T., Wisniewski, J., and Wu, C. (1997) Role of nucleosome remodeling factor NURF in transcriptional activation of chromatin. *Mol. Cell*, **1**, 141.

59. Pazin, M. J., Kamakaka, R. T., and Kadonaga, J. T. (1994) ATP-dependent nucleosome reconfiguration and transcriptional activation from preassembled chromatin templates. *Science*, **266**, 2007.
60. Alexiadis, V., Varga-Weisz, P. D., Bonte, E., Becker, P. B., and Gruss, C. (1998) *In vitro* chromatin remodelling by chromatin accessibility complex (CHRAC) at the SV40 origin of DNA replication. *EMBO J.*, **17**, 3428.
61. Stillman, B., Gerard, R. D., Guggenheimer, R. A., and Gluzman, Y. (1985) T antigen and template requirements for SV40 DNA replication *in vitro*. *EMBO J.*, **4**, 2933.
62. Ito, T., Bulger, M., Kobayashi, R., and Kadonaga, J. T. (1996) *Drosophila* NAP-1 is a core histone chaperone that functions in ATP-facilitated assembly of regularly spaced nucleosomal arrays. *Mol. Cell. Biol.*, **16**, 3112.
63. Smith, S. and Stillman, B. (1989) Purification and characterization of CAF-I, a human cell factor required for chromatin assembly during DNA replication *in vitro*. *Cell*, **58**, 15.
64. Trachtulcov, P., Janatov, I., Kohlwein, S. D., and Hasek, J. (2000) *Saccharomyces cerevisiae* gene ISW2 encodes a microtubule-interacting protein required for premeiotic DNA replication. *Yeast*, **16**, 35.
65. Shiratori, A., Shibata, T., Arisawa, M., Hanaoka, F., Murakami, Y., and Eki, T. (1999) Systematic identification, classification, and characterization of the open reading frames which encode novel helicase-related proteins in *Saccharomyces cerevisiae* by gene disruption and Northern analysis. *Yeast*, **15**, 219.
66. Deuring, R., Fanti, L., Armstrong, J. A., Sarte, M., Papoulas, O., Prestel, M., Daubresse, G., Verardo, M., Moseley, S. L., Berloco, M., *et al.* (2000) The ISWI chromatin remodeling protein is required for gene expression and the maintenance of higher order chromatin structure *in vivo*. *Mol. Cell*, **5**, 355.

7 | Histone acetyltransferase/ transcription co-activator complexes

SHELLEY L. BERGER, PATRICK A. GRANT, JERRY L. WORKMAN, and
C. DAVID ALLIS

1. Introduction

The precise organization of DNA in chromatin has important functional consequences. DNA-templated processes such as transcription, replication, repair, recombination, and segregation are influenced by the remarkable topological complexity of DNA in chromatin, imposed at the most fundamental level by histone proteins. For several decades it has been expected that modification of histone proteins would affect these processes. Among the well-known covalent modifications of core histones, the reversible acetylation of internal, often invariant, lysine residues in the amino-terminal domains has long been positively correlated to transcription (1; reviewed in 2–4). However, experimental data demonstrating a causative role of histone acetylation in gene activation remained elusive (reviewed in 5). To that end, many groups sought to identify the enzyme system(s) responsible for bringing about the steady-state balance of histone acetylation from a variety of tissues and cellular sources.

2. Histone acetyltransferase complexes

2.1 The elusive search for histone acetyltransferases

For several decades the long, and at times frustrating, search for enzymes capable of transferring acetyl groups onto histones from acetyl coenzyme A (acetyl CoA), referred to as histone acetyltransferases (or HATs), proceeded (4). However, progress was plagued by many factors. General instability of the activities in more purified forms, as well as low abundance of the enzymes in most starting material, stand out as being significant obstacles. In general, these activities had been grouped into two general classes based upon nuclear (type A) or cytoplasmic (type B) origin. Although arbitrary, this distinction fitted well with a large body of evidence that histone acetylation was linked to deposition- and transcription-related processes that were likely being catalysed by A- versus B-type HATs, respectively (reviewed in 6).

In hindsight, two developments dramatically influenced the explosive discovery of HATs in the mid to late 1990s. First, Sternglanz and co-workers doggedly pursued genetic screens for relevant HAT activities in budding yeast by assaying crude whole-cell extracts, from wild-type and mutant cells, for relative differences in bulk HAT activity in a conventional in-solution enzymatic assay. This approach ultimately proved successful, leading to the cloning and description of a HAT-encoding gene from *Saccharomyces cerevisiae*, *HAT1* (7). Close inspection of the amino-acid sequence of Hat1p, a B-type HAT purified biochemically soon thereafter by Parthun *et al.* (8), revealed short motifs found in other enzymes that bind and metabolize acetyl CoA, such as *N*-acetyltransferases.

Secondly, an in-gel HAT activity assay was developed to search for polypeptides, resolved in high-resolution SDS–PAGE gels containing mixtures of free histones as substrate, that contained catalytic activity upon careful renaturation (9). Application of this assay to crude macronuclear extracts from the ciliated protozoan, *Tetrahymena thermophila*, led to the discovery of a candidate catalytic type A HAT subunit, p55. Unexpectedly, the predicted sequence of p55 revealed remarkable similarity to a previously identified transcriptional co-activator in yeast, Gcn5p (10). Like Hat1p, ciliate and yeast Gcn5p display several highly conserved acetylation-related motifs, leading to the formal description of the GNAT (*Gcn5*-related *N*-*a*cetyl*t*ransferase) superfamily (11). These findings provided compelling evidence for an intimate link between histone acetylation and transcriptional output, and suggested a novel transcriptional regulatory scheme involving targeted acetylation. More specifically, it was envisioned that chromatin-modifying enzymes, such as type A HATs, are recruited to specific promoters through selective interactions with activator and co-activator proteins (6).

The above concept received rapid support by a remarkable series of subsequent findings, erupting in the close of the 1996 calendar year. The mammalian Gcn5 orthologue PCAF, CREB-binding protein (CBP), adenovirus E1A-binding protein p300, and TAF$_{II}$250, were each found to possess intrinsic HAT activity (reviewed in 12). Conversely, the discovery that a mammalian histone deacetylase (HDAC) was a homologue of the yeast co-repressor, Rpd3p (13) (see Chapter 8), gave rise to the hypothesis that regulated activation events might involve the exchange of complexes containing histone deacetylase function with those containing HAT activity. By the close of the millennium, over 20 distinct polypeptides had been reported with intrinsic HAT or acetyltransferase (AT) activity from a variety of sources, ranging from yeast to humans (see Table 1 of ref. 14). In parallel, the list of HDACs continues to grow at an equally rapid pace (see Chapter 8).

2.2 The diversity of histone acetyltransferases

The 'Gcn5p is a HAT' discovery (10) was made in an 'off-beat' organism *Tetrahymena*, a cousin to the freshwater *Paramecium* that most of us were exposed to in High School

Biology. The choice of this organism was fortunate since direct comparisons have subsequently shown that the ciliate Gcn5 is more readily able to refold following denaturation (e.g. in the in-gel HAT assay) than its yeast or human counterparts. For this reason NMR (15) and crystallographic (16) information were first collected on the catalytic domain of the *Tetrahymena* Gcn5. In turn, this information has provided key insights into structure and catalytic mechanism of the yeast and human Gcn5 family members (17). The extremely high degree of conservation between histones and histone-modifying activities (e.g. Gcn5 (18)) permits researchers working in this area to cross organismal boundaries to glean secrets from the unique biological strengths of each system.

This dizzying gold rush of HAT discoveries has been accompanied by unexpected connections between these activities and growth, development, and cellular transformation. For example, disruption of interactions between the human PCAF, a Gcn5 family member, and its co-activator p300/CBP, itself a HAT, by the product of the viral E1A oncogene, is required for E1A-mediated cellular transformation (19). Translocation of another putative acetyltransferase, MOZ, in-frame to CBP, is associated with specific subtypes of acute myeloid leukaemias (20, reviewed in ref. 21). Interestingly, MOZ is a MYST family member (14). The MYST family also includes Sas2p/Sas3p, MOF, Tip60, and Esa1p. The Sas2p/Sas3p proteins are implicated in transcriptional silencing in yeast (22). MOF is required for the hypertranscription of X-linked genes in *Drosophila* males (23, 24). Tip60 has been shown to interact with HIV-Tat protein in the two-hybrid system (25–27). *ESA1* is a yeast gene required for growth (28, 29). All of the above proteins possess intrinsic HAT activity, suggesting that the acetylation of histones, and/or a potentially vast array of other non-histone substrates (see below), is a fundamental aspect of biological regulation, much as is the case for protein phosphorylation. Understanding the nature, regulation, and specificity of these highly conserved chromatin-modifying activities is directly relevant to our understanding of both normal development and differentiation (reviewed in 30), as well as abnormal processes which lead to oncogenesis (reviewed in 21, 31).

2.3 Gcn5/PCAF-containing HAT complexes

As useful as the in-gel HAT assay was to identify some relevant catalytic activities such as Gcn5p, it had at least one significant limitation. Early experiments showed clearly that Gcn5p from any source was unable to acetylate nucleosomes efficiently, providing the first important clue that other factors, presumably dissociated from Gcn5p in the in-gel HAT assay, were required to permit this enzyme to act on chromatin substrates (32, 33). Gcn5p has since been identified as a component of numerous high molecular weight acetyltransferase/transcriptional co-activator complexes from both yeast and human cell extracts (Table 1) (reviewed in 12). The association of Gcn5p in multiprotein complexes is known to potentiate its HAT activity on chromatin substrates (32–35). The ability of Gcn5p to acetylate nucleosomes efficiently is important for the transcriptional stimulatory activity of these complexes from chromatin templates *in vitro* and *in vivo* (36–38).

Table 1 The SAGA group of histone acetyltransferase complexes

Subunit class/functions	Yeast SAGA	Human PCAF	Human TFTC	Human STAGA
Acetyltransferase	Gcn5	PCAF	Gcn5L	Gcn5L
ADA proteins; facilitate function of	Ada1			
transcription activators	Ada2	Ada2		
	Ada3	Ada3	Ada3	
	Ada5/Spt20			
TBP-group of Spt proteins; facilitate	Spt3	Spt3	Spt3	Spt3
TBP function	Spt7	Spt7		
	Spt8			
	Spt20/Ada5			
TAF$_{II}$ proteins: also found in TFIID,			TAF$_{II}$150	
facilitate nucleosome acetylation			TAF$_{II}$135	
interactions with TBP/activators?	TAF$_{II}$90	PAF65β	TAF$_{II}$100	
	TAF$_{II}$60	PAF65α	TAF$_{II}$80	
	TAF$_{II}$17	TAF$_{II}$31	TAF$_{II}$31	TAF$_{II}$31
	TAF$_{II}$25	TAF$_{II}$30	TAF$_{II}$30	
	TAF$_{II}$68	TAF$_{II}$20	TAF$_{II}$20	
PI-3 kinase related; activator interaction	Tra1	TRRAP	TRRAP	

Subunits that have been identified thus far for the yeast SAGA and each human complex are shown. The lack of a subunit listing in a complex does not indicate that it is not present in the complex. Only that it has not yet been identified. The left column lists the classes of proteins corresponding to each subunit. See text for references.

Yeast Gcn5p-dependent nucleosomal HAT complexes with sizes of 170–200 kDa, 800–900 kDa, and 1.8–2.0 MDa have been purified. In each of these complexes Gcn5p is associated with members of the Ada (*a*lteration/*d*eficiency in *a*ctivation) family of proteins (39). The 170–200 kDa complex (HAT-A2) (32) and 800–900 kDa (Ada) (40) complex consist of Gcn5p, Ada2p, Ada3p, and a number of unidentified subunits. The Ada complex is known to contain Ahc1p (see Table 2), which is apparently uniquely required for the integrity of this complex, but not for other Gcn5p complexes (40). The most notable Gcn5p-dependent HAT complex is the 1.8–2.0 MDa SAGA complex, which was found to contain at least four distinct groups of gene products previously implicated in transcription regulation (Table 1). The first of these are the Ada proteins Gcn5p (Ada4p), Ada1p, Ada2p, Ada3p, and Ada5p (Spt20p). The second group comprises all members of the TBP-related set of Spt proteins (see Chapter 4), except TBP (Spt15) itself (33). The third group within SAGA includes a subset of TBP-associated factors, TAF$_{II}$s (41). The final protein identified as a component of SAGA is the product of the essential gene *TRA1* (42, 43). Tra1p has been described as the homologue of the human transformation/transcription domain-associated protein TRRAP, a novel member of the ataxia telangiectasia mutated (ATM) family of phosphatidylinositol 3-kinases (44). Thus, the SAGA complex provides a link between histone acetylation by Gcn5p and the function of the Ada, Spt, Tra1, and certain TAF$_{II}$ proteins.

Complexes with homology to SAGA have also been described in human cells (Table 1). The cellular factor PCAF bears a high degree of homology to Gcn5p and also has intrinsic HAT activity (19). A PCAF complex has been purified from human cells and consists of approximately 20 polypeptides. This complex resembles SAGA, in that it contains counterparts of the yeast Ada2p, Ada3p, Spt3p, Spt7p, and Tra1p proteins, and a similar subset of $TAF_{II}s$ or TAF_{II}-related factors (45, 46). In mammalian cells long (L) and short forms of Gcn5 have been described and a complex containing hGcn5S and apparently indistinguishable subunits to the PCAF complex has also been described (45). This most likely represents a complex similar to the independently identified STAGA (Spt3–TAF_{II}31–Gcn5L acetyltransferase) complex (47) and TFTC (TBP-free TAF_{II}-containing complex) (48, 49) (Table 1). The identification of multiple HAT complexes from yeast and human extracts with common subunit compositions underscores a mechanism of transcriptional activation from chromatin via a conserved partnership between this group of co-activators and acetyltransferases.

2.4 Other 'co-activator' HAT complexes

In addition to the SAGA-related complexes, numerous other protein complexes that contribute to transcription have been found to contain HAT activity (Table 2). p300 and CBP (CREB-binding protein) are highly homologous global transcriptional co-activators which have been demonstrated to acetylate nucleosomal histones effectively (50, 51). Various cellular and viral factors target p300/CBP to modulate transcription and/or cell-cycle progression, including PCAF (reviewed in 4). Mammalian Gcn5 has also been demonstrated to interact with p300/CBP (52). PCAF and p300/CBP have also been found to interact with the proteins ACTR and SRC-1, two homologous co-activators, which bind nuclear hormone receptors directly and stimulate their transcriptional activities. Interestingly ACTR and SRC-1 have also been demonstrated to have HAT activity, indicating the recruitment of a complex containing

Table 2 Other transcription-related HATs

Complex	Acetyltransferase subunit	Transcription
	p300/CBP (human)	Global transcription adaptor
	SRC-1/ACTR (human)	Nuclear hormone receptor transcription adaptor
TFIID	$TAF_{II}250$ (human)	General transcription initiation factor for transcription by RNA polymerase II
	$TAF_{II}230$ (*Drosophila*)	
	$TAF_{II}130$ (yeast)	
TFIIIC	p90, p110, p220 (human)	General transcription initiation factor for transcription by RNA polymerase III
Elongator	Elp3p (yeast)	Transcription elongation
ADA	Gcn5p (yeast)	Unknown
NuA3	Sas3p (yeast)	Transcription elongation
NuA4	Esa1p (yeast)	Transcription adaptor? Activates transcription *in vitro*

The left column lists the complex, if any, in which a particular catalytic subunit (mddle column) is found. The right column lists the known function of the HAT/complex. See text for references.

multiple HATs by nuclear receptors, which may act cooperatively during activation of target genes (53, 54).

General transcription initiation factors for polymerase II and polymerase III transcription have been found to contain HAT activity. Human TAF$_{II}$250 and its yeast (yTaf$_{II}$130p/145p) and *Drosophila* (dTAF$_{II}$230) homologues also contain HAT activity (55) (Table 2). Each of these proteins is part of the evolutionarily conserved TFIID complex that consists of TBP and a number of TAF$_{II}$s (55). TBP plays a central role in transcription by recognizing a specific DNA sequence (TATA element) found in the core promoter of many genes, and is generally believed to be required for the assembly of other proteins in that region to initiate the process of gene transcription. The finding that TFIID contains a component possessing HAT activity suggests that acetylation by TFIID may be important at the core promoter to mediate transcription-factor interaction with nucleosomal DNA. Human TFIIIC is another multisubunit transcription factor that directly recognizes promoter elements and recruits other factors and RNA polymerase III to transcribe small structural RNAs. Human TFIIIC contains at least 9 subunits, three of which (220, 110, and 90 kDa) have intrinsic HAT activity (56, 57). This association of multiple HATs also suggests a critical role of acetylation in transcription initiation by RNA polymerase III.

Putative acetyltransferase domains have been identified within a group of proteins known as the MYST family, named for its founding members *MOZ*, *YBF2/Sas3p*, *Sas2*, and *Tip60*. The MYST protein Esa1p has been identified as the catalytic subunit of the NuA4 (*nu*cleosomal *a*cetyltransferase of histone H4) protein complex in yeast (58). This complex has an apparent molecular mass of 1.3 MDa and also contains the Tra1p component found in the SAGA-related family of HATs (58). The NuA3 (*nu*cleosomal *a*cetyltransferase of histone H3) complex has also been purified from yeast extracts and found to contain the MYST-related HAT Sas3p as well as Taf$_{II}$30p (59). Finally, the elongator complex, a major component of actively elongating RNA polymerase II holoenzyme, harbours the Elp3p acetyltransferase that acetylates all four core histones *in vitro* (60).

3.   Functions of HAT complexes and histone acetylation

3.1   Recruitment of HAT complexes to target genes

A long history of correlative studies has indicated an enrichment of acetylated histones on transcribed DNA sequences; however, more recent studies have begun to delineate the distribution of acetylated histones relative to transcriptional regulatory regions. Most of these advances have utilized the chromatin immunoprecipitation (CHIPs) technique, where histones are cross-linked to DNA, the DNA is sheared and then immunoprecipitated with anti-acetylated histone antibodies. Polymerase chain reaction (PCR) or blotting techniques are then used to determine whether a specific sequence is associated with the acetylated histones that have been immunoprecipitated. While early studies suggested that increased levels of histone acetylation may be spread over broad domains of 'active' chromatin, comprising several kilobase pairs (61), more recent experiments have also demonstrated more localized peaks of histone acetylation resulting from the action of HAT complexes. Mapping of Gcn5p-dependent acetylation of histone H3 near the *HIS3* promoter demonstrated that it was restricted to nucleosomes immediately adjacent to this promoter (36). Gcn5p-dependent histone H3 acetylation at the yeast *HO* promoter was mapped to within approximately 1 kb of the active promoter (62). Similarly, induction of the human interferon-β (IFN-β) promoter led to p300/CBP-dependent H3 and H4 acetylation of only 2–3 nucleosomes (i.e. 600 bp) within and surrounding the promoter (63).

Promoter proximal histone acetylation is consistent with the co-activator functions of some HAT complexes and suggests that they are recruited to promoters by DNA-binding transcriptional activators. A direct interaction with transcriptional activators has been demonstrated for several HAT complexes *in vitro*. The yeast SAGA and NuA4 complexes have been found to bind to a number of acidic transcriptional activation domains (64, 65). These interactions were found to target adjacent nucleosomes for acetylation and led to acetylation-dependent transcription activation from chromatin templates *in vitro* (64–67). In addition, as mentioned above, the mammalian p300/CBP and the pCAF complex interacts with numerous sequence-specific activators (e.g. CREB, nuclear hormone receptors, Pit1, MyoD) (68–70).

In addition to co-activators that function as HATs, other transcription-related proteins have been shown to have HAT activity (Table 2). This suggests additional pathways to recruit acetylation activities (Fig. 1). The observation that TFIID has a HAT activity is of particular interest since several TAFs are shared between TFIID and the SAGA complex (see above). Moreover, SAGA has been suggested to be able to substitute for TFIID in transcription in some instances and these two complexes appear to be structurally similar (49, 71, 72). Thus, SAGA and TFIID may have overlapping functions but contain distinct HAT activities, Gcn5p and Taf$_{II}$250p, respectively.

HAT activities also appear to play an important role in transcription elongation (Fig. 1). The elongator complex associates with the carboxy-terminal domain of the large subunit of elongating RNA polymerase II (60). The Elp3p subunit of the elongator complex has been shown to have HAT activity, indicating that elongating RNA polymerase II carries a histone acetyltransferase complex with it as it elongates through nucleosomes in the coding regions of transcribed genes. In addition, the Sas3p subunit of the NuA3 complex interacts with the Spt16p subunit of the yeast CP complex (59). The human homologue of the CP complex, FACT (*f*acilitates *a*ctivation of *c*hromatin *t*emplates), was shown to stimulate elongation of RNA polymerase II transcription through nucleosomes (73). By interacting with the CP complex, NuA3 may participate in this process. Interestingly, the CP complex has also been implicated in replication (74), raising the possibility that the NuA3 HAT complex might also function in replication.

3.2 Effects of histone acetylation

The mechanisms by which histone acetylation affects chromatin structure and transcription is not yet clear. Acetylation of lysines in histones reduces their positive

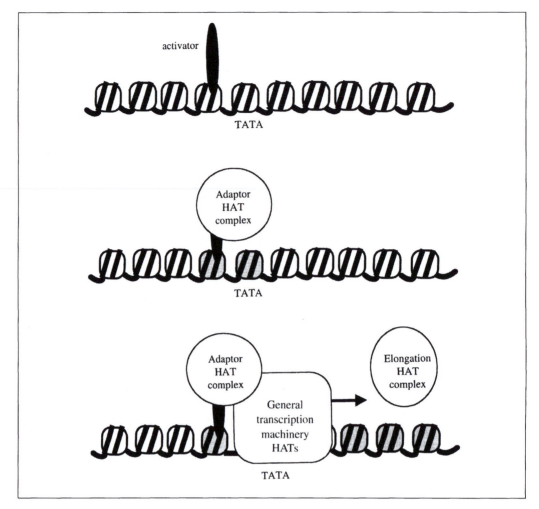

Fig. 1 Histone acetyltransferases function at several steps in transcription. The binding of a sequence-specific transcription activator to a promoter within a nucleosome array (top) has the ability to recruit HAT complexes that function as transcription adaptors. Both these complexes and the activators themselves recruit the general transcription machinery (bottom), which also includes HAT complexes (e.g. TFIID). Finally, transcription elongation is facilitated by additional complexes that acetylate histones (e.g. elongator, NuA3).

charge. This has been suggested to reduce the affinity or extent of histone–DNA interactions and effects along these lines have been measured (75–77). Consistent with this possibility is the fact that acetylation of histones has been shown to increase the *in vitro* binding of transcription factors to DNA contained in nucleosomes, which was otherwise suppressed by the histone tails (78–80). An alternative, and not mutually exclusive, model is that the histone tails are important in the packaging of arrays of nucleosomes into 30 nm chromatin fibres. Ultracentrifugation studies have measured a reduced compaction of nucleosome arrays containing acetylated histones (81, 82).

Another function of the histone tails and acetylation is to act as direct interaction sites for regulatory protein complexes. The unacetylated tails of histones H3 and H4 have been shown to bind the yeast Sir3p/Sir4p and Tup1p transcriptional repressors (83, 84), the *Drosophila* polycomb repressor (85), and the mammalian transducin-like enhancer of split repressor (86). The Sir and Tup proteins function as part of distinct repressor complexes at telomeres and other loci, including repressed mating-type genes (see Chapter 11). A striking finding revealed that bromodomains can bind acetyl-lysines within acetylated histone tails (87, 88). Bromodomains are found in a number of proteins in transcriptional regulatory complexes, including the $TAF_{II}250$ subunit of TFIID, the Gcn5p subunit of SAGA, the Swi2p subunit of SWI/SNF, and the Sth1p subunit of RSC (89). Thus, while the functions of the histone tails in recruiting other regulatory complexes is likely to be complex, the picture currently emerging is that the unacetylated tails may provide interaction sites for transcriptional repressors, while the acetylated tails may provide interaction sites for activating complexes.

4. Other substrates of histone acetyltransferases

4.1 HATs as FATs (factor acetyltransferases)

Soon after the initial identification of enzymes that acetylate histone substrates, the question arose whether the HATs modify proteins other than histones. A considerable number of non-histone substrates of acetyltransferases have now been identified (Table 3), indicating that many of these enzymes can also often be considered FATs (factor acetyltransferases). The first non-histone substrates identified *in vitro* were the general transcription factors, TFIIE and TFIIF (90), however, the relevance of their acetylation and whether it occurs *in vivo* has not been demonstrated. Most other non-histone substrates are DNA-binding transcriptional activators, and nearly all of these were tested initially because of a previous demonstration of a role of the acetyltransferases CBP/p300 or PCAF/Gcn5 as co-activators for their function. A second class comprises transcriptional co-activators themselves, which are also acetyltransferases. The third class of FAT substrates consists of non-histone chromatin components.

The underlying and very interesting question concerning each of the putative substrates (beyond establishing that they are genuine) is the effect of acetylation on function. There are two general possibilities, and each is reminiscent of similar questions for histones. Does acetylation alter interaction with other proteins, or, rather, does the charge alteration by acetylation change characteristics of the factors themselves, including effects on DNA binding or association with other factors? Examples of each of these mechanisms have been revealed in the investigation of the FAT substrates described below.

4.2 FAT substrates

The largest group of FAT substrates comprises classical transcriptional activators that bind to specific DNA sequences to induce gene expression. Acetylation of three

Table 3 Summary of factor acetyltransferase (FAT) substrates

FAT Substrate	Function *in vivo*	FAT enzyme *in vitro*	Effect of actylation
General transcription factors			
TFIIE	Required for basal transcription	PCAF, p300/CBP, TAF$_{II}$250	ND
TFIIF		PCAF, p300/CBP	ND
Transcriptional activators			
p53	Tumour suppressor	PCAF, p300/CBP	Increased DNA binding
GATA-1	Blood-cell differentiation	p300/CBP	DNA binding may be affected
c-Myb	Proto-oncogene product	p300/CBP	Increased DNA binding
EKLF	Globin gene expression	p300/CBP	Inhibitory domain modified
MyoD	Muscle differentiation	PCAF	Increased DNA binding
E2F	Cell-cycle control	PCAF	Increased DNA binding
dTCF	Developmental regulation	p300/CBP	Co-activator interaction disrupted
HIV Tat	HIV-1 transactivation	PCAF	Increased CDK9 binding
		p300/CBP	
Nuclear receptor co-activators			
ACTR	Transcriptional response to	p300/CBP	Receptor interaction disrupted
SRC-1	hormone signals	p300/CBP	ND
TIF2		p300/CBP	ND
Non-histone chromatin proteins			
HMG1/2	Chromatin component	ND	ND
HMG17	Nucleosome binding	ND	ND
HMG17	Nucleosome binding	PCAF	Nucleosome binding weakened
HMG I(Y)	Enhanceosome component	PCAF	Enhanceosome assembly[a]
		p300/CBP	Enhanceosome disruption

Non-histone substrates of acetyltransferases and their functions are listed in the left two columns. The HAT/FAT enzymes that have been found to acetylate them and the effect of acetylation on their activities is listed in the two right columns. See text for references.

[a] D. Thanos, personal communication.

ND, not determined.

---

activators has been shown clearly to cause increased DNA binding. The tumour suppressor p53 was among the first FAT substrates to be recognized (91). p53 was known to require both CBP/p300 and PCAF for optimal activation (91–93) and was shown to be acetylated by both of these enzymes *in vitro* (91, 94, 95). The acetylation occurred in an internal inhibitory domain to DNA binding, and resulted in increased DNA binding. Moreover, p53 acetylation increases following irradiation of cells resulting in DNA damage, which was previously known to turn on p53 function. Physiologically important acetylation has been documented for MyoD, a muscle cell differentiation factor, for E2F, a regulator of cell-cycle progression, and for GATA-1, an erythroid cell differentiation factor. As for p53, MyoD was a known CBP/p300- and PCAF-dependent activator (96, 97), and *in vitro* and *in vivo* studies indicated that PCAF, but not CBP/p300, acetylates MyoD (98). Acetylation resulted in increased DNA binding by MyoD *in vitro*, and mutation of the acetylated lysines to arginine resulted in a lack of *in vivo* transcriptional activation and myogenesis. E2F is also acetylated by PCAF, resulting in increased DNA binding, stabilization of the E2F protein, and increased activation ability *in vivo* (99). E2F has a dual role as a repressor, in a complex with Rb and a histone deacetylase (HDAC), and interestingly, the HDAC is capable of deacetylating E2F as well as histones. GATA-1 is acetylated by CBP/p300, also to enhance DNA binding, and mutation of the target lysines interferes with erythroid differentiation (100, 101). Another erythroid cell factor, EKLF, has also been shown to be acetylated by CBP/p300 (102).

Two transcription factors, TCF (a T-cell factor) and HIV Tat (which interacts with RNA to control transcriptional elongation) are acetylated to alter affinity for interacting regulatory factors. TCF interacts with a co-activator, Armidillo, to activate genes. CBP-mediated acetylation of TCF in its Armidillo-binding domain weakens activator–Armidillo interaction (103). In contrast, CBP and PCAF-mediated acetylation of Tat enhances transcriptional elongation through altering interactions with associated factors (104).

Transcription co-activators may also serve as FAT substrates. Transcriptional activation by nuclear hormone receptors involve an associated group of co-activators (and HATs), including CBP/p300, PCAF, and ACTR. In the first demonstration of modification of a HAT by a second HAT, CBP/p300 was shown to acetylate ACTR near a motif crucial for its interaction with hormone receptors, and acetylation interfered with ACTR–receptor interaction (105). Thus, for both TCF and ACTR, acetylation weakens docking with an interacting protein.

In addition to histones, most high-mobility group (HMG) chromatin-associated proteins can also be acetylated. HMGs are involved in higher-order chromatin structure and transcriptional enhanceosome assembly. Acetylation has been shown to regulate these functions. For example, HMG14 and 17 promote transcription by unfolding higher-order chromatin structure. PCAF-mediated acetylation of HMG17 reduces its interaction with nucleosomes (106). Importantly, the single site acetylated in HMG17 corresponds to the site modified *in vivo*. HMGI(Y) is an architectural component of transcription activator/co-activator complexes that assemble in the enhancers of certain promoters, such as that of the viral-inducible IFN-β gene. *In vitro* acetylation of factors in the enhancesome revealed that HMGI(Y) is specifically acetylated by both CBP/p300 and PCAF (107, 108). Both CBP/p300 and PCAF acetylation of HMGI(Y) is required for induction of IFN-β immediately following viral infection. In addition, acetylation by CBP/p300 is required for normal downregulation of the gene later in viral infection.

4.3 Additional important questions regarding factor acetylation

The evidence is now quite persuasive that FAT activity exists and several bona fide substrates have been identified. There are several outstanding questions about FATs and their substrates. First, do all HATs also function as FATs? To date FAT activity has been shown for only the Gcn5 family and CBP/p300 family. A second question is whether acetyltransferases act in a concerted mechanism to activate genes, incorporating both FAT and HAT activity. Thus, a cascade of acetylation is suggested by the current evidence: factor acetylation may enhance initial DNA binding by activators, as well as subsequent interaction of activators with co-activators, to assist recruitment to promoters. Next, histone acetylation would alter the chromatin around the basal promoter to stimulate binding of general factors and consequent transcriptional initiation.

5. Additional histone modifications: looking ahead

An extensive literature documents an elaborate collection of post-translational modifications (e.g. acetylation, phosphorylation, methylation, etc.) that occur on the 'tail' domains of these proteins (Fig. 2) (reviewed in 109). While these histone-tail domains have not been observed at atomic resolution, these covalent marks are believed to influence dynamic transitions that occur between decondensed versus condensed regions of the chromatin fibre in response to upstream signals. It remains unclear whether this diverse array of covalent modifications exert their effects mostly by regulating higher-order chromatin structure or by presenting a docking surface for, yet unidentified, downstream components (109). Despite these uncertainties, the importance of histone covalent modifications, notably acetylation, has become clear, particularly during the transcription process (3, 5, 110).

Cellular responses to environmental stimuli and nutritional fluctuations ultimately elicit defined nuclear responses in genetic programmes. Similarly, responses to DNA damage include the activation of repair pathways, control of cell-cycle checkpoints and apoptosis. All of these processes are regulated through signal transduction

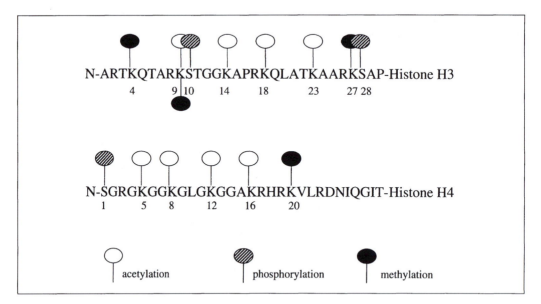

Fig. 2 Diagram of the amino-terminal tails of histones H3 (upper) and H4 (lower). The sequence of each is shown and further marked with the modifications that can be found at different amino acids. The code for the modifications indicated is at the bottom of the figure.

cascades that may terminate in chromatin-remodelling steps that are brought about, in part, by covalent histone modifications (111). Converging evidence, for example, suggests that histone H3 phosphorylation is also directly correlated with the induction of immediate-early genes such as *c-jun*, c-*fos*, and *c-myc* (112) reviewed in (113). Importance of this phosphorylation mark in H3 as a potentially relevant downstream signalling target is suggested by the finding that Rsk-kinase family members, known to be activated by growth factor and stress stimuli, also utilize H3 as a substrate *in vivo* and *in vitro* (114, 115). Moreover, mutations in Rsk-2 are closely associated with Coffin–Lowry syndrome in humans and result in a loss of epidermal growth factor (EGF)-stimulated H3 phosphorylation *in vivo* (115). Transcriptional activation in response to mitogenic and other stimuli is altered in Coffin–Lowry cells, suggesting a potential direct role for H3 phosphorylation in regulating gene transcription via a chromatin-remodelling step that remains to be clearly defined. Recent results, documenting co-activator-mediated H3 methylation (116), as well as mitogen-stimulated synergism between H3 phosphorylation and acetylation (30), make it likely that deciphering the 'language' of covalent histone modifications (109) will take some time.

Using histone acetylation as a paradigm, the literature strongly supports the value of combined biochemical and genetic approaches aimed at understanding the multisubunit enzymatic machinery that governs the steady-state balance of each histone modification. To fully understand the chromatin-based on/off 'switch', it will be necessary to define the components that catalyse and regulate each forward and reverse reaction. We will also need to know to what extent does one reaction, or one modification, influence another. What are the kinetics and intermediates formed in the pathway, and which are the physiologically relevant substrates? In using enzymatic assays to identify the relevant activities, careful attention will be needed to the form of the substrates (free histones, mononucleosomes, nucleosomal arrays, etc.) as well as the form of the enzyme (subunit or multisubunit complex) under examination. Recent history indicates a fascinating interplay of enzymatic activities that establish the 'history code' decorating the histone termini (109). For example, who would have ever predicted that Sir2p, a conserved silencing-associated protein in organisms ranging from yeast to humans (117), would be both a histone ribosylase (118) and a NAD-dependent histone deacetylase (119, 120). The next millennium is only beginning, it promises to be full of wonderful surprises.

6. Discussion

Comments from authors of other chapters raised several issues:

- One concerns the integrated functions of HAT complexes with other chromatin-remodelling complexes (see also Chapters 4 and 5). This is an emerging concept that will piece together many of the details of how these activities function together. Indeed, studies that examined the cell-cycle regulated *HO* promoter concluded that SWI/SNF function is a prerequisite for Gcn5 HAT function (62, 121). By contrast, a second study using a model promoter has revealed the opposite relationship: that Gcn5 function (and specifically the bromodomain of Gcn5) is required for chromatin remodelling and SWI/SNF recruitment (122).

- While Hat1p has been classified as a type B histone-deposition related HAT, it is quite possible that the type B HATs will ultimately be found to also participate in gene regulation.

- Some HAT complexes are very large and contain multiple subunits, suggesting that they will perform functions beyond acetylation, a fact that is already becoming apparent (123, 124).

- There is the potential for these HAT complexes to be dynamic, i.e. the HATs or their cofactors may associate or dissociate in response to cellular conditions, during development, etc. This may be important for the regulation of the activities, and could contribute an additional layer of subtlety to that provided by the different complexes themselves.

Acknowledgements

The authors wish to thank Peter Becker, Andrew Free, Toshio Tsukiyama, and Maria Vogelauer for comments on the manuscript and Ms Lorene Stitzer for assistance in its preparation.

References

1. Allfrey, V. G., Faulkner, R., and Mirsky, A. E. (1964) Acetylation and methylation of histones and their possible role in the regulation of RNA synthesis. *Proc. Natl Acad. Sci., USA*, **51**, 786.
2. Loidl, P. (1994) Histone acetylation: facts and questions. *Chromosoma*, **103**, 441.
3. Grunstein, M. (1997) Histone acetylation in chromatin structure and transcription. *Nature*, **389**, 349.
4. Workman, J. L. and Kingston, R. E. (1998) Alteration of nucleosome structure as a mechanism of transcriptional regulation. *Annu. Rev. Biochem.*, **67**, 545.
5. Struhl, K. (1998) Histone acetylation and transcriptional regulatory mechanisms. *Genes Develop.*, **12**, 599.
6. Brownell, J. E. and Allis, C. D. (1996) Special HATs for special occasions: linking histone acetylation to chromatin assembly and gene activation. *Curr. Opin. Genet. Devop.*, **6**, 176.
7. Kleff, S., Andrulis, E. D., Anderson, C. W., and Sternglanz, R. (1995) Identification of a gene encoding a yeast histone H4 acetyltransferase. *J. Biol. Chem.*, **270**, 24674.
8. Parthun, M. R., Widom, J., and Gottschling, D. E. (1996) The major cytoplasmic histone acetyltransferase in yeast: links to chromatin replication and histone metabolism. *Cell*, **87**, 85.
9. Brownell, J. E. and Allis, C. D. (1995) An activity gel assay detects a single, catalytically active histone acetyltransferase subunit in *Tetrahymena* macronuclei. *Proc. Natl Acad. Sci., USA*, **92**, 6364.
10. Brownell, J. E., Zhou, J., Ranalli, T., Kobayashi, R., Edmondson, D. G., Roth, S. Y., and Allis, C. D. (1996) *Tetrahymena* histone acetyltransferase A: a homolog to yeast Gcn5p linking histone acetylation to gene activation. *Cell*, **84**, 843.

11. Neuwald, A. F. and Landsman, D. (1997) Gcn5-related histone *N*-acetyltransferases belong to a diverse superfamily that includes the yeast Spt10 protein. *Trends Biochem. Sci.*, **22**, 154.

12. Brown, C. E., Lechner, I., Howe, I., and Workman, J. L. (2000) The many HATs of transcription coactivators [in-process citation]. *Trends Biochem. Sci.*, **25**, 15.

13. Taunton, J., Hassig, C. A., and Schreiber, S. L. (1996) A mammalian histone deacetylase related to the yeast transcriptional regulator Rpd3p. *Science*, **272**, 408.

14. Sterner, D. E. and Berger, S. L. (2000) Acetylation of histones and transcription-related factors. *Microbiol. Mol. Biol.*, **64**, 435.

15. Lin, Y., Fletcher, C. M., Zhou, J., Allis, C. D., and Wagner, G. (1999) Solution structure of the catalytic domain of Gcn5 histone acetyltransferase bound to coenzyme A. *Nature*, **400**, 86.

16. Trievel, R. C., Rojas, J. R., Sterner, D. E., Venkataramani, R. N., Wang, L., Zhou, J., Allis, C. D., Berger, S. L., and Marmorstein, R. (1999) Crystal structure and mechanism of histone acetylation of the yeast GCN5 transcriptional coactivator [see comments]. *Proc. Natl Acad. Sci., USA*, **96**, 8931.

17. Clements, A., Rojas, J. R., Trievel, R. C., Wang, L., Berger, S. L., and Marmorstein, R. (1999) Crystal structure of the histone acetyltransferase domain of the human PCAF transcriptional regulator bound to coenzyme A. *EMBO J.*, **18**, 3521.

18. Candau, R., Moore, P. A., Wang, L., Barlev, N., Ying, C. Y., Rosen, C. A., and Berger, S. L. (1996) Identification of human proteins functionally conserved with the yeast putative adaptors ADA2 and GCN5. *Mol. Cell. Biol.*, **16**, 593.

19. Yang, X.-J., Ogryzko, V. V., Nishikawa, J.-I., Howard, B. H., and Nakatani, Y. (1996) A p300/CBP-associated factor that competes with the adenoviral oncoprotein E1A. *Nature*, **382**, 319.

20. Borrow, J., Stanton, V. P. J., Andresen, J. M., Becher, R., Behm, F. G., Chaganti, R. S., Civin, C. I., Disteche, C., Dube, I., Frischauf, A. M., Horsman, D., *et al.* (1996). The translocation t(8;16) (p11;p13) of acute myeloid leukaemia fuses a putative acetyltransferase to the CREB-binding protein. *Nature Genet.*, **14**, 33.

21. Roth, S. Y. and Allis, C. D. (1996) Histone acetylation and chromatin assembly: a single escort, multiple dances? *Cell*, **87**, 5.

22. Reifsnyder, C., Lowell, J., Clarke, A., and Pillus, L. (1996) Yeast SAS silencing genes and human genes associated with AML and HIV-1 Tat interactions are homologous with acetyltransferases. *Nature Genet.*, **14**, 42.

23. Hilfiker, A., Hilfiker-Kleiner, D., Pannuti, A., and Lucchesi, J. C. (1997) mof, a putative acetyl transferase gene related to the Tip60 and MOZ human genes and to the SAS genes of yeast, is required for dosage compensation in *Drosophila*. *EMBO J.*, **16**, 2054.

24. Smith, E. R., Pannuti, A., Gu, W., Steurnagel, A., Cook, R. G., Allis, C. D., and Lucchesi, J. C. (2000) The drosophila MSL complex acetylates histone H4 at lysine 16, a chromatin modification linked to dosage compensation. *Mol. Cell. Biol.*, **20**, 312.

25. Kamine, J., Elangovan, B., Subramanian, T., Coleman, D., and Chinnadurai, G. (1996) Identification of a cellular protein that specifically interacts with the essential cysteine region of the HIV-1 Tat transactivator. *Virology*, **216**, 356.

26. Yamamoto, T. and Horikoshi, M. (1997) Novel substrate specificity of the histone acetyltransferase activity of HIV-1-Tat interactive protein Tip60. *J. Biol. Chem.*, **272**, 30595.

27. Kimura, A. and Horikoshi, M. (1998) Tip60 acetylates six lysines of a specific class in core histones *in vitro*. *Genes Cells*, **3**, 789.

28. Smith, E. R., Eisen, A., Gu, W., Sattah, M., Pannuti, A., Zhou, J., Cook, R.G., Lucchesi, J. C., and Allis, C.D. (1998) ESA1 is a histone acetyltransferase that is essential for growth in yeast. *Proc. Natl Acad. Sci., USA*, **95**, 3561.

29. Clarke, A. S., Lowell, J. E., Jacobson, S. J., and Pillus, L. (1999) Esa1p is an essential histone acetyltransferase required for cell cycle progression. *Mol. Cell. Biol.*, **19**, 2515.

30. Cheung, P., Tanner, K. G., Cheung, W. L., Sassone-Corsi, P., Denu, J. M., and Allis, C. D. (2000) Synergistic coupling of histone H3 phosphorylation and acetylation in response to epidermal growth factor stimulation. *Mol. Cell*, **5**, 905.

31. Jacobson, S. and Pillus, L. (1999) Modifying chromatin and concepts of cancer. *Curr. Opin. Genet. Develop.*, **9**, 175.

32. Ruiz-Garcia, A. B., Sendra, R., Pamblanco, M., and Tordera, V. (1997) Gcn5p is involved in the acetylation of histone H3 in nucleosomes. *FEBS Lett.*, **403**, 186.

33. Grant, P. A., Duggan, L., Côté, J., Roberts, S. M., Brownell, J. E., Candau, R., Ohba, R., Owen-Hughes, T., Allis, C. D., Winston, F., *et al.* (1997) Yeast Gcn5 functions in two multisubunit complexes to acetylate nucleosomal histones: characterization of an Ada complex and the SAGA (Spt / Ada) complex. *Genes Develop.*, **11**, 1640.

34. Pollard, K. J. and Peterson, C. L. (1997) Role of ADA/GCN5 products in antagonizing chromatin-mediated transcriptional repression. *Mol. Cell. Biol.*, **17**, 6212.

35. Syntichaki, P. and Thireos, G. (1998) The Gcn5–Ada complex potentiates the histone acetyltransferase activity of Gcn5. *J. Biol. Chem.*, **273**, 24414.

36. Kuo, M.-H., Zhou, J., Jambeck, P., Churchill, M. E. A., and Allis, C. D. (1998) Histone acetyltransferase activity of yeast Gcn5p is required for the activation of target genes *in vivo*. *Genes Develop.*, **12**, 627.

37. Wang, L., Liu, L., and Berger, S. L. (1998) Critical residues for histone acetylation by Gcn5, functioning in Ada and SAGA complexes, are also required for transcriptional function *in vivo*. *Genes Develop.*, **12**, 640.

38. Steger, D. J., Eberharter, A., John, S., Grant, P. A., and Workman, J. L. (1998) Purified histone acetyltransferase complexes stimulate HIV-1 transcription from preassembled nucleosomal arrays. *Proc. Natl Acad. Sci., USA.*, **95**, 12924.

39. Grant, P. A. and Berger, S. L. (1999) Histone acetyltransferase complexes. *Semin. Cell Develop. Biol.*, **10**, 169.

40. Eberharter, A., Sterner, D., Schieltz, D., Hassan, A., Yates III, J., Berger, S., and Workman, J. (1999) The ADA complex is a distinct histone acetyltransferase complex in *Saccharomyces cerevisiae*. *Mol. Cell. Biol.*, **19**, 6621.

41. Grant, P., Schieltz, D., Pray-Grant, M., Steger, D., Reese, J., Yates III, J., and Workman, J. (1998) A subset of TAFIIs are integral components of the SAGA complex required for nucleosome acetylation and transcription stimulation. *Cell*, **94**, 45.

42. Saleh, A., Schieltz, D., Ting, N., McMahon, S. B., Litchfield, D. W., Yates, J. R. 3rd, Lees-Miller, S. P., Cole, M. D., and Brandl, C. J. (1998) Tra1p is a component of the yeast Ada.Spt transcriptional regulatory complexes. *J. Biol. Chem.*, **273**, 26559.

43. Grant, P. A., Schieltz, D., Pray-Grant, M. G., Yates, J. R. R., and Workman, J. L. (1998) The ATM-related cofactor Tra1 is a component of the purified SAGA complex. *Mol. Cell*, **2**, 863.

44. McMahon, S. B., Van Buskirk, H. A., Dugan, K. A., Copeland, T. D., and Cole, M. D. (1998) The novel ATM-related protein TRRAP is an essential cofactor for the c-Myc and E2F oncoproteins. *Cell*, **94**, 363.

45. Ogryzko, V. V., Kotani, T., Zhang, X., Schlitz, R. L., Howard, T., Yang, X. J., Howard, B. H., Qin, J., and Nakatani, Y. (1998) Histone-like TAFs within the PCAF histone acetylase complex. *Cell*, **94**, 35.

46. Vassilev, A., Yamauchi, J., Kotani, T., Prives, C., Avantaggiati, M. L., Qin, J., and Nakatani, Y. (1998) The 400 kDa subunit of the PCAF histone acetylase complex belongs to the ATM superfamily. *Mol. Cell*, **2**, 869.

47. Martinez, E., Kundu, T. K., Fu, J., and Roeder, R. G. (1998) A human SPT3-TAFII31-GCN5-L acetylase complex distinct from transcription factor IID. *J. Biol. Chem.*, **273**, 23781. [Published erratum appears in *J. Biol. Chem.* (1998) **273**, (42), 27755.]

48. Wieczorek, E., Brand, M., Jacq, X., and Tora, L. (1998) Function of TAF(II)-containing complex without TBP in transcription by RNA polymerase II. *Nature*, **393**, 187.

49. Brand, M., Yamamoto, K., Staub, A., and Tora, L. (1999) Identification of TATA-binding protein-free TAFII-containing complex subunits suggests a role in nucleosome acetylation and signal transduction. *J. Biol. Chem.*, **274**, 18285.

50. Bannister, A. J. and Kouzarides, T. (1996) The CBP coactivator is a histone acetyltransferase. *Nature*, **384**, 641.

51. Ogryzko, V. V., Schiltz, R. L., Russanova, V., Howard, B. H., and Nakatani, Y. (1996) The transcriptional coactivators p300 and CBP are histone acetyltransferases. *Cell*, **87**, 953.

52. Xu, W., Edmondson, D. G., and Roth, S. Y. (1998) Mammalian GCN5 and P/CAF acetyltransferases have homologous amino-terminal domains important for recognition of nucleosomal substrates. *Mol. Cell. Biol.*, **18**, 5659.

53. Chen, H., Lin, R. J., Schiltz, R. L., Chakravarti, D., Nash, A., Nagy, L., Privalsky, M. L., Nakatani, Y., and Evans, R. M. (1997) Nuclear receptor coactivator ACTR is a novel histone acetyltransferase and forms a multimeric activation complex with P/CAF and CBP/p300. *Cell*, **90**, 569.

54. Spencer, T. E., Jenster, G., Burcin, M. M., Allis, C. D., Zhou, J., Mizzen, C. A., McKenna, N. J., Onate, S. A., Tsai, S. Y., Tsai, M.-J., *et al.* (1997) Steroid receptor coactivator-1 is a histone acetyltransferase. *Nature*, **389**, 194.

55. Mizzen, C. A., Yang, X.-J., Kokubo, T., Brownell, J. E., Bannister, A. J., Owen-Hughes, T., Workman, J. L., Wang, L., Berger, S. L., Kouzarides, T., *et al.* (1996). The TAFII250 subunit of TFIID has histone acetyltransferase activity. *Cell*, **87**, 1261.

56. Kundu, T. K., Wang, Z., and Roeder, R.G. (1999) Human TFIIIC relieves chromatin-mediated repression of RNA polymerase III transcription and contains an intrinsic histone acetyltransferase activity. *Mol. Cell. Biol.*, **19**, 1605.

57. Hsieh, Y. J., Kundu, T. K., Wang, Z., Kovelman, R., and Roeder, R. G. (1999) The TFIIIC90 subunit of TFIIIC interacts with multiple components of the RNA polymerase III machinery and contains a histone-specific acetyltransferase activity. *Mol. Cell. Biol.*, **19**, 7697.

58. Allard, S., Utley, R., Savard, J., Clark, A., Grant, P., Brandl, C., Pillus, L., Workman, J., and Cote, J. (1999) NuA4, an essential transcription adaptor/histone H4 acetyltransferase complex containing Esa1p and the ATM-related cofactor Tra1p. *EMBO J.*, **18**, 5108.

59. John, S., Howe, L., Tafrov, S. T., Grant, P. A., Sternglanz, R., and Workman, J. L. (2000) The Something About Silencing protein, Sas3, is the catalytic subunit of NuA3, a yTAF(II)30-containing HAT complex that interacts with the Spt16 subunit of the yeast CP (Cdc68/Pob3)-FACT complex. *Genes Develop.*, **14**, 1196.

60. Wittschieben, B. O., Otero, G., de Bizemont, T., Fellows, J., Erdjument-Bromage, H., Ohba, R., Li, Y., Allis, C. D., Tempst, P., and Svejstrup, J. Q. (1999) A novel histone acetyltransferase is an integral subunit of elongating RNA polymerase II holoenzyme. *Mol. Cell*, **4**, 123.

61. Hebbes, T. R., Clayton, A. L., Thorne, A. W., and Crane-Robinson, C. (1994) Core histone hyperacetylation co-maps with generalized DNase I sensitivity in the chicken β-globin chromosomal domain. *EMBO J.*, **13**, 1823.

62. Krebs, J. E., Kuo, M.-H., Allis, C. D., and Peterson, C. L. (1999) Cell cycle-regulated histone acetylation required for expression of the yeast *HO* gene. *Genes Develop.*, **13**, 1412.

63. Parekh, B. S. and Maniatis, T. (1999) Virus infection leads to localized hyperacetylation of histones H3 and H4 at the IFN-beta promoter. *Mol. Cell*, **3**, 125.

64. Utley, R. T., Ikeda, K., Grant, P. A., Côté, J., Steger, D. J., Eberharter, A., John, S., and Workman, J. L. (1998) Transcriptional activators direct histone acetyltransferase complexes to nucleosomes. *Nature*, **394**, 498.

65. Wallberg, A. E., Neely, K. E., Gustafsson, J. A., Workman, J. L., Wright, A. P., and Grant, P. A. (1999) Histone acetyltransferase complexes can mediate transcriptional activation by the major glucocorticoid receptor activation domain. *Mol. Cell. Biol.*, **19**, 5952.

66. Ikeda, K., Steger, D. J., Eberharter, A., and Workman, J. L. (1999) Activation domain-specific and general transcription stimulation by native histone acetyltransferase complexes. *Mol. Cell. Biol.*, **19**, 855.

67. Vignali, M., Steger, D., Neely, K., and Workman, J. (2000) Distribution of acetylated histones resulting from Gal4-VP16 recruitment of SAGA and NuA4 complexes. *EMBO J.*, **19**, 2629.

68. Puri, P. L., Sartorelli, V., Yang, X.-J., Hamamori, Y., Ogryzko, V. V., Howard, B. H., Kedes, L., Wang, J. Y. J., Graessmann, A., Nakatani, Y., *et al.* (1997) Differential role of p300 and PCAF acetyltransferases in muscle differentiation. *Mol. Cell*, **1**, 35.

69. Korzus, E., Torchia, J., Rose, D. W., Xu, L., Kurokawa, R., McInerney, E. M., Mullen, T. M., Glass, C. K., and Rosenfeld, M. G. (1998) Transcription factor-specific requirements for coactivators and their acetyltransferase functions. *Science*, **279**, 703.

70. Xu, L., Lavinsky, R. M., Dasen, J. S., Flynn, S. E., McInerney, E. M., Mullen, T. M., Heinzel, T., Szeto, D., Korzus, E., Kurokawa, R., *et al.* (1998) Signal-specific co-activator domain requirements for Pit-1 activation. *Nature*, **395**, 301.

71. Wieczorek, E., Brand, M., Jacq, X., and Tora, L. (1998) Function of TAF(II)-containing complex without TBP in transcription by RNA polymerase II [see comments]. *Nature*, **393**, 187.

72. Brand, M., Leurent, C., Mallouh, V., Tora, L., and Schultz, P. (1999) Three-dimensional structures of the TAFII-containing complexes TFIID and TFTC. *Science*, **286**, 2151.

73. Orphanides, G., LeRoy, G., Chang, C. H., Luse, D. S., and Reinberg, D. (1998) FACT, a factor that facilitates transcript elongation through nucleosomes. *Cell*, **92**, 105.

74. Wittmeyer, J. and Formosa, T. (1997) The *Saccharomyces cerevisiae* DNA polymerase alpha catalytic subunit interacts with Cdc68/Spt16 and with Pob3, a protein similar to an HMG1-like protein. *Mol. Cell. Biol.*, **17**, 4178.

75. Simpson, R. T. (1978) Structure of chromatin containing extensively acetylated H3 and H4. *Cell*, **13**, 691.

76. Ausio, J. and van Holde, K. E. (1986) Histone hyperacetylation: its effects on nucleosome conformation and stability. *Biochemistry*, **25**, 1421.

77. Norton, V. G., Marvin, K. W., Yau, P., and Bradbury, E. M. (1990) Nucleosome linking number change controlled by acetylation of histones H3 and H4. *J. Biol. Chem.*, **265**, 19848.

78. Lee, D. Y., Hayes, J. J., Pruss, D., and Wolffe, A. P. (1993) A positive role for histone acetylation in transcription factor access to nucleosomal DNA. *Cell*, **72**, 73.

79. Vettese-Dadey, M., Grant, P. A., Hebbes, T. R., Crane-Robinson, C., Allis, C. D., and Workman, J. L. (1996) Acetylation of histone H4 plays a primary role in enhancing transcription factor binding to nucleosomal DNA in vitro. *EMBO J.*, **15**, 2508.

80. Vitolo, J. M., Thiriet, C., and Hayes, J. J. (2000) The H3–H4 N-terminal tail domains are the primary mediators of transcription factor IIIA access to 5S DNA within a nucleosome. *Mol. Cell. Biol.*, **20**, 2167.

81. Garcia-Ramirez, M., Rocchini, C., and Ausio, J. (1995) Modulation of chromatin folding by histone acetylation. *J. Biol. Chem.*, **270**, 17923.

82. Tse, C., Sera, T., Wolffe, A. P., and Hansen, J. C. (1998) Disruption of higher-order folding by core histone acetylation dramatically enhances transcription of nucleosomal arrays by RNA polymerase III. *Mol. Cell. Biol.*, **18**, 4629.

83. Hecht, A., Laroche, T., Strahl-Bolsinger, S., Gasser, S. M., and Grunstein, M. (1995) Histone H3 and H4 N-termini interact with SIR3 and SIR4 proteins: A molecular model for the formation of heterochromatin in yeast. *Cell*, **80**, 583.

84. Edmondson, D. G., Smith, M. M., and Roth, S. Y. (1996) Repression domain of the yeast global repressor Tup1 interacts directly with histones H3 and H4. *Genes Develop.*, **10**, 1247.

85. Breiling, A., Bonte, E., Ferrari, S., Becker, P. B., and Paro, R. (1999) The *Drosophila* polycomb protein interacts with nucleosomal core particles *in vitro* via its repression domain. *Mol. Cell. Biol.*, **19**, 8451.

86. Palaparti, A., Baratz, A., and Stifani, S. (1997) The Groucho/transducin-like enhancer of split transcriptional repressors interact with the genetically defined amino-terminal silencing domain of histone H3. *J. Biol. Chem.*, **272**, 26604.

87. Dhalluin, C., Carlson, J. E., Zeng, L., He, C., Aggarwal, A. K., and Zhou, M. M. (1999) Structure and ligand of a histone acetyltransferase bromodomain. *Nature*, **399**, 491.

88. Ornaghi, P., Ballario, P., Lena, A. M., Gonzalez, A., and Filetici, P. (1999) The bromodomain of Gcn5p interacts *in vitro* with specific residues in the N terminus of histone H4. *J. Mol. Biol.*, **287**, 1.

89. Winston, F. and Allis, C. D. (1999) The bromodomain: a chromatin-targeting module? *Nature Struct. Biol.*, **6**, 601.

90. Imhof, A., Yang, X. J., Ogryzko, V. V., Nakatani, Y., Wolffe, A. P., and Ge, H. (1997) Acetylation of general transcription factors by histone acetyltransferases. *Curr. Biol.*, **7**, 689.

91. Gu, W. and Roeder, R. (1997) Activation of p53 sequence-specific DNA binding by acetylation of the p53 C-terminal domain. *Cell*, **90**, 595.

92. Scolnick, D. M., Chehab, N. H., Stavridi, E. S., Lien, M. C., Caruso, L., Moran, E., Berger, S. L., and Halazonetis, T. D. (1997) CREB-binding protein and p300/CBP-associated factor are transcriptional coactivators of the p53 tumor suppressor protein. *Cancer Res.*, **57**, 3693.

93. Lill, N. L., Grossman, S. R., Ginsberg, D., DeCaprio, J., and Livingston, D. M. (1997) Binding and modulation of p53 by p300/CBP coactivators. *Nature*, **387**, 823.

94. Sakaguchi, K., Herrera, J. E., Saito, S., Miki, T., Bustin, M., Vassilev, A., Anderson, C. W., and Appella, E. (1998) DNA damage activates p53 through a phosphorylation-acetylation cascade. *Genes Develop.*, **12**, 2831.

95. Liu, L., Scolnick, D. M., Trievel, R. C., Zhang, H. B., Marmorstein, R., Halazonetis, T. D., and Berger, S. L. (1999) p53 sites acetylated in vitro by PCAF and p300 are acetylated *in vivo* in response to DNA damage. *Mol. Cell. Biol.*, **19**, 1202.

96. Puri, P. L., Sartorelli, V., Yang, X. J., Hamamori, Y., Ogryzko, V. V., Howard, B. H., Kedes, L., Wang, J. Y., Graessmann, A., Nakatani, Y., *et al.* (1997) Differential roles of p300 and PCAF acetyltransferases in muscle differentiation. *Mol. Cell*, **1**, 35.

97. Puri, P. L., Avantaggiati, M. L., Balsano, C., Sang, N., Graessmann, A., Giordano, A., and Levrero, M. (1997) p300 is required for MyoD-dependent cell cycle arrest and muscle-specific gene transcription. *EMBO J.*, **16**, 369.

98. Sartorelli, V., Puri, P. L., Hamamori, Y., Ogryzko, V., Chung, G., Nakatani, Y., Wang, J. Y. J., and Kedes, L. (1999) Acetylation of MyoD directed by PCAF is necessary for the execution of the muscle program. *Mol. Cell*, **4**, 725.

99. Martínez-Balbás, M. A., Bauer, U.-M., Nielsen, S. J., Brehm, A., and Kouzarides, T. (2000) Regulation of E2F1 activity by acetylation. *EMBO J.*, **19**, 662.

100. Boyes, J., Byfield, P., Nakatani, Y., and Ogryzko, V. (1998) Regulation of activity of the transcription factor GATA-1 by acetylation. *Nature*, **396**, 594.

101. Hung, H. L., Lau, J., Kim, A. Y., Weiss, M. J., and Blobel, G. A. (1999) CREB-Binding protein acetylates hematopoietic transcription factor GATA-1 at functionally important sites. *Mol. Cell. Biol.*, **19**, 3496.

102. Zhang, W. and Bieker, J. J. (1998) Acetylation and modulation of erythroid Krüppel-like factor (EKLF) activity by interaction with histone acetyltransferases. *Proc. Natl Acad. Sci., USA*, **95**, 9855.

103. Waltzer, L. and Bienz, M. (1998) *Drosophila* CBP represses the transcription factor TCF to antagonize Wingless signalling. *Nature*, **395**, 521.

104. Kiernan, R. E., Vanhulle, C., Schiltz, L., Adam, E., Xiao, H., Maudoux, F., Calomme, C., Burny, A., Nakatani, Y., Jeang, K.T., *et al.* (1999) HIV-1 Tat transcriptional activity is regulated by acetylation. *EMBO J.*, **18**, 6106.

105. Chen, H., Lin, R. J., Xie, W., Wilpitz, D., and Evans, R. M. (1999) Regulation of hormone-induced histone hyperacetylation and gene activation via acetylation of an acetylase. *Cell*, **98**, 675.

106. Herrera, J. E., Sakaguchi, K., Bergel, M., Trieschmann, L., Nakatani, Y., and Bustin, M. (1999) Specific acetylation of chromosomal protein HMG-17 by PCAF alters its interaction with nucleosomes. *Mol. Cell. Biol.*, **19**, 3466.

107. Munshi, N., Merika, M., Yie, J., Senger, K., Chen, G., and Thanos, D. (1998) Acetylation of HMG I(Y) by CBP turns off IFN beta expression by disrupting the enhanceosome. *Mol. Cell*, **2**, 457.

108. Parekh, B. S. and Maniatis, T. (1999) Virus infection leads to localized hyperacetylation of histones H3 and H4 at the IFN promoter. *Mol. Cell*, **3**, 125.

109. Strahl, B. D. and Allis, C. D. (2000) The language of covalent histone modifications. *Nature*, **403**, 41.

110. Howe, L., Brown, C. E., Lechner, T., and Workman, J.L. (1999) Histone acetyltransferase complexes and their link to transcription. *Crit. Rev. Eukaryot. Gene Expr.*, **9**, 231.

111. Mizzen, C., Kuo, M. H., Smith, E., Brownell, J., Zhou, J., Ohba, R., Wei, Y., Monaco, L., Sassone-Corsi, P., and Allis, C. D. (1998) Signaling to chromatin through histone modifications: how clear is the signal? *Cold Spring Harb. Symp. Quant. Biol.*, **63**, 469.

112. Mahadevan, L. C., Willis, A. C., and Barratt, M. J. (1991) Rapid histone H3 phosphorylation in response to growth factors, phorbol esters, okadaic acid, and protein synthesis inhibitors. *Cell*, **65**, 775.

113. Thomson, S., Mahadevan, L. C., and Clayton, A. L. (1999) MAP kinase-mediated signalling to nucleosomes and immediate-early gene induction [see comments]. *Semin. Cell Develop. Biol.*, **10**, 205.

114. Thomson, S., Clayton, A. L., Hazzalin, C. A., Rose, S., Barratt, M. J., and Mahadevan, L. C. (1999) The nucleosomal response associated with immediate-early gene induction is mediated via alternative MAP kinase cascades: MSK1 as a potential histone H3/HMG-14 kinase. *EMBO J.*, **18**, 4779.

115. Sassone-Corsi, P., Mizzen, C. A., Cheung, P., Crosio, C., Monaco, L., Jacquot, S., Hanauer, A., and Allis, C. D. (1999) Requirement of Rsk-2 for epidermal growth factor-activated phosphorylation of histone H3. *Science*, **285**, 886.

116. Chen, D., Ma, H., Hong, H., Koh, S. S., Huang, S. M., Schurter, B. T., Aswad, D. W., and Stallcup, M. R. (1999) Regulation of transcription by a protein methyltransferase. *Science*, **284**, 2174.

117. Braunstein, M., Rose, A. B., Holmes, S. G., Allis, C. D., and Broach, J. R. (1993) Transcriptional silencing in yeast is associated with reduced nucleosome acetylation. *Genes Develop.*, **7**, 592.

118. Tanny, J. C., Dowd, G. J., Huang, J., Hilz, H., and Moazed, D. (1999) An enzymatic activity in the yeast Sir2 protein that is essential for gene silencing. *Cell*, **99**, 735.

119. Imai, S., Armstrong, C. M., Kaeberlein, M., and Guarente, L. (2000) Transcriptional silencing and longevity protein Sir2 is an NAD- dependent histone deacetylase [in process citation]. *Nature*, **403**, 795.

120. Landry, J., Sutton, A., Tafrov, S., Heller, R., Stebbins, J., Pillus, L., and Sternglanz, R. (2000) The silencing protein SIR2 and its homologs are NAD-dependent protein deacetylases. *Proc. Natl Acad. Sci., USA*, **97**, 5807.

121. Cosma, M. P., Tanaka, T., and Nasmyth, K. (1999) Ordered recruitment of transcription and chromatin remodeling factors to cell cycle and developmentally regulated promoters. *Cell*, **97**, 299.

122. Syntichaki, P., Topalidou, I., and Thireos, G. (2000) The Gcn5 bromodomain co-ordinates nucleosome remodelling. *Nature*, **404**, 414.

123. Sterner, D. E., Grant, P. A., Roberts, S. M., Duggan, L., Belotserkovskaya, R., Pacella, L. A., Winston, F., Workman, J. L., and Berger, S. L. (1999) Functional organization of the yeast SAGA complex: distinct components involved in structural integrity, nucleosome acetylation, and TATA-binding protein interaction. *Mol. Cell. Biol.*, **19**, 86.

124. Dudley, A. M., Rougeulle, C., and Winston, F. (1999) The Spt components of SAGA facilitate TBP binding to a promoter at a post-activator-binding step *in vivo*. *Genes Develop.*, **13**, 2940.

8 | Histone deacetylation: mechanisms of repression

ANDREW FREE, MICHAEL GRUNSTEIN, ADRIAN BIRD, and
MARIA VOGELAUER

1. Introduction and general remarks

The preceding chapter introduced the concept of modification of the amino-terminal tails of the core histones by acetylation, and described the range of acetyltransferase enzymes that can be targeted to certain regions of chromatin to undertake this modification *in vivo*. Acetylation of the histone tails is correlated with the relief of transcriptional repression, and a corollary of this is that deacetylation of modified histone tails will lead to transcriptional repression. It is therefore not surprising that eukaryotic cells contain a similar range of histone deacetylase enzymes to reverse the modifications carried out by the acetyltransferases, and that these enzymes also reside in multiprotein complexes which can be targeted to specific regions of chromatin by a variety of transcriptional repressors.

This chapter will describe the deacetylase enzymes and their complexes, and will then look in detail at the regions of the chromosomes of both yeast and higher eukaryotic cells which are hypoacetylated (i.e. relatively deficient in acetylated core histones). We then consider how the deacetylases may be targeted to these transcriptionally repressed regions by specific repressor proteins, and also examine the question of whether deacetylation is restricted to gene promoters or may be a common feature of longer stretches of chromatin encompassing the bodies of genes. Finally, we examine how a distinct repression mechanism, modification of DNA by cytosine methylation, makes use of histone deacetylation to silence transcription.

2. Histone deacetylases and their complexes

2.1 Histone deacetylase enzymes in yeast and mammals

In recent years, the identification of a growing family of proteins associated with histone deacetylase activity has highlighted the role of deacetylation of histone tails in transcriptional repression. There are two main families of deacetylase enzymes,

Table 1 Histone deacetylases and their complexes from yeast and higher eukaryotes

Deacetylase enzymes

Class I histone deacetylases		Class II histone deacetylases	
Yeast	**Mammals**	**Yeast**	**Mammals**
RPD3	HDAC1	HDA1	HDAC4
	HDAC2	HOS1	HDAC5
	HDAC3	HOS2	HDAC6
		HOS3	HDAC7

SIR2 NAD-dependent histone deacetylase

Deacetylase complexes (mammals)

Sin3/HDAC core complex (histone deacetylase-B in yeast)	Mi-2/NuRD core complex
Sin3	Mi-2
HDAC1	HDAC1
HDAC2	HDAC2
RbAp46	RbAp46
RbAp48	RbAp48
SAP30	MBD3
SAP18	MTA2
	p66[a]

Other complexes and interactions
Histone deacetylase-A (yeast): HDA1, HDA2, and HDA3
HOS1–HOS2? (yeast)
HDAC3–HDAC4–RbAp48 (mammals)
HDAC3–HDAC5 (mammals)
HDAC7–Sin3 (mammals)

[a] Found in *Xenopus* Mi-2 complex only.

with members in eukaryotes from yeast to mammals (see Table 1). In the first class, the founding members from yeast and mammals, respectively, are the RPD3 and HDAC1 enzymes, which are closely related. Interestingly, RPD3 was initially identified genetically as a transcriptional repressor (1, 2) and only later suggested to possess deacetylase activity following biochemical identification of its mammalian homologue (3). The related mammalian enzymes HDAC2 (4) and HDAC3 (5) form the class I deacetylase family along with HDAC1 and RPD3; HDAC1/2 and RPD3 appear to be members of similar complexes in mammalian and yeast cells, and most information has been gathered about this family of deacetylases.

A second yeast deacetylase, HDA1, is structurally distinct from RPD3 but more closely related to the yeast enzymes HOS1, HOS2, and HOS3 (6). These proteins are also similar structurally to the mammalian enzymes HDAC4, HDAC5 (also called mHDA1) and HDAC6 (mHDA2), which form the class II deacetylases (7). Very recently, a further member of the family, mammalian HDAC7, has been identified (8). Mammalian class II deacetylases are larger (902–1215 amino acids) than the class I enzymes, and, in the cases of HDACs 4, 5, and 7, contain a well-conserved carboxy-terminal deacetylase domain. In contrast, HDAC6 includes two catalytic domains near the amino terminus, which have been postulated to form an intramolecular dimer.

Less is known about the *in vivo* association partners and functions of the class II deacetylases (see below).

Although the families of deacetylases described above are quite diverse, they all contain a well-conserved stretch of 60–70 amino acids which has homology to regions in both acetylpolyamine amidohydrolase (6) and the acetate utilization protein AcuC from *Bacillus subtilis*. This conserved domain contains two pairs of histidine residues separated by 37 amino acids which, when mutated individually or in combination to alanine, eliminate the histone deacetylase activity of RPD3 *in vitro*, and its repression activity *in vivo* (9). Therefore it is likely that this domain constitutes a core deacetyl-ase domain, and that the conserved histidine residues are key catalytic residues within it. As noted above, the human HDAC6 protein contains two highly homo-logous central catalytic domains, which may have arisen by an internal duplication event and which have been shown by mutagenesis of the conserved histidine residues to be independently active (7).

2.2 Histone deacetylases are members of multiprotein complexes

Like histone acetyltransferases (Chapter 7), histone deacetylase enzymes have been identified as members of multiprotein complexes (Table 1), which probably serve to modulate and target their function *in vivo*. In yeast cells, RPD3 is found within a 600 kDa complex known as histone deacetylase-B, which also contains the co-repressor protein SIN3 (10, 11). The Sin3 complex has been more exhaustively studied in mammalian cells, and found to comprise at least seven subunits which include two class I deacetylases, HDAC1 and HDAC2. Sin3 is a large protein which contains several *paired amphipathic helix* (PAH) domains postulated to be involved in protein–protein interactions (12), and is likely to act as a molecular scaffold for assembly of the other proteins involved in the complex (11). Among these proteins is a related pair of factors, RbAp46 and RbAp48, named for their association with retinoblastoma protein through which they were originally identified (13). RbAp46 and RbAp48 are interesting because they can bind directly to helix 1 of histone H4 *in vitro* (14), and at least *Xenopus* RbAp48 can bind H4 *in vivo* (15). They are also found in HAT complexes (14, 16) and the Mi-2/NuRD complex (see below), and may there-fore constitute core histone-binding subunits for these complexes. However, helix 1 of H4 is normally buried within the nucleosomal structure; thus it is not surprising that the assembled Sin3 complex can deacetylate free core histones but not assembled nucleosomes *in vitro* (17). It is likely that additional factors are required to open up the chromatin *in vivo* to allow the Sin3 complex to bind to and deacetylate the histones. It is possible that the DNA-binding factors which recruit the Sin3 complex to specific sites in chromatin may perform such a role. The Sin3 complex also contains additional core factors SAP18 and SAP30, the role of which is less clear (17, 18). SAP30 may be involved in the recruitment of the Sin3 complex by a specific subset of NCoR co-repressor complexes (19).

A second major mammalian histone deacetylase complex containing the HDAC1 and HDAC2 deacetylases has also been identified. This is the *nucleosome remodel-ling histone deacetylase* (NuRD) or Mi-2 complex, which has at its core the large Mi-2 protein, originally identified as an autoantigen in dermatomyositis, a human con-nective tissue disease. Intriguingly, Mi-2 is a member of the Snf2 family of ATPases associated with chromatin remodelling, and possesses nucleosome-stimulated ATPase activity, suggesting that it may be involved in remodelling nucleosomes to make them accessible to the deacetylase activity (20). In *Xenopus* oocyte extracts, the Mi-2 complex was identified as the major peak of deacetylase activity obtained following biochem-ical fractionation (20), although histone deacetylase activity appears to reside mainly in smaller complexes in the intact oocyte nucleus (21); it is possible that the Mi-2 com-plex assembles at later stages in *Xenopus* development or during purification from oocytes. Mi-2/NuRD from mammalian cells has been affinity purified and shown to contain a total of seven subunits (22). Aside from Mi-2 itself and HDAC1/2, these include RbAp46 and RbAp48, which presumably perform a similar histone-targeting function as in the Sin3 complex. Thus the association of HDAC1 and HDAC2 with a major core protein and with these histone-binding subunits is a key common feature of the Sin3 and Mi-2/NuRD complexes. The other subunits of the complex identified in mammals are the MTA2 protein (related to *metastasis-associated* protein 1; also known as MTA1-like) and MBD3, which is a member of the *methyl-CpG-binding domain* protein family (see below). In *Xenopus*, an additional unknown polypeptide, p66, has been found to be associated with the Mi-2 complex (23). It seems likely that taken together, the Sin3 and Mi-2/NuRD complexes may form the major functional forms of the HDAC1 and HDAC2 deacetylases in mammalian cells.

In contrast to RPD3/HDAC1 and HDAC2, much less is known about the associ-ation partners of the other deacetylase enzymes (Table 1). In yeast, HDA1 is part of a 350 kDa complex known as histone deacetylase-A, which exhibits much greater sensitivity to the deacetylase inhibitor trichostatin A (TSA) than the histone deacetylase-B/Sin3 complex (6, 10). The yeast HOS1 and HOS2 enzymes have also been reported to be found in multiprotein complexes following purification, while HOS3 is unusual in that it can form a homodimer with intrinsic deacetylase activity without the need for accessory factors, and is also TSA insensitive (24). Of the mammalian class II enzymes which are related to these deacetylases, HDAC4 can be co-immunoprecipitated with the class I enzyme HDAC3 and with RbAp48, while HDAC5 has been shown to interact with at least HDAC3 (7). The recent discovery of HDAC7 also led to evidence that this deacetylase can directly interact with Sin3, a property not ascribed to other class II deacetylases (8). It is apparent that molecules such as Sin3, Mi-2 and RbAp46/48 are likely to be crucial to the correct assembly and targeting of many histone deacetylase activities *in vivo*.

2.3   The SIR2 protein: a unique NAD-dependent histone deacetylase

Recently, a very different histone deacetylase has been identified in the guise of the yeast SIR2 protein. SIR2 is a component of heterochromatin in *Saccharomyces cerevisiae* (see Chapter 10), and is involved in transcriptional silencing of silent mating-type loci, telomeric regions, and ribosomal RNA genes (rDNA). It can also suppress recombination within this rDNA, and acts as a longevity factor in yeast cells. The mating-type and telomeric silencing functions described above require chromatin binding of the SIR3 protein, which may require deacetylation of lysine 16 of the H4. Interestingly, it has also been shown that overexpression of SIR2 leads to histone hypoacetylation (25), although the protein has no sequence similarity to the deacetylases described above. None the less, recent work (26–28) shows that SIR2 does indeed have *in vitro* histone deacetylase activity. This activity requires the cofactor NAD, which may provide a link between deacetylase activity and the cellular energy or oxidation state. In fact NAD had previously been identified as the cofactor for a SIR2-catalysed ADP ribosylation reaction (29), but mutations that remove this activity while leaving the deacetylase activity intact have little effect on the silencing function of SIR2, whereas deacetylase-negative mutants are silencing-defective (26). It is interesting to note that the SIR2 protein family is evolutionarily widespread, being found in organisms from bacteria to humans (30), although presumably at least the eubacterial proteins will not function as histone deacetylases *in vivo*, as histone substrates are not present.

3. Heterochromatin and hypoacetylation

3.1 Yeast models of heterochromatin (telomeric, mating-type, and ribosomal DNA silencing)

In the yeast *S. cerevisiae* certain genomic loci resemble the heterochromatin of more complex eukaryotes (see Chapter 10). Regions at the silent mating loci *HML* and *HMR* and at telomeres are condensed, replicate late in S phase, and associate in foci at the nuclear periphery. Moreover, these regions, together with the ribosomal DNA (rDNA) locus, have the ability to silence genes transposed there, and do so epigenetically. Many of the *cis*- and *trans*-elements necessary to establish and maintain the repressive chromatin structure at these loci have been identified, and models predicting how these factors interact and function have been proposed and are rapidly evolving. Although there are some differences, the various yeast heterochromatic loci share many structural and functional characteristics.

Silencing at *HML* and *HMR* is established by a combination of DNA-binding sequences for RAP1, ABF1, and ORC (origin recognition complex), forming the so-called 'silencers'. Variants of these silencers bracket both *HML* and *HMR*, where the interaction of these proteins with the SIR (silent information regulator) proteins, SIR1, SIR2, SIR3, and SIR4, establishes the repressive chromatin structure. Once the structure has been nucleated, SIR1 is no longer necessary for its maintenance. At telomeres SIR1 is not necessary for transcriptional silencing, but a similar mechanism is involved in the formation of heterochromatin. For example, at the right telomere of chromosome VI, RAP1 interacts with the telomeric (TG_{1-3}) repeats at the very end (~300 bp) of the chromosome and recruits SIR proteins to this site (Fig. 1A). It has

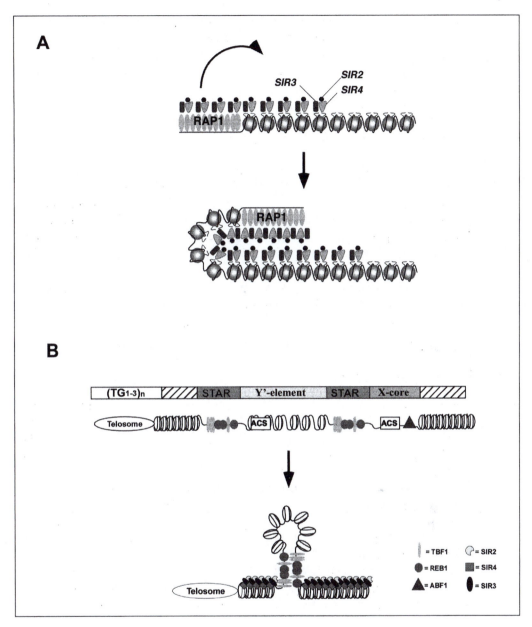

Fig. 1 Yeast telomeric heterochromatin. (A) Spreading of SIR proteins from a RAP1 nucleation site induces the yeast telomere to fold back. (B) Heterochromatin at yeast telomeres is discontinuous. Subtelomeric anti-silencing regions (STARs) act like boundaries that delimit stretches of transcriptionally active chromatin from surrounding heterochromatin. Dimerization of REB1 and TBF1 could lead to the formation of a loop of more open chromatin. ARS consensus sequences (ACS) in combination with ABF1 contribute to re-establishing heterochromatin.

been shown that SIR3 and SIR4 proteins interact directly with histones H3 and H4 *in vitro* (31), and this interaction may be modulated by the acetylation state of the histone N-termini. Moreover, SIR3–SIR4–SIR2 interact in that order (32, 33). Since silencing from the right telomere of chromosome VI occurs in a continuous fashion (34), it has been proposed that SIR proteins spread from the RAP1 initiation sites by interacting with the histone H3 and H4 N-termini to silence adjacent chromatin (31, 32). The SIR protein complex is also thought to induce folding back of the chromosome end (Fig. 1A; 32), analogous to the loop structures seen at human telomeres.

The flow of SIR proteins from a single RAP1 cluster into the subtelomeric regions is thought to be disrupted by insulator elements named STARs (*sub*telomeric *anti*-silencing *r*egions) within the subtelomeric X- and Y'-elements (Fig. 1B; 35). The STAR elements function as barriers to the formation of yeast heterochromatin, and a subtelomeric reporter gene is derepressed when positioned between two STARs. STAR elements all contain multimeric binding sites for two proteins, TBF1 (*TT*AGGG *r*epeat-*b*inding *f*actor *1*) and REB1 (*r*RNA *e*nhancer-*b*inding protein *1*). Due to these elements, chromatin at most telomeres may be imagined as short stretches of continuous heterochromatin separated from each other by islands of a more open structure, delimited by STARs (Fig.1B) (35). Generally, these less-silenced regions coincide with the Y'-elements, and a model has been proposed that views them as domains that escape heterochromatin by forming an isolated loop that extrudes out of the bulk compact structure (Fig. 1B) (36).

The silent chromatin structure at the rDNA locus is quite different from the other loci. There are approximately 150 copies of tandemly arranged rDNA genes (9.1 kb per copy) at the locus. Here, transcriptional silencing is SIR2-dependent, but SIR3- and SIR4-independent (37, 38). The association of SIR2 with the rDNA locus depends instead on the presence of NET1 (39). These two proteins, together with CDC14, associate in the RENT (*r*egulator of *n*ucleolar silencing and *t*elophase exit) complex that localizes to the nucleolus.

An important feature of heterochromatin in both yeast and more complex organisms is the hypoacetylation of its histones. It is now clear that all acetylatable lysines examined of all four core histones are hypoacetylated at the *HM* loci and at telomeres (N. Suka *et al.*, submitted for publication). The hypoacetylation of lysine K16 in H4 seems to be particularly important. This residue is found in that portion of H4 (residues 16–29) which interacts with SIR3 (40); substitutions in this site that mimic its acetylation strongly prevent silencing (41). Interestingly, K16-Ac is the only acetylated site in monoacetylated histone H4 in euchromatin, a feature that may prevent promiscuous SIR3 interactions in euchromatin (42).

How acetylation in heterochromatin is prevented is unclear. Disruption of any of the five related histone deacetylases of yeast (*RPD3*, *HDA1*, *HOS1*, *HOS2*, and *HOS3*) does not affect telomeric histone acetylation (N. Suka *et al.*, submitted for publication). However, the recent evidence that SIR2 is an NAD-dependent histone deacetylase allows one to imagine a scenario in which SIR2 deacetylates K16 in histone H4, allowing SIR3 binding. The resulting compacted structure would allow the subsequent 'passive' protection of other histone sites from acetylation (Fig. 2). It must be stressed

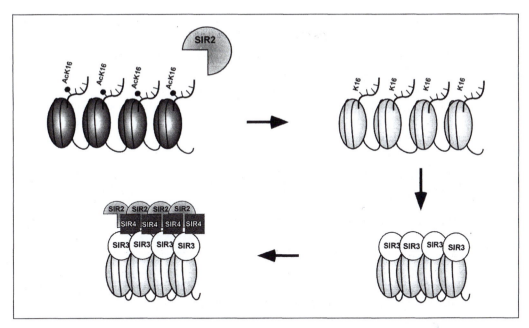

Fig. 2 Establishment of heterochromatin by SIR proteins. SIR2 could deacetylate telomeric histones. Deacetylation of lysine 16 of histone H4 might then stabilize the interaction between histones and SIR3, favouring the spreading of the SIR2/SIR3/SIR4 complex and resulting in the formation of a higher-order chromatin structure.

that the SIR2 deacetylase activity, but not the mono-ADP-ribosylation activity, has been shown to be necessary for silencing *in vivo* (26). Nevertheless, it is clear that two novel post-translational activities have been added to the function of SIR2, and that one or both of these activities provides the signal that allows silencing to be inherited epigenetically during DNA replication.

3.2 Higher eukaryotic heterochromatin and hypoacetylation

As in yeast, inactive regions of higher eukaryotic chromosomes are also organized into a distinct heterochromatic structure, and can silence euchromatic genes which are juxtaposed with them. Again, the heterochromatic regions exhibit hypoacetylation relative to the bulk chromatin, and are associated with distinct proteins which are likely to contribute to their transcriptionally silent state. James and Elgin (43) identified the first such protein, HP1 (heterochromatin protein 1) in *Drosophila*, where it is found associated with pericentric heterochromatin, the heterochromatic regions of the small fourth chromosome and, to a lesser extent, telomeric heterochromatin and specific euchromatic locations (44). In mammals, three HP1 proteins have been identified in both mouse (mHP1α, M31/MOD1, and M32/MOD2) and human (HP1^{Hsα}, HP1^{Hsβ}, and HP1^{Hsγ}), and of these at least M31/MOD1 and HP1^{Hsβ} can be localized to heterochromatic regions *in vivo* (reviewed in 45). The HP1 family proteins contain both an N-terminal 'chromo domain', with structural similarity to archaebacterial histone-like proteins, and a C-terminal 'chromo shadow domain'; both domains can target HP1 to heterochromatin. Both of these domains contain hydrophobic pockets which are thought to mediate protein–protein interactions, and indeed several proteins have been found to interact with HP1 family members, including SU(VAR)3–7, which co-localizes with HP1 on *Drosophila* metaphase and polytene chromosomes, ORC proteins, lamin receptors (which may target hetero-chromatic regions to the nuclear membrane), and histone H1-like proteins.

Using human HP1α as a bait in a yeast two-hybrid screen, a protein known as TIF1β was also shown to be an HP1-interacting factor (46). This protein belongs to a large group of *transcriptional intermediary factors* (TIFs; also known as co-activators or co-repressors) which mediate the function of many transcription factors, often by remodelling chromatin structure. TIF1β itself may function as a co-repressor for the numerous KRAB domain-containing zinc-finger proteins. Two recent studies (47, 48) showed that TIF1β (also called KAP-1) can be found co-localized with HP1 in pericentric heterochromatin, as well as in punctate euchromatic domains, and can be co-immunoprecipitated with HP1 from nuclear extracts. The deacetylase inhibitor TSA was found partially to relieve silencing by both TIF1β and all HP1 family members (48), suggesting a role for histone deacetylation in heterochromatic silencing and in TIF1β-dependent silencing. Therefore, as is the case in yeast, there may exist a functional relationship between heterochromatin-mediated silencing and histone deacetylation in mammalian cells.

Several possible levels at which hypoacetylation could be involved in hetero-chromatic silencing exist. First, hypoacetylated histones may contribute directly to the altered chromatin structure in heterochromatic regions. By itself, this altered chromatin structure could define a heterochromatic region. Alternatively, by analogy with the situation in *S. cerevisiae*, where the SIR proteins interact with deacetylated histone tails and form large protein complexes, HP1 proteins could form a multi-meric higher-order structure which is dependent upon hypoacetylated chromatin. Indeed, it is known that the various HP1 proteins can interact with each other and self-associate (46). It is not difficult to envisage that such a repressive structure could be disrupted by the reacetylation of histone tails following TSA treatment.

It is also possible that HP1 family members may target HDAC complexes to the regions of chromatin where HP1 is already bound. A precedent is found in recent observations on the Ikaros DNA-binding proteins. Ikaros proteins are necessary for determination of the lymphoid cell lineage, and are associated with transcriptionally silent genes in centromeric heterochromatin domains (49). They are also involved in the repositioning of silent genes into heterochromatic domains in the nuclei of T cells following cell activation (50). Kim *et al.* (51) have shown that a major fraction of Ikaros (and the related Aiolos) protein in T-cell nuclei is associated with the Mi-2/NuRD histone deacetylase complex, and that the Ikaros–Mi-2/NuRD complex is active in chromatin remodelling and histone deacetylation. Moreover, Ikaros recruits the Mi-2/NuRD complex to heterochromatic regions following T-cell activation. This may be a means by which the silent state in heterochromatin is established. It is poss-ible that targeted deacetylation and the assembly of HP1 complexes on chromatin which is already hypoacetylated may act in concert to ensure that heterochromatic silencing is complete and efficient.

3.3 Hypoacetylation of the inactive X chromosome in mammals

One of the best-studied examples of heterochromatic silencing is the mammalian inactive X chromosome. Female mammals achieve X-chromosome dosage compensation by inactivating one of the two copies of the chromosome in their cells (see Chapter 12). The chromatin of the inactive X chromosome (Xi) exhibits many characteristics of heterochromatin, being late replicating, extensively cytosine-methylated at CpG dinucleotides, and enriched in histone variants, in this case the H2A variant, macroH2A. It also displays a marked hypoacetylation of histones H3 and H4, which can be detected by immunocytology (52, 53). This hypoacetylation, unlike DNA methylation, is a molecular aspect of X-inactivation shared by both marsupials and eutherian mammals (54), and may thus represent an evolutionarily ancient component of the inactivation mechanism. Although the vast majority of genes on Xi are transcriptionally silent, there are certain loci which escape inactivation. Key among these is the XIST gene, which encodes an RNA product essential for the X-inactivation mechanism; XIST is transcribed from Xi but not from the active X chromosome (Xa). There is also a subset of X-linked genes which is transcribed from both Xi and Xa (55); examples are the ZFX and SMCX loci. Intriguingly, cytological analysis of Xi with anti-acetylated H4 antibodies identified three strongly immunofluorescent bands within the generally unlabelled, hypoacetylated bulk of the chromosome; two of these bands were found in cytogenetic regions known to contain Xi-expressed genes (52).

More recent studies have attempted to define the role of histone hypoacetylation in the mechanism of X-inactivation. Keohane *et al.* (56) followed the course of deacetylation of Xi during the early development of female mammalian cells. Differentiation of embryonic stem cells in culture leads to X-inactivation, which is manifested as the appearance of a single late-replicating X chromosome on day 2 following differentiation. In contrast, H4 deacetylation only appears on Xi at day 4, and reaches a maximum at day 6. This timing is similar to the appearance of hypoacetylated pericentric heterochromatin, and suggests that deacetylation may be required for maintenance, but not initiation, of the inactivation process (56, 57). However, it seems possible that changes in the acetylation pattern of chromatin upstream of the XIST gene may play a functional role in the later stages of X-inactivation. A region extending 120 kb upstream of XIST remains hyperacetylated during the early stages of the inactivation process, consistent with the requirement for XIST transcription for inactivation, but this hyperacetylation is lost by day 7 of differentiation, when inactivation is complete. However, treatment of differentiating cells with TSA inhibited formation of a normal inactive X, whereas mutant cells in which XIST is partially deleted and not initially hyperacetylated were less affected by TSA (58). It is possible that deacetylation of the XIST promoter is necessary for the completion of X-inactivation. A recent study (59) has employed chromatin immunoprecipitation methods to analyse H4 acetylation in the bodies and promoters of genes on both Xi and Xa. It was found that the bodies of the genes exhibited a moderate level of acetylation, similar on both chromosomes, whereas the promoters of X-inactivated genes were hyperacetylated on Xa and hypoacetylated on Xi, correlating with both their methylation status and transcriptional activity. In contrast, genes which escape X-inactivation had hyperacetylated promoters on both X chromosomes, while the *XIST* promoter was again found to be hyperacetylated only on Xi. Thus it seems that the differences in histone acetylation between Xi and Xa may occur primarily in gene promoters; the mechanisms by which histone deacetylases might be selectively targeted to these promoters on Xi, but not Xa, are as yet unknown.

4. Deacetylation and repression of euchromatic genes

4.1 Recruitment of HDAC complexes to yeast promoters

Nucleosomes, when positioned over critical promoter elements such as the TATA-box, can repress transcription initiation (60). This effect can be modulated by the controlled acetylation/deacetylation of histone N-terminal tails (61). Repression of gene activity occurs in many cases through the recruitment of a histone deacetylase-containing complex to the repressor sequence adjacent to the TATA-box. For example, in yeast the DNA-binding repressor UME6 recruits a large multiprotein complex (greater than 600 kDa), which includes the adaptor SIN3 and the catalytic subunit RPD3, to an upstream repressor element (URS) of the *INO1* gene (62). The effect of RPD3 is highly localized, as shown using *ch*romatin *i*mmuno*p*recipitation (CHIP) with antibodies raised against individual sites of acetylation (63, 64). It results in deacetylation of histone H4 lysines K5 and K12 in a domain that includes approximately two nucleosomes surrounding the TATA element of *INO1*. The importance of the RPD3 catalytic activity is illustrated by the observation that repression is relieved in strains bearing catalytically inactive alleles of RPD3 (11). Moreover, RPD3 can repress transcription when artificially tethered to a promoter by a SIN3–LexA fusion protein (62).

The chromatin immunoprecipitation assay used to determine the requirement for and specific activity of RPD3 depends on the specificity of the antibodies for specific sites of acetylation. It has recently been found that antibodies believed to be specific for individual sites of histone acetylation can immunoprecipitate significant DNA even when the site of acetylation is mutated (N. Suka *et al.*, submitted for publication). Thus, a new set of antibodies has been generated against specific sites of acetylation in the histone tails; these antibodies are only capable of chromatin immunoprecipitation in the presence of the acetylation site in question. Using this new set of reagents it has been discovered that RPD3 and SIN3 are required for the hypoacetylation at the *INO1* promoter of most histone H4 sites of acetylation (N. Suka *et al.*, submitted for publication), leading to the conclusion that RPD3 is required for the hypoacetylation of all core histones at *INO1*. This helps to explain the strong repressor activity of RPD3.

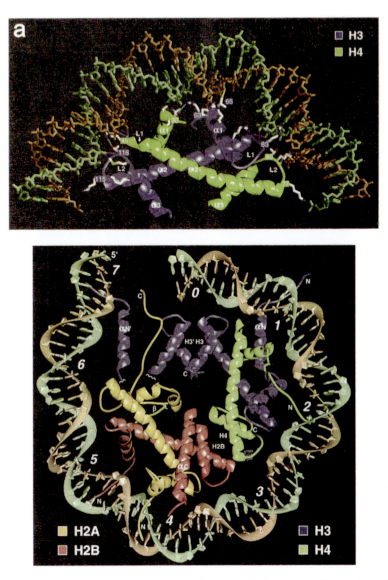

Plate 1 Histone-DNA organization. (a) The histone-fold (α1-L1-α2-L2-α3 structural elements) pair for H3 and H4 is shown bound to DNA. Amino-acid side chains that make hydrogen bonds or hydrophobic interactions with the DNA backbones, and arginine side-chains inserted into the minor groove are shown. Histone main-chain to DNA phosphate hydrogen bonds are shown in magenta. (b) One-half of the nucleosome core is shown for clarity with the view down the DNA superhelix axis and with pseudo-dyad axis of the whole structure aligned vertically. The central base pair through which the dyad passes is labelled). Each successive label (1–7) represents one further DNA double-helix turn, progressing from the DNA centre. Histone regions that are part of the other half are shown as primed. The four-helix bundles are labelled as H3′ H3 and H2B H4; histone-fold extensions of H3 and H2B as αN′, αN, and αC, respectively; the interface between the H2A docking domain and the H4 C-terminus as β; and N- and C-terminal tail regions as N or C. (Reproduced by permission from *Nature* **389**, 251–260, copyright 1997 Macmillan Magazines Ltd.)

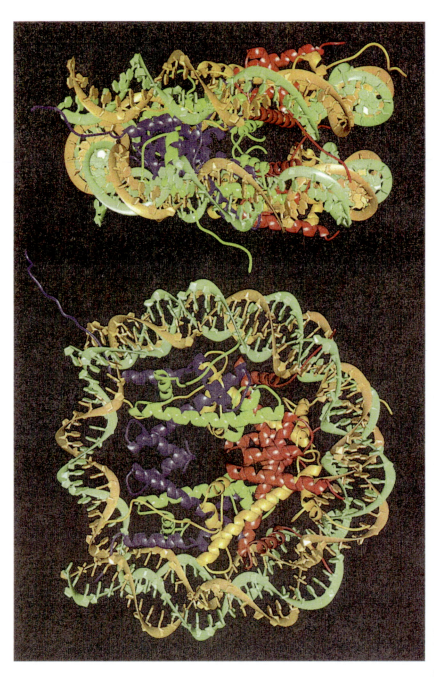

Plate 2 The nucleosome core is viewed (left) down the DNA superhelix axis, and (right) orthogonal to the superhelix axis. For both views, the molecular pseudo-twofold axis is aligned vertically, with the DNA centre at the top. Ribbon traces are shown for the 147 bp DNA phosphodiester backbones (brown and turquoise) and eight histone protein main chains (H3, blue; H4, green; H2A, yellow; H2B, red). The mean radius of the DNA superhelix is 42 Å, and the mean spereation of DNA double helices or pitch of the superhelix is 24 Å. (Reproduced by permission from *Nature* **389**, 251–260, copyright 1997 Macmillan Magazines Ltd.)

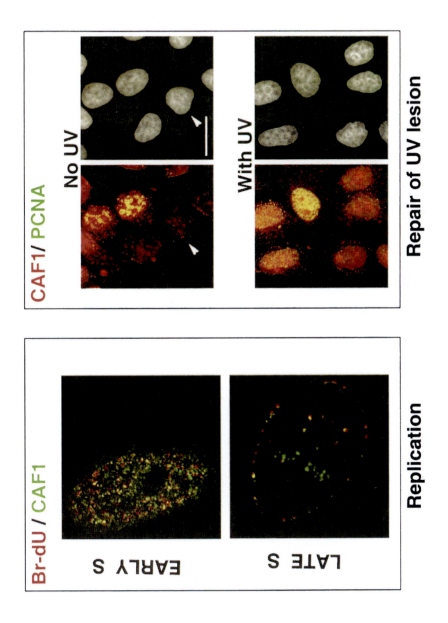

Plate 3 Localization of CAF-1 during the cell cycle and upon UV irradiation in human cells. (a) CAF-1 is associated with early and late S-phase replication foci. DNA replication sites in an asynchronous population of HeLa cells were labelled by a pulse of bromodeoxyuridine (BrdU, red); CAF-1 p150 subunit was simultaneously immunolocalized (green). Early and late replicating cells were identified by their BrdU-staining pattern (for details, see Chapter 2, ref. 65). (b) Localization of Caf-1 and PCNA after UV irradiation. In unirradiated HeLa cells ('No UV', top), the merging (yellow) of the CAF-1 p60 (red) and PCNA (green) signals was detected in S-phase cells by immunoflourescence. After irradiation, the signal corresponding to the chromatin-bound fraction of both CAF-1 p60 and PCNA increases in all cells ('with UV', bottom). DAPI staining (in grey, right-hand panels) displays all the nuclei, including those with no detectable signal for either CAF-1 p60 or PCNA (indicated by the arrow heads) (for details, see Chapter 2, ref.90)

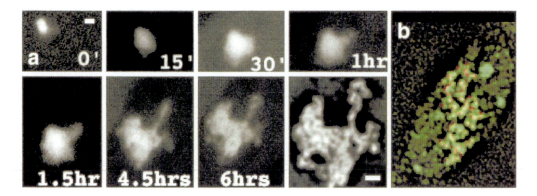

Plate 4 (a) An amplified, c. 90 Mb chromosome region containing lac operator repeats adjacent to DHFR cDNA transgenes, interspersed with large blocks of co-amplified genomic DNA, was targeted with a GFP-lac repressor–VP16 acidic activation domain fusion protein (Chapter 14, ref. 4). A straightening and unfolding of a 90–100 nm fibre was observed over a 6 hour time period after microinjection of the fusion protein in live cells. Unfolding involves the entire chomosome region including the genomic DNA sequences. (b) Striking increase in RNA synthesis at the chromosome site determined by BrUTP incorporation followed by immunostaining (green). Sites of maximum RNA accumulation overlap partially with lac repressor-VP16-GFP signal (red).

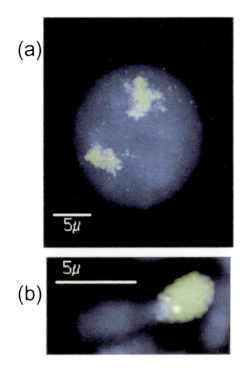

Plate 5 Chromosome probes. FISH using a comples probe to the entire short arm of human chromosome 11 (HSA11p) (green) combined with a cosmid specific to a locus in 11p12 (red) in human lymphoblastoid cells. The territorial nature of the interphase chromosome is illustrated (a) as compared with the more compact mitotic chromosome (b).

Yeast contains five putative histone deacetylases: RPD3, HDA1, HOS1, HOS2, and HOS3 (6). The HDA1-containing complex (histone deacetylase-A) contains three proteins, HDA1, HDA2 and HDA3. TUP1 is a general repressor in yeast affecting pathways involved in mating, stress response, oxygen and glucose utilization, and response to DNA damage (65). When HDA1 or TUP1 is disrupted *in vivo*, the nucleosomes of TUP1-regulated genes are hyperacetylated specifically at histones H3 and H2B, leading to the suggestion that TUP1 recruits HDA1 deacetylase (J. Wu *et al.*, submitted for publication). The cause of this specificity may be the unique availability of H3 and H2B in the nucleosome structure. Both histones share common features not only in their histone-fold domain, when compared to H4 and H2A, but also in the way their N-terminal tails protrude from the nucleosomal particle. The N-termini of histones H3 and H2B pass through channels in the superhelix formed by the minor grooves of the nucleosomal DNA, while the H4 and H2A tails pass over the gyres of the superhelix (66; Chapter 1). Whether these structural differences determine histone specificity remains to be determined.

As it has previously been shown that TUP1 interacts preferentially with the deacetylated forms of H3 and H4 *in vitro* (67), it is possible that TUP1 interacts with the hypoacetylated histone tails to repress gene activity. Recent work suggests that TUP1 spreads from the promoter of *STE6* along the coding region chromatin (68), although it remains to be determined whether TUP1 spreads along histone tails *in vivo*. It should not be assumed that acetylation and deacetylation are merely a means of altering the charge on the histone N-termini, thus regulating interactions of the tails with appropriate factors: when arginines or glycines are used to replace lysines at the N-termini of histones H4 and H3, telomeric silencing is disrupted (69).

In summary, the promoter-specific recruitment of histone deacetylases by repressors is a common and important feature of eukaryotic transcription repression. Different histone deacetylases show distinct promoter targeting and distinct histone substrate specificity. Analysis of the remaining yeast deacetylases related to HDA1 and RPD3 may yet uncover other roles and specificities. HOS3, for example, has been shown to form a homodimer with intrinsic histone deacetylase activity when expressed in *E. coli*. Its lack of accessory factors suggests a unique form of targeting, perhaps even via direct DNA binding by HOS3 itself (24). Whether this occurs, and how the other yeast histone deacetylases are targeted to genes or other domains, remains to be determined.

4.2 Recruitment of HDAC complexes to mammalian promoters

Once again, the precedents established for the targeting of histone deacetylase complexes to yeast promoters appear also to hold true in higher eukaryotic systems. There is an ever-growing list of transcriptional repressors which mediate all or part of their repressive effect through the recruitment of histone deacetylase complexes to their target gene promoters (see ref. 70 for review and references). The list includes repressors that interact with the Sin3 complex (e.g. Mad/Max and Mxi/Max heterodimers, MeCP2 and homeodomain proteins), with the Mi-2/NuRD complex (e.g. *Drosophila* Hunchback and MBD2), or with both complexes (the Ikaros family proteins). There are also factors that have been shown to target deacetylases but where the nature of the HDAC complex is unknown (e.g. retinoblastoma protein). In certain cases a histone deacetylase which is not part of a classical complex may be targeted in isolation: for instance the muscle-specific *m*yocyte *e*nhancer *f*actor-2 (MEF2) can recruit HDAC4 to DNA and hence silence reporter genes in a TSA-sensitive manner (71). It remains to be seen whether other proteins are also involved in the HDAC4–MEF2 repressor complex.

The repressor proteins which appear to target histone deacetylases as a primary means of transcriptional silencing cover a wide range of biological functions. A key group of such proteins is the nuclear hormone receptors, which in the absence of cognate hormone act as transcriptional repressors at specific promoters. When converted to an activated conformation by hormone binding, they instead recruit activator complexes and activate transcription (see Chapter 9). In the repressive state, a group of unliganded receptors has been shown to interact with the co-repressor proteins NCoR (*n*uclear hormone receptor *co*-*r*epressor) and SMRT (*s*ilencing *m*ediator of *r*etinoid and *t*hyroid hormone receptor) (72). These co-repressors in turn recruit the Sin3 deacetylase complex to hormone receptor-regulated promoters. A similar mechanism is involved in repression by the *Drosophila* ecdysone receptor (73), although the hormone ecdysone is not found in mammals. Thus there is evolutionary conservation of the silencing mechanism, involving recruitment of the Sin3 complex by hormone receptors. A very different protein which also represses transcription via deacetylase recruitment is the tumour suppressor, p53. Recent work (74) shows that p53-dependent repression of the target genes *Map4* and *stathmin* can be relieved by TSA treatment, and that p53 can be co-immunoprecipitated with the Sin3 complex. It has also been shown by chromatin immunoprecipitation that both p53 and Sin3 are associated with the *Map4* promoter *in vivo*, and that inactivation of a temperature-sensitive p53 results in increased H3 acetylation at the *Map4* promoter. Deacetylase recruitment has also been implicated in heterochromatic silencing by Ikaros proteins, and may also be involved in X-chromosome inactivation. Thus is seems clear that histone deacetylases are indeed 'silencers for hire' (70).

4.3 Non-promoter acetylation and deacetylation

The preceding two sections dealt with the importance of histone deacetylation at promoters for transcriptional repression. However, promoter-specific targeting of acetyltransferases and deacetylases may play a minor role in the overall balance of acetylation and deacetylation in the cell. Changes in the acetylation state of chromatin associated with the major DNA-dependent processes of replication and transcription may have a dominant effect. Acetylation of cellular histones occurs soon after their synthesis, which may be due in part to HAT1, a histone acetyltransferase that acetylates K5 and K12 of free histone H4, but which is unable to modify whole nucleosomes. This enzyme is associated with a regulatory subunit, HAT2 in yeast and

RbAp46 in mammalian cells, which may target the enzyme to its substrate (14, 75, 76). CAC3 in yeast and RbAp48 in mammals are related to these targeting proteins and may have similar functions (3, 77). Interestingly, both CAC3 and RbAp48 are also subunits of a chromatin assembling factor, CAF-I, arguing that they target other proteins to histones as well.

The human CAF-I complex has been shown to assemble nucleosomes specifically on newly replicated DNA *in vitro*. H4 associated with CAF-I is acetylated mainly at K5, K8, or K12, or a combination of these sites. To test the importance of the histone N-termini, truncations were made in the H3 and H4 N-termini of yeast. It was shown that deletions in either the H3 or H4 N-termini had little effect on nucleosome assembly; however, the loss of both tails disrupted nucleosome assembly *in vitro* and *in vivo* (78). This finding led to an examination of the role of the acetylated lysine residues in the absence of the redundant histone tail. Mutations made in the H4 N-terminus (in the absence of the H3 tail) demonstrated that the simultaneous mutation of the three H4 acetylatable lysines (K5, K8, and K12) was required to inhibit chromatin assembly *in vitro* and *in vivo* (79). Since the histone N-termini are not essential for CAF-I-mediated nucleosome assembly *in vitro* (B. Stillman, personal communication) it may be that histone acetylation is necessary only for factors that are redundant with CAF-I in assembly. Such factors could include ASF1 (*anti-silencing function protein 1*), particularly since ASF1 is found associated with acetylated histone H4.

Transcriptional processes may also have a major impact on the state of chromatin acetylation on a larger scale, and vice versa. Histone acetylation during transcription elongation could help open up chromatin structure over the coding region, allowing the advance of RNA polymerase. Recently, several factors associated with histone acetylation have been reported to stimulate transcription elongation on chromatin templates in *S. cerevisiae* (80, 81). One of them, the so-called Elongator, associates preferentially with the phosphorylated form of RNA polymerase II (81). This complex contains three subunits, ELP1, ELP2, and ELP3. Of these, ELP3 possesses intrinsic HAT activity and has similarity to the histone acetyltransferase GCN5 (Chapter 7). *In vitro*, ELP3 can acetylate all four core histones. Deletion of *ELP3* delays activation of *PHO5*, indicating its functional importance *in vivo*. While it has not been shown that ELP3 acetylates histones as RNA polymerase traverses the coding region, the possibility of widespread histone modification throughout the gene exists.

Many yeast genes, such as *PHO5*, have a peak of histone acetylation centred on the promoter. It is possible that the hypoacetylation of the coding regions is passive, due to the absence of acetylation activity on these domains, or, alternatively, there may be an active deacetylation of coding regions. It has recently been discovered that deletion of *RPD3* and *HDA1* results in increased acetylation over an extended region surrounding and including the *PHO5* gene (M. Vogelauer *et al.*, submitted for publication). This argues for a widespread background activity of histone acetylation and deacetylation, upon which targeted promoter acetylation and deacetylation occurs. These widespread modifications may ensure that histones acetylated during transcription elongation are returned to their hypoacetylated state after transcription is complete; the function of acetylases and deacetylases in this manner would provide a rapid turnover of acetyl groups on nucleosomes. This may be necessary to ensure a quick response of the transcription apparatus to changes in histone acetylation state.

In conclusion, the processes of DNA replication and transcription are likely to contribute substantially to the overall pattern of histone acetylation within the cell. Newly assembled nucleosomes contain acetylated histones, which may become deacetylated following chromatin assembly. Transcription of genes, activated by the targeting of histone acetyltransferases to their regulatory regions, could induce acetylation of the coding region as the transcription complex proceeds through the gene. A ubiquitous general background of acetylation/deacetylation would then return histones to a default state of acetylation. The localized hypo- or hyperacetylation of promoters, while functionally important, may constitute a small fraction of the total work carried out by histone deacetylases and acetyltransferases.

5. DNA methylation and chromatin structure

5.1 Repression mediated by DNA methylation

A common modification of higher eukaryotic genomic DNA is the methylation of the C-5 position of cytosine in the dinucleotide sequence CpG. In the vertebrates, about 60–90% of all CpGs are methylated, and as a result the CpG dinucleotide is underrepresented in the genome due to the spontaneous deamination of methylated cytosine to thymine. Short regions of DNA where the CpG dinucleotide frequency approaches that statistically expected are known as CpG islands; these regions are generally unmethylated, and constitute the bulk of the unmethylated fraction of CpGs in the genome. CpG islands are frequently associated with functional promoters or origins of replication.

There is substantial historical evidence that the methylation of CpG sequences, particularly within CpG islands, correlates with transcriptional repression (reviewed in 82). *In vitro*-methylated reporter gene constructs are repressed upon transfection into cultured cells, and treatment with the demethylating agent 5-azacytidine can activate both endogenous genes and retroviruses that have been silenced by methylation. More recent work shows that DNA methylation is essential for mouse development and important for X-chromosome inactivation and genomic imprinting, and that the phenomenon of methylation-dependent gene silencing may function in plants and fungi as well as in vertebrate animals.

There are several ways in which DNA methylation may lead to transcriptional repression. First, transcription factors or other components of the transcription machinery may be unable to bind to methylated DNA; many transcription factors contain CpG dinucleotides in their recognition sequences, and indeed some of these can no longer bind to cytosine-methylated DNA. However, other factors (such as Sp1) bind in a manner unaffected by methylation, and this mechanism may only play a role in a limited number of specific cases. Secondly, methylation of the DNA may influence chromatin structure directly. Dense CpG methylation can change the preference of core histones for certain sequences (83), and there have been reports that histone H1 binds preferentially to methylated DNA (84). However, the latter point is contentious (see ref. 85), and repression of methylated templates microinjected into *Xenopus* oocytes can occur in the absence of somatic forms of H1 (86).

It is thought that the major means by which methylation represses transcription is via a family of specific methyl-CpG-binding proteins. The first such protein to be characterized in detail was MeCP2 (87), which can bind to a single methyl-CpG base pair via a *methyl-CpG-binding domain* (MBD) (88). Subsequently, four further proteins containing MBD domains have been identified in mice and humans, named MBD1–4 (89). Of these, all except MBD3 can bind specifically to methylated DNA in a mobility-shift assay, although MBD4 is thought to be a mCpG–TpG mismatch-specific glycosylase rather than a transcriptional repressor (90). However, MeCP2, MBD1, and MBD2 have all been shown to function as methylation-dependent transcriptional repressors *in vitro* (91–93). Until recently, though, the means by which MBD proteins might repress transcription remained unclear.

5.2 Chromatin and methylation-dependent repression

It is well known that DNA methylation correlates with changes in the chromatin structure of the methylated regions. Artificially methylated DNA adopts a nuclease-resistant conformation when integrated into the genome (94), while unmethylated CpG islands are associated with nuclease-hypersensitive chromatin (95). Notably, highly acetylated histones H3 and H4 are also found at unmethylated CpG islands. When a methylated herpes simplex virus thymidine kinase (HSV TK) reporter gene was injected into rodent cells, it was transcriptionally repressed only after a delay of 8 hours, unless it was first reconstituted into chromatin (96), suggesting that chromatin formation is essential for methylation-dependent repression. Indeed, it is observed that assembly of chromatin on an *in vitro*-methylated reporter construct injected into *Xenopus* oocytes occurs simultaneously with the loss of transcriptional activity from the reporter (86).

As methyl-CpG-binding proteins are the presumptive mediators of the majority of methylation-dependent repression, their interaction with chromatin is of interest. The founder member of the MBD family, MeCP2, is known to be able to bind to chromatin assembled on methylated DNA in oocyte extracts, and in doing so can displace the linker histone H1 (91). A more detailed study suggests that MeCP2 can bind to a *Xenopus* 5S rDNA mononucleosome and protect mCpGs both within the linker DNA and within the nucleosome itself (97). Those mCpGs situated where the major groove faces outwards appear to be preferentially protected. Thus it is apparent that at least the best-characterized methyl-DNA-binding protein can interact with methylated sequences in chromatin, and that chromatin is important for the mechanism of methylation-dependent silencing. Clues to the nature of this connection came from observations that the HDAC inhibitor sodium butyrate can relieve repression of a methylated episomal reporter (98), and that both butyrate and TSA can activate silent, methylated plant rRNA genes (99). A subsequent study indicated that histone acetylation is lower on methylated versus unmethylated reporter genes *in vivo*; TSA can specifically relieve the methylation-dependent repression and alter the DNAase I insensitivity of the chromatin (100). These results set the stage for the discovery that several of the MBD proteins could associate with histone deacetylase complexes to target them to methylated regions of the genome.

5.3 MeCP2, MBD2, and MBD1 are HDAC-dependent silencers

The first methyl-DNA-binding protein to be shown to interact with histone deacetylase complexes was MeCP2. This protein contains a *t*ranscriptional *r*epression *d*omain (TRD) C-terminal to its MBD domain (91), and this region of the protein when fused to glutathione S-transferase (GST) was shown to be able to co-localize components of the Sin3–HDAC complex (Table 1) to glutathione–Sepharose beads (101). Anti-Sin3 antibodies could also co-immunoprecipitate MeCP2, and vice versa. It appears that the direct interaction is between MeCP2 and the Sin3 protein, as *in vitro* translated HDACs did not bind MeCP2 strongly. A deletion analysis of Sin3 identified two regions that contribute to its interaction with MeCP2, one within its central HDAC-interaction domain, and the second at its C-terminus (101). A second study (102) showed that the *Xenopus* MeCP2 protein can also be co-immunoprecipitated with Sin3. More than 90% of MeCP2 co-fractionates with Sin3 and with a major peak of histone deacetylase activity in oocyte extracts subjected to three column purification steps, suggesting a highly stable association under these conditions. Consistent with the *in vitro* association between MeCP2 and a deacetylase complex, reporter gene repression by a transfected MeCP2 TRD construct targeted to the gene was found to be relieved by the deacetylase inhibitor, TSA; this suggests that localization of a functional deacetylase to the target promoter is an important factor in MeCP2-mediated repression (101) (Fig. 3). It is noteworthy, however, that complete de-repression by TSA was not obtained, suggesting that the MeCP2 TRD may also act as a more direct silencer of transcription. Again, parallel experiments in the *Xenopus* system confirmed that a significant proportion of MeCP2-mediated silencing is deacetylase dependent (102).

MBD2 is another member of the MBD protein family which binds specifically to methylated DNA *in vitro* and associates with methylated chromosomal regions *in vivo* (89). A recent study showed that MBD2 is a potent methylation-dependent transcriptional repressor *in vivo* (93). At certain promoters this repression could be relieved by TSA, and, like MeCP2, MBD2 was found to be associated with HDAC activity in HeLa nuclear extracts. HDAC1, HDAC2, and the histone-binding protein RbAp48 could all be co-immunoprecipitated by an anti-MBD2 antiserum, while anti-HDAC1 antibodies could precipitate MBD2. Most interestingly, it was found that MBD2 was a component of the MeCP1 complex, which was previously identified as a methyl-CpG-binding activity implicated in transcriptional repression (103). MeCP1 binding activity could be immunodepleted by anti-HDAC1 antibodies (93), suggesting that MeCP1 may be an important histone deacetylase-containing complex within

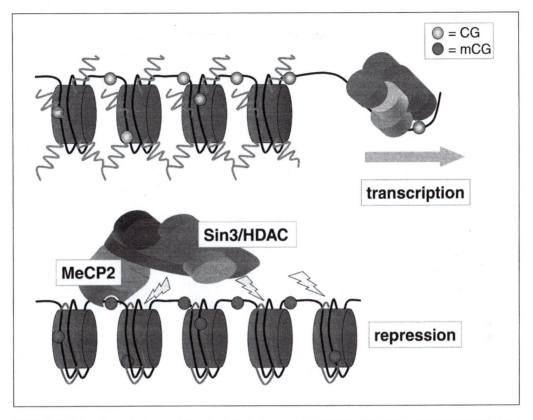

Fig. 3 Methylation-dependent deacetylase targeting silences transcription. Chromatin containing unmethylated CpG sites has hyperacetylated histone tails and is competent for transcription (above). Methylation of CpG sites (mCG; below) attracts methyl-CpG-binding proteins such as MeCP2 which target the Sin3 histone deacetylase complex, causing deacetylation of the histone tails and transcriptional repression.

the cell. Thus it seems that MBD2 is a second transcriptional repressor involved in targeting histone deacetylase complexes to methylated DNA.

MBD1 is a third MBD family member which has recently been shown to function as a transcriptional repressor (92, 104). Previously, MBD1 was thought to be a component of MeCP1, but this now seems unlikely as more specific antisera against MBD1 do not immunodeplete the MeCP1 complex and, unlike MBD2 and MeCP1, MBD1 is not immunodepleted by anti-HDAC1 antibodies (93, 104). However, a transcriptional repression domain can be identified at the C-terminus of MBD1, and repression by this domain can be partially relieved by TSA (104). This suggests that MBD1-dependent silencing does involve histone deacetylation, although MBD1 does not appear to associate with HDAC1 or other known deacetylase complexes. Consistent with this, MBD1 is found to co-localize with regions of low H4 acetylation within pericentromeric heterochromatin. It will be interesting to determine which proteins associate with MBD1 *in vivo*.

There may also be a connection between the major histone deacetylase complex

Mi-2/NuRD and DNA methylation. It was recently shown that a subunit of the Mi-2/NuRD complex is the MBD family member MBD3 (22, 23), which is homologous throughout its length to the MBD2b protein (89). However, unlike other MBD family members, mammalian MBD3 shows little preference for methylated DNA sequences *in vitro*, and does not localize to methylated sequences *in vivo* (89), although the *Xenopus* homologue can bind preferentially to methylated DNA (23). Likewise, the intact mammalian Mi-2/NuRD complex shows no significant affinity for methylated DNA *in vitro* (22). However, a second component of the complex, MTA2 (Table 1), has also been suggested to have a methyl-DNA-binding activity in *Xenopus* (23). A further complication is that the MBD2 protein, a component of the MeCP1 deacetylase complex (see above), can also interact with Mi-2/NuRD and may target it to methylated DNA. Therefore it is not clear whether the Mi-2/NuRD complex is not normally involved in methylation-dependent repression *in vivo*, but can be recruited in special circumstances by MBD2, or whether it is intrinsically implicated in repressing methylated DNA via its core components MBD3 and MTA2. Further study of the *in vivo* situation is required to resolve these questions.

5.4 Other links between DNA methylation and histone deacetylation

A recent study has established a further provocative connection between DNA methylation and histone deacetylases. Work in Kouzarides' laboratory (105) has shown that the enzyme that maintains methylation at CpG sites and ensures the copying of methylation patterns following DNA replication, *DNA methyltransferase 1* (Dnmt1), can also interact with deacetylase activity *in vitro*, and that methyltransferase activity can be purified from nuclear extracts by association with HDAC1. A repression domain within Dnmt1 was identified which functions at least in part by recruiting histone deacetylase activity, and which has homology to the repression domain of the *human trithorax* protein (HRX). It was proposed by these authors that Dnmt1 may target HDACs directly to the replication fork, where Dnmt1 is localized, and thus deacetylate newly synthesized chromatin on methylated DNA without the need for intermediary methyl-CpG-binding proteins. However, it is also possible that the interaction may be a means of targeting Dnmt1 to chromatin which is being deacetylated, thus bringing in a second level of repression to enhance the silencing of specific loci. Whatever the nature of the methyltransferase–deacetylase connection, it is clear that two repression mechanisms, deacetylation and methylation, function in a coordinated manner in mammals to ensure that chromatin remains stably repressed.

6. Final comments

This chapter has described a diverse array of silencing processes in eukaryotic cells to which deacetylation of histone tails contributes. Indeed, to our knowledge there are no silencing systems in which histone deacetylation has been proven to play no role.

However, it is not clear that histone deacetylation is causative for silencing in all cases, and certainly true that in many cases there are other deacetylase-independent mechanisms which contribute significantly to the silencing of a particular locus or stretch of chromatin. Perhaps the deacetylation of histones is best regarded as a layer of repression which can be added to other repressive activities. Because of the involvement of deacetylation in such a variety of systems, a large family of deacetylase enzymes has evolved, and these can be targeted by different repressor proteins via interactions with common intermediary proteins. Many, if not all, of the deacetylases function in complexes, which may be dynamic and contribute themselves to the regulation of deacetylase activity.

We have discussed the reversibility of histone acetylation and deacetylation due to the activities of deacetylase and acetyltransferase enzymes, but silencing itself is often a very stable phenomenon. One function that may contribute to the stability of silencing is DNA methylation in higher eukaryotes, which could lead to the continual targeting of deacetylase enzymes to the silenced regions, as well as contributing deacetylase-independent silencing. Many organisms in which stable silencing exists lack DNA methylation; in these cases other systems such as the Polycomb proteins in *Drosophila* or the Sir proteins in *S. cerevisiae* could contribute the stability. However, the utility of histone deacetylation in silencing, whether as a key factor or as an additional 'top-up' repression mechanism, is apparent.

References

1. Vidal, M., Buckley, A. M., Hilger, F. and Gaber, R. F. (1990) Direct selection for mutants with increased K$^+$ transport in *Saccharomyces cerevisiae*. *Genetics*, **125**, 313.
2. Vidal, M. and Gaber, R. F. (1991) RPD3 encodes a second factor required to achieve maximum positive and negative transcriptional states in *Saccharomyces cerevisiae*. *Mol. Cell. Biol.*, **11**, 6317.
3. Taunton, J., Hassig, C. A., and Schreiber, S. L. (1996) A mammalian histone deacetylase related to the yeast transcriptional regulator Rpd3p. *Science*, **272**, 408.
4. Yang, W. M., Inouye, C., Zeng, Y., Bearss, D., and Seto, E. (1996) Transcriptional repression by YY1 is mediated by interaction with a mammalian homolog of the yeast global regulator RPD3. *Proc. Natl Acad. Sci., USA*, **93**, 12845.
5. Emiliani, S., Fischle, W., Van Lint, C., Al-Abed, Y., and Verdin, E. (1998) Characterization of a human RPD3 ortholog, HDAC3. *Proc. Natl Acad. Sci., USA*, **95**, 2795.
6. Rundlett, S. E., Carmen, A. A., Kobayashi, R., Bavykin, S., Turner, B. M., and Grunstein, M. (1996) HDA1 and RPD3 are members of distinct yeast histone deacetylase complexes that regulate silencing and transcription. *Proc. Natl Acad. Sci., USA*, **93**, 14503.
7. Grozinger, C. M., Hassig, C. A., and Schreiber, S. L. (1999) Three proteins define a class of human histone deacetylases related to yeast Hda1p. *Proc. Natl Acad. Sci., USA*, **96**, 4868.
8. Kao, H. Y., Downes, M., Ordentlich, P., and Evans, R. M. (2000) Isolation of a novel histone deacetylase reveals that class I and class II deacetylases promote SMRT-mediated repression. *Genes Develop.*, **14**, 55.
9. Kadosh, D. and Struhl, K. (1998) Histone deacetylase activity of Rpd3 is important for transcriptional repression *in vivo*. *Genes Develop.*, **12**, 797.

10. Carmen, A. A., Rundlett, S. E., and Grunstein, M. (1996) HDA1 and HDA3 are components of a yeast histone deacetylase (HDA) complex. *J. Biol. Chem.*, **28**, 15837.

11. Kasten, M. M., Dorland, S., and Stillman, D. J. (1997) A large protein complex containing the yeast Sin3p and Rpd3p transcriptional regulators. *Mol. Cell. Biol.*, **17**, 4852.

12. Halleck, M. S., Pownall, S., Harder, K. W., Duncan, A. M., Jirik, F. R., and Schlegel, R. A. (1995) A widely distributed putative mammalian transcriptional regulator containing multiple paired amphipathic helices, with similarity to yeast SIN3. *Genomics*, **26**, 403.

13. Qian, Y. W. and Lee, E. Y. (1995) Dual retinoblastoma-binding proteins with properties related to a negative regulator of Ras in yeast. *J. Biol. Chem.*, **270**, 25507.

14. Verreault, A., Kaufman, P. D., Kobayashi, R., and Stillman, B. (1998) Nucleosomal DNA regulates the core-histone-binding subunit of the human Hat1 acetyltransferase. *Curr. Biol.*, **8**, 96.

15. Vermaak, D., Wade, P. A., Jones, P.L., Shi, Y. B., and Wolffe, A. P. (1999) Functional analysis of the SIN3-histone deacetylase RPD3-RbAp48-histone H4 connection in the *Xenopus* oocyte. *Mol. Cell. Biol.*, **19**, 5847.

16. Imhof, A. and Wolffe, A. P. (1999) Purification and properties of the *Xenopus* Hat1 acetyl-transferase: association with the 14–3–3 proteins in the oocyte nucleus. *Biochemistry*, **38**, 13085.

17. Zhang, Y., Sun, Z. W., Iratni, R., Erdjument-Bromage, H., Tempst, P., Hampsey, M., and Reinberg, D. (1998) SAP30, a novel protein conserved between human and yeast, is a component of a histone deacetylase complex. *Mol. Cell*, **1**, 1021.

18. Zhang, Y., Iratni, R., Erdjument-Bromage, H., Tempst, P., and Reinberg, D. (1997) Histone deacetylases and SAP18, a novel polypeptide, are components of a human Sin3 complex. *Cell*, **89**, 357.

19. Laherty, C. D., Billin, A. N., Lavinsky, R. M., Yochum, G. S., Bush, A. C., Sun, J. M., Mullen, T. M., Davie, J. R., Rose, D. W., Glass, C. K., *et al.* (1998) SAP30, a component of the mSin3 corepressor complex involved in N-CoR-mediated repression by specific transcription factors. *Mol. Cell*, **2**, 33.

20. Wade, P. A., Jones, P. L., Vermaak, D., and Wolffe, A. P. (1998) A multiple subunit Mi-2 histone deacetylase from *Xenopus laevis* cofractionates with an associated Snf2 superfamily ATPase. *Curr. Biol.*, **8**, 843.

21. Ryan, J., Llinas, A. J., White, D. A., Turner, B. M., and Sommerville, J. (1999) Maternal histone deacetylase is accumulated in the nuclei of *Xenopus* oocytes as protein complexes with potential enzyme activity. *J. Cell Sci.*, **112**, 2441.

22. Zhang, Y., Ng, H.-H., Erdjument-Bromage, H., Tempst, P., Bird, A., and Reinberg, D. (1999) Analysis of the NuRD subunits reveals a histone deacetylase core complex and a connection with DNA methylation. *Genes Develop.*, **13**, 1924.

23. Wade, P. A., Gegonne, A., Jones, P. L., Ballestar, E., Aubry, F., and Wolffe, A. P. (1999) Mi-2 complex couples DNA methylation to chromatin remodelling and histone deacetylation. *Nature Genet.*, **23**, 62.

24. Carmen, A. A., Griffin, P. R., Calaycay, J. R., Rundlett, S. E., Suka, Y., and Grunstein, M. (1999) Yeast HOS3 forms a novel trichostatin A-insensitive homodimer with intrinsic histone deacetylase activity. *Proc. Natl Acad. Sci., USA*, **96**, 12356.

25. Braunstein, M., Sobel, R. E., Allis, C. D., Turner, B. M., and Broach, J. R. (1996) Efficient transcriptional silencing in *Saccharomyces cerevisiae* requires a heterochromatin histone acetylation pattern. *Mol. Cell. Biol.*, **16**, 4349.

26. Imai, S., Armstrong, C. M., Kaeberlein, M., and Guarente, L. (2000) Transcriptional silencing and longevity protein Sir2 is an NAD-dependent histone deacetylase. *Nature*, **403**, 795.

27. Landry, J., Sutton, A., Tafrov, S. T., Heller, R. C., Stebbins, J., Pillus, L., and Sternglanz, R. (2000) The silencing protein Sir2 and its homologs are NAD-dependent protein deacetylases. *Proc. Natl Acad. Sci., USA*, **97**, 5807.

28. Smith, J. S., Brachmann, C. B., Celic, I., Kenna, M. A., Muhammad, S., Starai, V. J., Avalos, J., Escalante-Semerena, J. C., Grubmeyer, C., Wolberger, C., *et al.* (2000) A phylogenetically conserved NAD⁺-dependent protein deacetylase activity in the Sir2 protein family. *Proc. Natl Acad. Sci., USA*, **97**, 6658.

29. Frye, R. A. (1999) Characterization of five human cDNAs with homology to the yeast SIR2 gene: Sir2-like proteins (sirtuins) metabolize NAD and may have protein ADP-ribosyltransferase activity. *Biochem. Biophys. Res. Commun.*, **24**, 273.

30. Brachmann, C. B., Sherman, J. M., Devine, S. E., Cameron, E. E., Pillus, L., and Boeke, J. D. (1995) The SIR2 gene family, conserved from bacteria to humans, functions in silencing, cell cycle progression, and chromosome stability. *Genes Dev.*, **9**, 2888.

31. Hecht, A., Laroche, T., Strahl-Bolsinger, S., Gasser, S. M., and Grunstein, M. (1995) Histone H3 and H4 N-termini interact with SIR3 and SIR4 proteins: a molecular model for the formation of heterochromatin in yeast. *Cell*, **80**, 583.

32. Strahl-Bolsinger, S., Hecht, A., Luo, K., and Grunstein, M. (1997) SIR2 and SIR4 interactions differ in core and extended telomeric heterochromatin in yeast. *Genes Develop.*, **11**, 83.

33. Moazed, D., Kistler, A., Axelrod, A., Rine, J., and Johnson, A. D. (1997) Silent information regulator protein complexes in *Saccharomyces cerevisiae*: a SIR2/SIR4 complex and evidence for a regulatory domain in SIR4 that inhibits its interaction with SIR3. *Proc. Natl Acad. Sci., USA*, **94**, 2186.

34. Renauld, H., Aparicio, O. M., Zierath, P. D., Billington, B. L., Chhablani, S. K., and Gottschling, D. E. (1993) Silent domains are assembled continuously from the telomere and are defined by promoter distance and strength, and by SIR3 dosage. *Genes Develop.*, **7**, 1133.

35. Fourel, G., Revardel, E., Koering, C. E., and Gilson, E. (1999) Cohabitation of insulators and silencing elements in yeast subtelomeric regions. *EMBO J.*, **18**, 2522.

36. Pryde, F. E. and Louis, E. J. (1999) Limitations of silencing at native yeast telomeres. *EMBO J.*, **18**, 2538.

37. Bryk, M., Banerjee, M., Murphy, M., Knudsen, K. E., Garfinkel, D. J., and Curcio, M. J. (1997) Transcriptional silencing of Ty1 elements in the RDN1 locus of yeast. *Genes Develop.*, **11**, 255.

38. Smith, J. S. and Boeke, J. D. (1997) An unusual form of transcriptional silencing in yeast ribosomal DNA. *Genes Develop.*, **11**, 241.

39. Straight, A. F., Shou, W., Dowd, G. J., Turck, C. W., Deshaies, R. J., Johnson, A. D., and Moazed, D. (1999) Net1, a Sir2-associated nucleolar protein required for rDNA silencing and nucleolar integrity. *Cell*, **97**, 245.

40. Grunstein, M. (1997) Molecular model for telomeric heterochromatin in yeast. *Curr. Opin. Cell Biol.*, **9**, 383.

41. Johnson, L. M., Kayne, P. S., Kahn, E. S., and Grunstein, M. (1990) Genetic evidence for an interaction between SIR3 and histone H4 in the repression of the silent mating loci in *Saccharomyces cerevisiae. Proc. Natl Acad. Sci., USA*, **87**, 6286.

42. Clarke, D. J., O'Neill, L. P., and Turner, B. M. (1993) Selective use of H4 acetylation sites in the yeast *Saccharomyces cerevisiae. Biochem. J.*, **294**, 557.

43. James, T. C. and Elgin, S. C. (1986) Identification of a nonhistone chromosomal protein associated with heterochromatin in *Drosophila melanogaster* and its gene. *Mol. Cell. Biol.*, **6**, 3862.

44. James, T. C., Eissenberg, J. C., Craig, C., Dietrich, V., Hobson, A., and Elgin, S. C. (1989) Distribution patterns of HP1, a heterochromatin-associated nonhistone chromosomal protein of *Drosophila*. *Eur. J. Cell Biol.*, **50**, 170.

45. Wallrath, L. L. (1998) Unfolding the mysteries of heterochromatin. *Curr. Opin. Genet. Dev.*, **8**, 147.

46. Le Douarin, B., Nielsen, A. L., Garnier, J. M., Ichinose, H., Jeanmougin, F., Losson, R., and Chambon, P. (1996) A possible involvement of TIF1 alpha and TIF1 beta in the epigenetic control of transcription by nuclear receptors. *EMBO J.*, **15**, 6701.

47. Ryan, R. F., Schultz, D. C., Ayyanathan, K., Singh, P. B., Friedman, J. R., Fredericks, W. J., and Rauscher, F. J. 3rd (1999) KAP-1 corepressor protein interacts and colocalizes with heterochromatic and euchromatic HP1 proteins: a potential role for Kruppel-associated box-zinc finger proteins in heterochromatin-mediated gene silencing. *Mol. Cell. Biol.*, **19**, 4366.

48. Nielsen, A. L., Ortiz, J. A., You, J., Oulad-Abdelghani, M., Khechumian, R., Gansmuller, A., Chambon, P., and Losson, R. (1999) Interaction with members of the heterochromatin protein 1 (HP1) family and histone deacetylation are differentially involved in transcriptional silencing by members of the TIF1 family. *EMBO J.*, **18**, 6385.

49. Brown, K. E., Guest, S. S., Smale, S. T., Hahm, K., Merkenschlager, M., and Fisher, A. G. (1997) Association of transcriptionally silent genes with Ikaros complexes at centromeric heterochromatin. *Cell*, **91**, 845.

50. Brown, K. E., Baxter, J., Graf, D., Merkenschlager, M., and Fisher, A. G. (1999) Dynamic repositioning of genes in the nucleus of lymphocytes preparing for cell division. *Mol. Cell*, **3**, 207.

51. Kim, J., Sif, S., Jones, B., Jackson, A., Koipally, J., Heller, E., Winandy, S., Viel, A., Sawyer, A., Ikeda, T., *et al.* (1999) Ikaros DNA-binding proteins direct formation of chromatin remodeling complexes in lymphocytes. *Immunity*, **10**, 345.

52. Jeppesen, P. and Turner, B. M. (1993) The inactive X chromosome in female mammals is distinguished by a lack of histone H4 acetylation, a cytogenetic marker for gene expression. *Cell*, **74**, 281.

53. Boggs, B. A., Connors, B., Sobel, R. E., Chinault, A. C., and Allis, C. D. (1996) Reduced levels of histone H3 acetylation on the inactive X chromosome in human females. *Chromosoma*, **105**, 303.

54. Wakefield, M. J., Keohane, A. M., Turner, B. M., and Graves, J. A. (1997) Histone underacetylation is an ancient component of mammalian X chromosome inactivation. *Proc. Natl Acad. Sci., USA*, **94**, 9665.

55. Brown, C. J., Carrel, L., and Willard, H. F. (1997) Expression of genes from the human active and inactive X chromosomes. *Am. J. Hum. Genet.*, **60**, 1333.

56. Keohane, A. M., O'Neill, L. P., Belyaev, N. D., Lavender, J. S., and Turner, B. M. (1996) X-Inactivation and histone H4 acetylation in embryonic stem cells. *Dev. Biol.*, **180**, 618.

57. Keohane, A. M., Lavender, J. S., O'Neill, L. P., and Turner, B. M. (1998) Histone acetylation and X inactivation. *Dev. Genet.*, **22**, 65.

58. O'Neill, L. P., Keohane, A. M., Lavender, J. S., McCabe, V., Heard, E., Avner, P., Brockdorff, N., and Turner, B. M. (1999) A developmental switch in H4 acetylation upstream of Xist plays a role in X chromosome inactivation. *EMBO J.*, **18**, 2897.

59. Gilbert, S. L. and Sharp, P. A. (1999) Promoter-specific hypoacetylation of X-inactivated genes. *Proc. Natl Acad. Sci., USA*, **96** 13825.

60. Han, M. and Grunstein, M. (1988) Nucleosome loss activates yeast downstream promoters *in vivo*. *Cell*, **55**, 1137.

61. Struhl, K. (1999) Fundamentally different logic of gene regulation in eukaryotes and prokaryotes. *Cell*, **98**, 1.
62. Kadosh, D. and Struhl, K. (1997) Repression by Ume6 involves recruitment of a complex containing Sin3 corepressor and Rpd3 histone deacetylase to target promoters. *Cell*, **89**, 365.
63. Rundlett, S. E., Carmen, A. A., Suka, N., Turner, B. M., and Grunstein, M. (1998) Transcriptional repression by UME6 involves deacetylation of lysine 5 of histone H4 by RPD3. *Nature*, **392**, 831.
64. Kadosh, D. and Struhl, K. (1998) Targeted recruitment of the Sin3-Rpd3 histone deacetylase complex generates a highly localized domain of repressed chromatin *in vivo*. *Mol. Cell. Biol.*, **18**, 5121.
65. Keleher, C. A., Redd, M. J., Schultz, J., Carlson, M., and Johnson, A. D. (1992) Ssn6–Tup1 is a general repressor of transcription in yeast. *Cell*, **68**, 709.
66. Luger, K., Mader, A. W., Richmond, R. K., Sargent, D. F., and Richmond, T. J. (1997) Crystal structure of the nucleosome core particle at 2.8 Å resolution. *Nature*, **389**, 251.
67. Edmondson, D. G., Smith, M. M., and Roth, S. Y. (1996) Repression domain of the yeast global repressor Tup1 interacts directly with histones H3 and H4. *Genes Develop.*, **10**, 1247.
68. Ducker, C. E. and Simpson, R. T. (2000) The organized chromatin domain of the repressed yeast a cell-specific gene STE6 contains two molecules of the corepressor Tup1p per nucleosome. *EMBO J.*, **19**, 400.
69. Thompson, J. S., Hecht, A., and Grunstein, M. (1993) Histones and the regulation of heterochromatin in yeast. *Cold Spring Harbor Symp. Quant. Biol.*, **58**, 247.
70. Ng, H.-H. and Bird, A. (2000) Histone deacetylases: silencers for hire. *Trends Biochem. Sci.*, **25**, 121.
71. Miska, E. A., Karlsson, C., Langley, E., Nielsen, S. J., Pines, J., and Kouzarides, T. (1999) HDAC4 deacetylase associates with and represses the MEF2 transcription factor. *EMBO J.*, **18**, 5099.
72. Xu, L., Glass, C. K., and Rosenfeld, M. G. (1999) Coactivator and corepressor complexes in nuclear receptor function. *Curr. Opin. Genet. Develop.*, **9**, 140.
73. Tsai, C. C., Kao, H. Y., Yao, T. P., McKeown, M., and Evans, R. M. (1999) SMRTER, a *Drosophila* nuclear receptor coregulator, reveals that EcR-mediated repression is critical for development. *Mol. Cell*, **4**, 175.
74. Murphy, M., Ahn, J., Walker, K. K., Hoffman, W. H., Evans, R. M., Levine, A. J., and George, D. L. (1999) Transcriptional repression by wild-type p53 utilizes histone deacetylases, mediated by interaction with mSin3a. *Genes Develop.*, **13**, 2490.
75. Parthun, M. R., Widom, J., and Gottschling, D. E. (1996) The major cytoplasmic histone acetyltransferase in yeast: links to chromatin replication and histone metabolism. *Cell*, **87**, 85.
76. Kleff, S., Andrulis, E. D., Anderson, C. W., and Sternglanz, R. (1995) Identification of a gene encoding a yeast histone H4 acetyltransferase. *J. Biol. Chem.*, **270**, 24674.
77. Ahmad, A., Takami, Y., and Nakayama, T. (1999) WD repeats of the p48 subunit of chicken chromatin assembly factor-1 required for *in vitro* interaction with chicken histone deacetylase-2. *J. Biol. Chem.*, **274**, 16646.
78. Ling, X., Harkness, T. A., Schultz, M. C., Fisher-Adams, G., and Grunstein, M. (1996) Yeast histone H3 and H4 amino termini are important for nucleosome assembly *in vivo* and *in vitro*: redundant and position-independent functions in assembly but not in gene regulation. *Genes Develop.*, **10**, 686.

79. Ma, X. J., Wu, J., Altheim, B. A., Schultz, M. C., and Grunstein, M. (1998) Deposition-related sites K5/K12 in histone H4 are not required for nucleosome deposition in yeast. *Proc. Natl Acad. Sci., USA*, **95**, 6693.

80. Hartzog, G. A., Wada, T., Handa, H., and Winston, F. (1998) Evidence that Spt4, Spt5, and Spt6 control transcription elongation by RNA polymerase II in *Saccharomyces cerevisiae*. *Genes Develop.*, **12**, 357.

81. Wittschieben, B. O., Otero, G., de Bizemont, T., Fellows, J., Erdjument-Bromage, H., Ohba, R., Li. Y., Allis, C. D., Tempst, P., and Svejstrup, J. Q. (1999) A novel histone acetyltransferase is an integral subunit of elongating RNA polymerase II holoenzyme. *Mol. Cell*, **4**, 123.

82. Bird, A. P. and Wolffe, A. P. (1999) Methylation-induced repression–belts, braces, and chromatin. *Cell*, **99**, 451.

83. Davey, C., Pennings, S., and Allan, J. (1997) CpG methylation remodels chromatin structure *in vitro*. *J. Mol. Biol.*, **267**, 276.

84. McArthur, M. and Thomas, J. O. (1996) A preference of histone H1 for methylated DNA. *EMBO J.*, **15**, 1705.

85. Campoy, F. J., Meehan, R. R., McKay, S., Nixon, J., and Bird, A. (1995) Binding of histone H1 to DNA is indifferent to methylation at CpG sequences. *J. Biol. Chem.*, **270**, 26473.

86. Kass, S. U., Landsberger, N., and Wolffe, A.P. (1997) DNA methylation directs a time-dependent repression of transcription initiation. *Curr. Biol.*, **7**, 157.

87. Lewis, J. D., Meehan, R. R., Henzel, W. J., Maurer-Fogy, I., Jeppesen, P., Klein, F., and Bird, A. (1992) Purification, sequence, and cellular localization of a novel chromosomal protein that binds to methylated DNA. *Cell*, **69**, 905.

88. Nan, X., Meehan, R. R., and Bird, A. (1993) Dissection of the methyl-CpG binding domain from the chromosomal protein MeCP2. *Nucleic Acids Res.*, **21**, 4886.

89. Hendrich, B. and Bird, A. (1998) Identification and characterization of a family of mammalian methyl-CpG binding proteins. *Mol. Cell. Biol.*, **18**, 6538.

90. Hendrich, B., Hardeland, U., Ng, H.-H., Jiricny, J., and Bird, A. (1999) The thymine glycosylase MBD4 can bind to the product of deamination at methylated CpG sites. *Nature*, **401**, 301.

91. Nan, X., Campoy, F. J., and Bird, A. (1997) MeCP2 is a transcriptional repressor with abundant binding sites in genomic chromatin. *Cell*, **88**, 471.

92. Fujita, N., Takebayashi, S., Okumura, K., Kudo, S., Chiba, T., Saya, H., and Nakao, M. (1999) Methylation-mediated transcriptional silencing in euchromatin by methyl-CpG binding protein MBD1 isoforms. *Mol. Cell. Biol.*, **19**, 6415.

93. Ng, H.-H., Zhang, Y., Hendrich, B., Johnson, C. A., Turner, B. M., Erdjument-Bromage, H., Tempst, P., Reinberg, D., and Bird, A. (1999) MBD2 is a transcriptional repressor belonging to the MeCP1 histone deacetylase complex. *Nature Genet.*, **23**, 58.

94. Keshet, I., Lieman-Hurwitz, J., and Cedar, H. (1986) DNA methylation affects the formation of active chromatin. *Cell*, **44**, 535.

95. Tazi, J. and Bird, A. (1990) Alternative chromatin structure at CpG islands. *Cell*, **60**, 909.

96. Buschhausen, G., Wittig, B., Graessmann, M., and Graessmann, A. (1987) Chromatin structure is required to block transcription of the methylated herpes simplex virus thymidine kinase gene. *Proc. Natl Acad. Sci., USA*, **84**, 1177.

97. Chandler, S. P., Guschin, D., Landsberger, N., and Wolffe, A. P. (1999) The methyl-CpG binding transcriptional repressor MeCP2 stably associates with nucleosomal DNA. *Biochemistry*, **38**, 7008.

98. Hsieh, C. L. (1994) Dependence of transcriptional repression on CpG methylation density. *Mol. Cell. Biol.*, **14**, 5487.

99. Chen, Z. J. and Pikaard, C. S. (1997) Epigenetic silencing of RNA polymerase I transcription: a role for DNA methylation and histone modification in nucleolar dominance. *Genes Dev.*, **11**, 2124.

100. Eden, S., Hashimshony, T., Keshet, I., Cedar, H., and Thorne, A.W. (1998) DNA methylation models histone acetylation. *Nature*, **394**, 842.

101. Nan, X., Ng, H.-H., Johnson, C. A., Laherty, C. D., Turner, B. M., Eisenman, R. N., and Bird, A. (1998) Transcriptional repression by the methyl-CpG-binding protein MeCP2 involves a histone deacetylase complex. *Nature*, **393**, 386.

102. Jones, P. L., Veenstra, G. J., Wade, P. A., Vermaak, D., Kass, S. U., Landsberger, N., Strouboulis, J., and Wolffe, A. P. (1998) Methylated DNA and MeCP2 recruit histone deacetylase to repress transcription. *Nature Genet.*, **19**, 187.

103. Meehan, R. R., Lewis, J. D., McKay, S., Kleiner, E. L., and Bird, A. P. (1989) Identification of a mammalian protein that binds specifically to DNA containing methylated CpGs. *Cell*, **58**, 499.

104. Ng, H.-H., Jeppesen, P., and Bird, A. (2000) Active repression of methylated genes by the chromosomal protein MBD1. *Mol. Cell. Biol.*, **20**, 1394.

105. Fuks, F., Burgers, W. A., Brehm, A., Hughes-Davies, L., and Kouzarides, T. (2000) DNA methyltransferase Dnmt1 associates with histone deacetylase activity. *Nature Genet.*, **24**, 88.

9 | Developmental regulation of chromatin function and gene expression

ALAN P. WOLFFE AND MICHELLE CRAIG BARTON

1. Introduction

Previous chapters have defined chromatin structure and discussed multiple mechanisms for altering this structure with subsequent consequences for gene regulation. In this chapter, we will discuss gene regulation linking chromatin structure composition and perturbation to activation or repression of gene expression during differentiation and development. Developmentally regulated processes are likely universal in exploiting the previously outlined mechanisms of chromatin assembly during replication, targeted remodelling of established chromatin structure, modification of histones, and other mechanisms as yet unknown. In this chapter, we will focus on two very different examples of developmental gene regulation that have been well defined at the level of chromatin structure and function. One is the determinative expression of the β-globin gene family during red blood cell differentiation as a prototype of progression through defined stages of development, each stage marked by distinct chromosomal structure and functional regulation. The other example is thyroid hormone receptor control of gene regulation, a primary regulatory switch initiating a cascade of events marked by dramatic morphological alterations during amphibian metamorphosis. Once the switch is activated during early embryogenesis, changes in chromatin composition and functional properties occur (reviewed in 1, 2). These transitions contribute directly to repression of gene families (3) and impose constraints on signal transduction pathways that determine cell fate (4).

2. Chromatin structure and gene regulation during red cell development

Erythrocyte (red blood cell) development and expression of globin genes is an excellent paradigm for analysis of chromatin structure reorganization at defined develop-

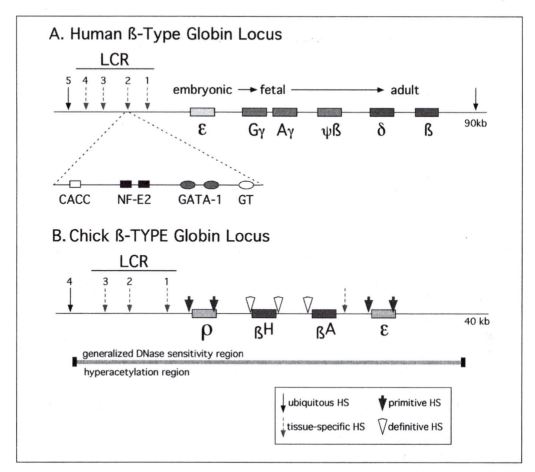

Fig. 1 Multiple, developmentally regulated genes lie within the β-type globin locus. (A) The approximately 90 kb human β-globin locus is arranged in order of developmental expression timing. Embryonic (ε), fetal (Gγ and Aγ) and adult (δ and β) β-type globin genes are expressed at specific stages of development in erythroid cells. The genes lie downstream of a red-cell-specific cluster of DNAase I hypersensitive sites known as the LCR (HS1 through HS4, dashed arrows). Upstream of these tissue-specific hypersensitive sites is a ubiquitous hypersensitive site (HS5, solid arrow). Multiple *trans*-acting factors bind within each HS site; an expanded map is shown for HS2. These proteins are not unique to LCR HS sites and also act at promoters and enhancers of the downstream genes. (B) The chick β-globin locus encompasses two primitive genes, ρ and ε, and two definitive genes, β^H and β^A, within 40 kb. A region of generalized DNAase sensitivity and hyperacetylation occurs across the locus (broad, shaded line below). Within this domain, developmental- (broad closed arrows, primitive; and broad open arrows, definitive) and tissue-specific (thin, dashed arrows) hypersensitive sites have been mapped. The 5′ tissue-specific HS cluster defines an LCR region, and a ubiquitous HS site (thin, solid arrow) acts as a locus insulator. All genes shown with the globin locus are transcribed from the 5′ (left) to 3′ (right) direction.

mental stages. The β-globin genes are arranged in a genomic cluster, and encode variant forms of β-type globin protein expressed specifically and progressively during red cell development. Human β-type globin genes display a 5′ to 3′ ordered arrangement of embryonic, fetal, and adult genes within the locus that correlates with timing of developmental expression (Fig. 1A). The change in globin gene expression patterns between stages of development is known as globin switching. Human β-globins display a three-part switch from embryonic to fetal to adult, while the avian β-globins switch simply from embryonic to adult (reviewed in 5). The avian (chick) β-type globin locus may have arisen from duplication of a gene locus comprising one primitive (fetal) and one definitive (adult) globin gene, perhaps accounting for the lack of conservation in the physical order of genes and developmental expression timing (Fig. 1B) (reviewed in 6).

The availability of developmentally staged, primary red cells for biochemical and chromatin structure analyses facilitated the use of chick β-type globin regulation as a model for developmental regulation of chromatin structure and gene expression. This system has a solid foundation of *in vivo* studies mapping chromatin structural changes associated with transcription regulation, which has been followed by continued development of *in vitro* approaches that recapitulate chromatin structure and function. Despite the differences in gene 5′ to 3′ order and expression timing between mammalian and chick globin genes, the lessons of developmentally regulated chromatin alterations and gene expression can, generally, be related to both systems.

2.1 Tissue-specific expression and chromatin structure of the globin locus

2.1.1 Locus-wide potentiation of gene expression

Tissue-specific, e.g. red-cell-specific, modification of chromatin structure has been defined across the chick β-globin locus by a greater sensitivity to DNAase I digestion compared to both flanking genomic DNA and globin loci in non-expressing tissues (Fig. 1B) (7). Regions of increased accessibility are approximately five- to tenfold more sensitive to digestion than flanking regions of DNA. This broad, locus-wide, generalized DNAase I sensitivity is distinct from the highly localized perturbations in chromatin structure known as hypersensitive (HS) sites. Sequence-specific interactions between *trans*-acting factors and DNA can generate these chromatin structure alterations that are hypersensitive to DNAase I digestion, rather than the generalized, moderate sensitivity observed across the globin locus in red cells (Fig. 1B).

Histone modification, as discussed in Chapter 7, may play an intrinsic role in establishing a chromatin structure that is relatively sensitive to probes such as DNAase I treatment. Analysis of globin locus chromatin structure has correlated changes in histone acetylation with tissue-specific gene expression (7, 8). Using the technique of *ch*romatin *i*mmuno*p*recipitation (CHIP) analysis to compare specific genomic regions where acetylated versus unacetylated histones were incorporated into chromatin, Crane-Robinson and colleagues developed a map of histone acetylation spanning the β-type globin locus in red blood cells. Regions of chromatin flanking the globin locus, as well as globin loci in non-expressing cells, lacked histone acetylation. As with the generalized DNAase sensitivity described above, these experiments reveal chromatin structural alterations along the entirety of the 'transcriptionally competent' globin locus (Fig. 1B). Transcriptional competence is distinct from transcriptional activity as it defines genes with a potential for expression (possessed by all globin genes in red blood cells), as opposed to genes actively engaged in transcription. These results imply an ordered sequence of events in which tissue-specific alterations in chromatin occur across the entire globin locus prior to activation of individual genes within the locus.

2.1.2 Activation of gene expression is marked by localized perturbations of chromatin structure

Highly localized, DNA sequence-specific changes in chromatin structure concomitant with developmentally regulated β-type globin gene activation have been defined within the red cell tissue-specific, potentiated chromatin. Two fetal or primitive red cell genes, ρ and ε, are expressed maximally between 3 and 5 days of fetal development in primitive red cells. As development progresses, ρ and ε fetal gene expression declines and is replaced by definitive red cell expression of two genes, β^H (hatching) and β^A (adult) within 10 days of embryonic development (Fig. 1B). Stage-specific gene activation within the globin locus is marked by localized alterations in chromatin structure at the promoter or enhancer of each gene (9). Compared to surrounding DNA, these sites of altered chromatin structure may be 50- to 100-fold more sensitive to structural probes such as DNAase I, generating hypersensitive (HS) cleavage sites. As shown in Fig. 1B, three classes of HS sites lie within the globin locus:

(1) tissue-specific sites that appear in all red cells;

(2) developmentally specific sites that flank actively expressing genes within the locus; and

(3) boundary-defining sites ubiquitous to all cell types (discussed in Chapter 13).

2.1.3 A cluster of tissue-specific hypersensitive sites defines a locus control region

A cluster of distal HS sites lies at the far 5' end of the globin locus, at least five in mammalian globin loci and four in the chick (named HS1–5 or HS1–4, respectively; Fig. 1) (10, 11). These hypersensitive sites are nuclease-accessible at all stages of red cell development. The most 5' HS site is a ubiquitous site that may demarcate the 5' boundary of the globin locus (see Chapter 13), while the other HS sites appear only in red blood cells. This cluster of tissue-specific hypersensitive sites has been termed a locus control region (LCR), based on the exciting observation that introduction of this region into the mouse genome conferred robust tissue-specific, integration-site-independent, copy-number-dependent activation of flanking globin transgenes. The LCR's link to developmental timing and control of gene expression was suggested by numerous studies of transgenic mouse models as well as analysis of human patients with natural mutations in this region. Patients with LCR mutations are thalassaemic, displaying altered developmental expression of β-type globin genes within the locus, changes in DNA replication timing of the globin locus, as well as formation of a repressive chromatin structure over a broad domain (reviewed in 11).

Whether this cluster of hypersensitive sites functions by altering chromatin structure over the entire locus, potentiating it for activation, or enhances expression from individual genes within the locus based on their complement of locally bound regulatory proteins, has been the subject of intense investigation. Dissection of LCR action by assessing the contribution of individual hypersensitive regions to gene activation has not been straightforward. A model of the LCR as a complex, combinatorial regulatory element has arisen from a number of studies identifying each site as an enhancer of gene expression with a limited range of individual function. Fraser, Grosveld, and colleagues have proposed a dynamic model of LCR activation of a single gene within the globin locus at a given time, controlling rapid switching between fetal and adult expression (12, 13).

These studies must now be placed in the context of recent, surprising evidence from the Groudine laboratory that ablation of the LCR from its natural location within the mouse genome, using homologous recombination, does not inactivate the globin locus, but does affect relative levels of individual gene expression (14). Overall chromatin structure of the endogenous globin locus was unchanged, and developmental timing of gene expression was normal. An affect on transcription rate rather than gene silencing was the primary consequence of LCR deletion. These results suggest that, rather than a dominant regulatory role for the LCR in establishing an active chromatin locus, multiple regulatory elements spaced throughout the globin locus or flanking the locus, or both, contribute to chromatin structure regulation and gene expression during development.

Further studies are needed to begin to understand regulatory control of chromatin structure and transcription regulation across an entire gene locus (reviewed in 15). Larger deletions 5' of the previously defined LCR might recapitulate the effects on transcription and chromatin structure observed in human thalassaemic deletion mutations, if these thalassaemias are due to negative regulatory elements now juxtaposed next to the globin locus, along with loss of the LCR (16). Additionally, recent analyses of intergenic, transcribed (non-coding) genes within the globin locus suggest that transcription-linked chromatin structure remodelling may occur at specific developmental stages. These data must now be incorporated into locus regulatory models (17).

2.2 Developmentally regulated chromatin structure and globin expression

Red cell nuclei isolated at defined stages of chick embryonic development have been employed as a substrate for DNAase I hypersensitivity and restriction enzyme accessibility analyses. These studies of chick globin chromatin structure revealed sites of preferential cleavage, indicating discrete perturbations in chromatin generated by *trans*-acting proteins binding to DNA (Fig. 1B) (18). Identification of promoters and enhancers, as well as interacting regulatory proteins, has focused on these regions. In primitive red cell nuclei isolated from 5-day-old embryos, chromatin structure perturbations occur both 5' and 3' of ρ and ε genes. At later stages of development, when primitive genes are not expressed, these sites are absent, with the exception of the shared β–ε enhancer. DNAase cleavage of definitive-stage red cell nuclei, from 10–14 days of development, reveals hypersensitive sites 5′ of both the $β^H$ and $β^A$ coding regions. One of the hypersensitive sites lying between the $β^A$ and ε genes is an enhancer region shared by both genes. It is bound by a dynamic collection of regulatory proteins during both primitive and definitive stages of development, and is structurally accessible to enzymatic probes throughout development. Chromatin accessibility at the $β^A$ promoter appears only in definitive red cells. A particularly useful marker for this 'open' chromatin structure is the release of a 114 bp fragment by MspI restriction enzyme digestion of actively expressing $β^A$ red cell chromatin. MspI sites lie proximal to the transcription start site of the $β^A$ gene, and accessibility is repressed when the $β^A$ gene is silent. Active expression of $β^A$ is associated with disruption or removal of a nucleosome that leaves this region open to MspI cleavage (19).

2.3 DNA-binding proteins that perturb chromatin structure

Investigations mapping chromatin accessibility patterns *in vivo* set the stage for biochemical analyses that could be linked to developmental regulation of gene expression. Determining the specific *trans*-acting proteins and *cis*-regulatory elements that generate a hypersensitive site or region has been an active area of study for both mammalian and chick globin loci. Characterization of *cis*-acting DNA-binding elements relied on mutagenesis and transfection studies using primary red cells and immortalized erythroleukaemia cell lines. The DNA-binding proteins that interact with globin promoters and enhancers are a collection that includes ubiquitous, tissue-specific, and developmentally restricted proteins. Biochemical isolation of DNA-binding proteins from developmentally staged red cell nuclear extracts has proceeded on the basis of protein binding to key *cis*-regulatory elements, followed by functional transcription analyses. The erythroid-specific GATA-1 protein is a zinc-finger DNA-binding protein and original member of the diverse GATA family of transcription activators (reviewed in 5). GATA-1 protein interacts with all globin genes, often at both promoter and enhancer. EKLF (*erythroid Kruppel-like factor*), is an essential erythroid protein that has been implicated in targeting chromatin remodelling to the active $β^A$ gene (20, 21). Along with NF-E2, GATA-1 and EKLF factors bind within hypersensitive sites comprising the LCR element, as well as at most globin promoters and enhancers (Fig. 1A) (5). Additionally, repressors of transcription have been described for both human and chick globin genes. The *pal*indromic binding factor (PAL) associates with the chick $β^A$ gene promoter in non-expressing tissues as well as in silenced adult red cell nuclei (22). GATA-1 and YY-1 proteins, which can be either activators or repressors of transcription, bind the human ε promoter in fetal and adult stages of development to silence ε gene expression (23).

Although chromatin accessibility of the $β^A$ gene enhancer is not developmentally regulated, *trans*-acting proteins that bind the enhancer region are present stage-specifically. The switch between $β^A$ and ε gene expression relies on enhancer activation shifting from one gene promoter to the other. Although this switch has been correlated with binding of developmentally restricted proteins at both enhancer and promoter, and changes in chromatin structure at the promoter of β^A, the mechanism of this switch is not completely understood. Many of the proteins binding the promoter and enhancer regions are identical, leading to models of enhancer–promoter communication based on protein–protein interaction (24). Competition for shared enhancer regulation of either primitive or definitive promoters may be established by both structural changes and changes in activator and repressor proteins bound to DNA.

2.4 Linking biochemical analyses with physiological significance

The development of *in vitro* systems that mimicked genomic structure and transcription has laid the groundwork for assessing the functional consequences of chromatin structural changes in globin regulation. Two *in vitro* systems have been used to assemble physiologically spaced nucleosomes on globin DNA templates. *Xenopus laevis* egg/oocyte and *Drosophila* embryonic extracts are rich sources of maternal histones and nucleosome assembly factors (25, 26). Addition of cloned DNA to these extracts initiates a stagewise assembly of chromatin that can be 'programmed' to reconstitute *in vivo* patterns of chromatin structure and function. Programming is achieved by addition of red cell nuclear extracts or purified DNA-binding proteins either prior to or during chromatin assembly. Most of the attention has focused on the adult chick β-globin gene, due to the relative ease of obtaining red cell nuclear extracts at the definitive stage of development versus earlier stages.

The first step in establishing a physiologically relevant *in vitro* system was to assemble globin DNA into chromatin in the presence of definitive red cell extracts, monitoring this *in vitro* assembled chromatin for structural motifs mapped *in vivo* (described above). Using both DNAase hypersensitivity analysis and determination of *Msp*I restriction enzyme accessibility at the promoter, Emerson and Felsenfeld showed that globin DNA assembly in *Xenopus* oocyte extracts plus staged red cell nuclear proteins established chromatin structure that appeared identical to the appropriate stage of development *in vivo* (27, 28). This important link to landmarks of developmental function *in vivo* paved the way for further *in vitro* investigation of chromatin structure/function and mechanisms of developmental activation.

With the development of an *in vitro* system that exhibited globin chromatin structure similar to *in vivo* nuclear DNA, the next step was to link structure and function at the level of transcriptional activation. Similar to the above approach, globin DNA templates were assembled into chromatin using *Xenopus* egg extracts. The soluble fraction of *Xenopus* egg extract assembles physiologically spaced nucleosomes onto globin DNA, which in the presence of definitive stage red cell extracts, resembles the chromatin structure of expressing β^A genes. Identification of specific proteins responsible for generating accessible, transcriptionally competent chromatin structure proceeded with deletion of putative regulatory sites and mutagenesis of the globin

DNA template in parallel with biochemical isolation of these factors. Additionally, these studies built upon the previous identification of a *cis*-acting promoter element, known as the developmental-stage selector, to which adult-expressed activator proteins bound (29), and the reported essential role of GATA-1 in assembly of the preinitiation complex at the globin promoter (30).

The crude nuclear extract of definitive-stage red cells could be replaced functionally by two proteins: GATA-1, an erythroid-specific factor, and NF-E4, a protein expressed in red cells at the definitive stage of development. In the presence of GATA-1 and NF-E4 proteins, chromatin assembly established a transcriptionally competent globin template in the absence of ongoing transcription. With the addition of exogenous nucleotides and a source of RNA polymerase II and general transcription factors, this transcriptionally competent template was actively expressed. In the absence of red cell nuclear extract or either of the two proteins GATA-1 and NF-E4, chromatin-assembled globin DNA was repressed for transcription (31). Thus, the accessible chromatin structure correlated with active expression *in vivo* was linked directly to functional transcription *in vitro*.

2.5 Mechanisms for establishing active chromatin structure

An important study by Boyes and Felsenfeld explored the mechanism by which a DNAase HS site could be generated (32). This study focused on a tissue-specific HS site in the shared β^A and ϵ enhancer, and employed integrated stable constructs of the entire β^A gene with specific erythroid and ubiquitous factor enhancer binding sites mutated. Restriction enzyme accessibility and DNAase hypersensitivity analyses revealed a direct correlation between the number of erythroid proteins bound to DNA and the probability of generating a region of perturbed chromatin structure through disruption of nucleosome binding. Though erythroid protein binding was additive, generation of hypersensitivity was an all-or-none phenomenon. Previous studies using non-physiological templates in transient transfection and *in vitro* transcription analyses supported the additive nature of transactivator binding and regulation of globin expression. Imposing chromatin structure on DNA conferred a threshold to activation at the level of chromatin structure derepression, marked by generation of hypersensitive sites.

Competition between nucleosome assembly and the collection of ubiquitous and erythroid-specific proteins that bind at promoters and enhancers of globin genes is one model for gene activation or repression during development. The equilibrium between the two processes may be shifted by increasing concentration or numbers of *trans*-acting proteins binding DNA at critical sites. Outside influences, or epigenetic processes, may affect the competition as well. Targeted chromatin remodelling complexes may facilitate destabilizing nucleosomes bound to globin DNA and enhance transactivator binding (see below). Additionally, the physical process of DNA replication may set the stage for competition by passage of a replication fork and disruption of DNA–protein complexes (33). The two pathways to activated chromatin, DNA replication, and ATP-dependent chromatin remodelling, are shown in Fig. 2.

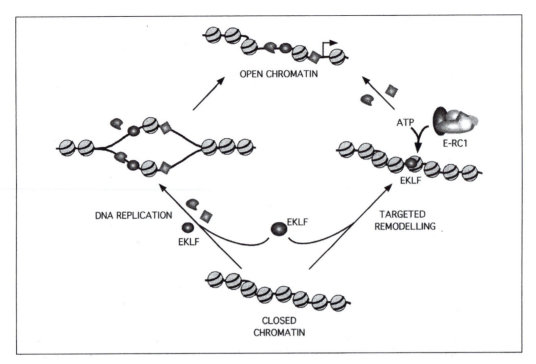

Fig. 2 Activation of β-globin gene expression relies on transactivator binding and chromatin structure remodelling. The erythroid Kruppel-like factor (EKLF) is essential for β-globin transcription. (Right) EKLF binding to the nucleosome-assembled promoter targets E-RC1 interaction and chromatin remodelling in the presence of ATP. Chromatin remodelling may precede DNA binding of other transactivating proteins. (Left) Passage of a replication fork during globin DNA replication may mediate binding of EKLF and transactivators to the globin promoter and enhancer in parallel, or in concert, with E-RC1 remodelling activity.

Both means of gaining accessibility depend on the presence of *trans*-acting factors, including ubiquitous, tissue- and developmental-specific proteins, and are not mutually exclusive.

2.5.1 Gaining accessibility to repressed chromatin by DNA replication

The *in vivo* correlation between replication timing and the activity state of globin genes is well known, with later replication associated with repressed globin transcription, including certain thalassaemias (34). An *in vitro* model of DNA replication and its potential role in altering gene regulatory programming has been developed by coupling synthetic nuclei assembly with *in vitro* transcription analysis. Cloned DNA incubated in *Xenopus* egg cytoplasmic extracts not only forms physiologically spaced chromatin, but also assembles functional nuclear organelles in the presence of maternal stores of membrane vesicles (35). These synthetic nuclei perform regulated nuclear transport (36), and replicate DNA in a semi-conservative fashion during a single cell cycle (37). The entire chick β-globin locus was assembled into synthetic nuclei in the presence or absence of definitive red cell nuclear extracts. After nuclear assembly and DNA replication, *in vitro* transcription was performed to assess whether globin gene chromatin was activated or repressed for expression. In the absence of 11-day red cell proteins, β-globin gene transcription was repressed. The established chromatin-repressed state could be altered *in vitro* by DNA replication in the presence of 11-day red cell nuclear extract. Promoter accessibility to *Msp*I digestion and globin gene expression occurred under these conditions. Transcription activation was dependent on the physical process of DNA replication, as a DNA polymerase inhibitor, aphidicolin, prevented any change in the repressed state (38).

This *in vitro* study supported a role for DNA replication in mediating competition between nucleosome assembly and transactivator binding, but more extensive investigations are needed to determine whether specific HS site formation is affected and what other factors act in chromatin structure perturbation. Several inducible genes have been reported that can be activated independently of DNA replication (39). However, cell-cycle and replication timing during differentiation may play a significant role in developmentally regulated gene expression. DNA replication could open the field for competitive transactivator binding to chromatin, acting in synergy with erythroid- and developmental-specific proteins that target chromatin remodelling complexes or modification of nucleosomes.

2.5.2 Chromatin remodelling in globin gene activation

Key regulatory factors, *cis*-acting elements, and correlative patterns of chromatin structure and active expression at specific stages of development have been described for the human β-globin gene locus (reviewed in 5). Using nuclear extracts of established erythroleukaemia cell lines and purified erythroid proteins, Armstrong and Emerson developed an *in vitro* transcription system with human β-globin DNA, chromatin-assembled by incubation in *Drosophila* embryonic extracts (20). These investigators showed that the essential erythroid-specific factor, EKLF, required a co-activating protein complex to establish an active chromatin structure at the human adult β-globin promoter. Open chromatin, marked within the promoter by a DNAase hypersensitive site, appeared as a requisite precursor to β-globin gene expression. Biochemical fractionation of mouse erythroleukaemia (MEL) cells, which express definitive-stage β-globin genes, yielded a complex of proteins that displayed ATP-dependent chromatin remodelling at the promoter, the *e*rythroid-*r*emodelling complex 1 (E-RC1).

Analysis of the proteins that comprised E-RC1 showed that many or most of the previously identified mammalian SWI/SNF subunits were present (see Chapter 5). Although the activation domain of EKLF was required for globin gene expression, DNA binding of EKLF, even in the absence of the transactivator domain, established the DNAase hypersensitive site associated with transcriptionally competent human β-globin chromatin. DNA-bound EKLF protein most likely targets the remodelling complex E-RC1 to the β-globin promoter. Whether other SWI/SNF chromatin remodelling complexes can be targeted by EKLF remains to be determined. Erythroid-specific NF-E2 protein has been shown to interact with ubiquitous chromatin remodelling complexes, such as NURF, CHRAC, or ACF, present in *Drosophila* embryonic extracts (see Chapter 6 and ref. 40), to disrupt chromatin structure within the

HS2 site at the LCR (41). These general remodelling complexes were unable to establish active chromatin structure at the β-globin promoter in the presence of EKLF protein. Thus, specific activators or repressors of gene function, restricted to certain stages of development or tissues, may interact with a subset of chromatin remodelling complexes to target localized reorganization of promoter or enhancer chromatin.

3. Amphibian metamorphosis and thyroid hormone receptor

Amphibian metamorphosis provides an almost unique opportunity to study the regulatory circuits responsible for transforming the structure and function of particular differentiated tissues and organs. This transformation involves both the programmed death of larval cells through apoptotic mechanisms and the rapid proliferation and differentiation of adult cell types (reviewed in 42).

The process of metamorphosis depends upon *thyroid hormone* (TH) and *thyroid hormone receptor* (TR). The biologically active forms of TH are 3, 5, 3'-triiodothyronine (T_3) and 3, 5, 3', 5'-tetraiodothyronine (T_4, also known as thyroxine). If production of these hormones in tadpoles is prevented by thyroidectomy, then metamorphosis does not occur. Conversely, if TH is added to tadpoles (by addition to their water) prior to the development of a functional thyroid gland, then metamorphosis can be induced precociously.

3.1 Thyroid hormone receptor

TRs are members of the large family of nuclear/steroid hormone receptors. Based on comparison between these related proteins, at least five common structural subdomains can be identified, which are named A/B through F, from amino (N)-terminal to carboxy (C)-terminal. An N-terminal A/B domain contributes to transcriptional activation for some steroid receptors. DNA-binding properties depend on two zinc fingers in the C domain. The D domain interacts with both positive and negative transcriptional regulators, and the E domain has a role in dimerization, binds hormone, and helps activate transcription. The TR has a ligand-dependent transactivation F domain located at the very C-terminus. This domain is also required for the release of a co-repressor necessary for transcriptional silencing that associates with the D domain (43).

TR recognizes *thyroid response elements* (TREs) which minimally contain an AGGTCA sequence. Although TRs can associate with TREs as monomers and homodimers, they bind with much higher affinities as heterodimers with *retinoid-X receptors* (RXRs), the receptors for 9-*cis* retinoic acid. The formation of a TR–RXR heterodimer results in distinct requirements for the orientation and spacing of the two AGGTCA 'half sites' recognized by each component of the heterodimer. TR–RXR heterodimers also confer the appropriate specificity of gene regulation in responsive cells, suggesting that they represent the physiologically relevant form of the receptor (44).

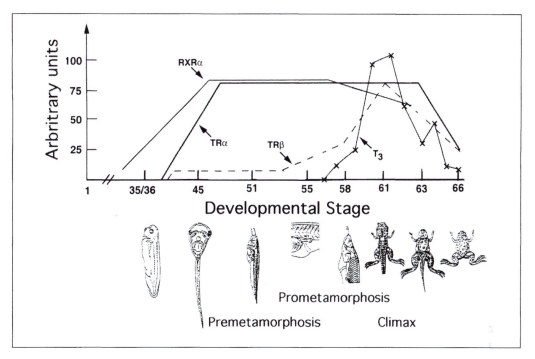

Fig. 3 Correlation of the levels of endogenous TH and the mRNAs of TRα, TRβ and RXRα genes with stages in *Xenopus laevis* metamorphosis. The plasma concentrations (arbitrary units) of thyroid hormone T_3 (crosses), and mRNA levels for TRα (solid line), TRβ (broken line), and RXRα (thin line) during development are based on published data (reviewed in ref. 5). The upregulation of TRβ around stage 55 correlates temporally with increased plasma T_3 levels; lower levels of T_3 are likely present at earlier stages. In addition, the upregulation of TRα and RXRα genes after tadpole hatching (stages 35/36) correlates with repression of TH-response genes in premetamorphic tadpoles. These tadpoles are now competent to respond to exogenous TH, which causes precocious expression of TH-response genes.

There are several genes encoding different variants of TR and RXR in *Xenopus*. The TRα, RXRα, and RXRγ receptor genes are all activated shortly after tadpole hatching (Fig. 3, stage 36) and high levels of expression are maintained (from stage 48) until the completion of metamorphosis (stage 66). In contrast, the TRβ genes are only weakly expressed in premetamorphic tadpoles, but are strongly activated during metamorphosis (from stage 55 to 61); TRβ gene expression is then substantially reduced (stage 66) (reviewed in 42).

3.2 Gene control through modification of chromatin targeted by the thyroid hormone receptor

A key early event in initiating the most dramatic changes in gene expression during metamorphosis is the dramatic increase in transcription of the TRβ gene itself in response to the first appearance of TH. We have investigated the regulation of the TRβ gene in detail in the expectation that regulatory features might be generalized to other TR–RXR responsive genes. The assembly of minichromosomes within the *Xenopus* oocyte nucleus has been used to examine the role of chromatin in both transcriptional silencing and activation of the *Xenopus* TRβA promoter by TR–RXR (45).

Microinjection of either single-stranded or double-stranded DNA templates into the *Xenopus* oocyte nucleus offers the opportunity for examination of chromatin assembly pathways, either coupled or uncoupled to DNA synthesis, and their influence on gene regulation (46). By injecting template DNA and then staging the injection times of mRNAs encoding transcriptional regulatory proteins, one has the potential to examine the mechanisms of transcription factor-mediated activation of promoters within a chromatin environment. In particular, it is possible to discriminate between pre-emptive mechanisms in which transcription factors bind during chromatin assembly to activate transcription, and post-replicative mechanisms in which transcription factors gain access to their recognition elements after they have been assembled into mature chromatin structures. The TRβA promoter has a substantial level of basal transcriptional activity even following assembly into chromatin (45). TR–RXR heterodimers bind constitutively within the minichromosome, independently of whether the receptor is synthesized before or after chromatin assembly. Rotational positioning of the TRE (*t*hyroid *r*esponse *e*lement) DNA-binding site on the surface of the histone octamer allows the specific association of the TR–RXR heterodimer *in vitro*. The unliganded TR–RXR effectively represses basal transcription of the TRβA gene. The coupling of chromatin assembly to the replication process augments transcriptional repression by unliganded TR–RXR without influencing the final level of transcriptional activity in the presence of thyroid hormone (45–47).

The molecular mechanisms by which the unliganded TR–RXR makes use of chromatin in order to augment transcriptional repression involve the two proteins, Sin3 and a histone deacetylase (48). Sin3 and histone deacetylases are cofactors in several systems in which transcriptional repression has a major role (see Chapter 8). Unliganded (hormone-free) thyroid hormone receptors and retinoic acid receptors bind NCoR (*n*egative *co*-*r*epressor) or SMRT (*s*ilencing *m*ediator for *r*etinoid and *t*hyroid receptors) co-repressor proteins (49) (Fig. 4). NCoR/SMRT interacts with Sin3 and recruits a histone deacetylase. All of the transcriptional repression conferred by the unliganded thyroid hormone receptor in *Xenopus* oocytes can be alleviated by the inhibition of histone deacetylase using trichostatin A, indicative of an essential role for deacetylation in establishing transcriptional repression in a chromatin environment (50).

The addition of thyroid hormone to the chromatin-bound receptor leads to the disruption of chromatin structure (51). Chromatin disruption is not restricted to the receptor binding site; it involves the reorganization of chromatin structure and the targeted recruitment of the acetyltransferases SRC1, P/CAF, and p300/CBP (52) (Fig. 5). It is possible to separate chromatin disruption from productive recruitment of the basal transcription machinery *in vivo* by deletion of regulatory elements essential for transcription initiation at the start site and by the use of transcriptional inhibitors (51). Therefore, chromatin disruption is an independent hormone-regulated function

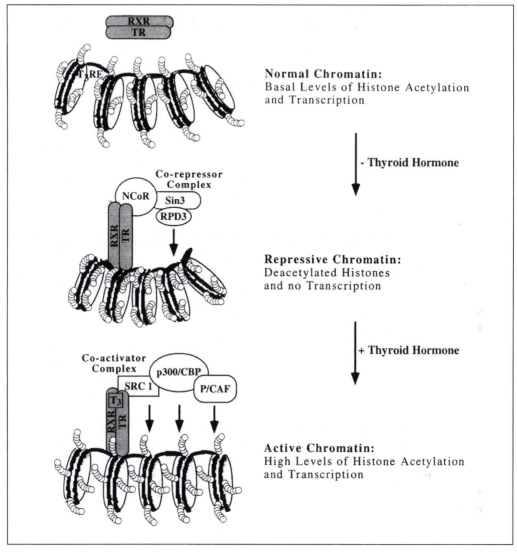

Fig. 4 A working model for the function of the thyroid hormone (T$_3$) receptor in chromatin. Normal chromatin has a basal level of histone acetylation and transcriptional activity. The binding of the thyroid hormone receptor (TR–RXR) to a thyroid response element (TRE) on a positioned nucleosome in the absence of T$_3$ leads to the recruitment of a co-repressor complex (NCoR, Sin3, RPD3) to direct histone deacetylation and transcriptional repression. The binding of the TR–RXR to a TRE in the presence of ligand, T$_3$, leads to the recruitment of the co-activator complex (SRC1, p300/CBP, P/CAF) that directs histone acetylation and facilitates transcription.

targeted by DNA-bound thyroid hormone receptor. It is remarkable just how effectively the various functions of the thyroid hormone receptor are mediated through the recruitment of enzyme complexes that modify chromatin. These results provide compelling evidence for the productive utilization of structural transitions in chromatin as a regulatory principle in gene control.

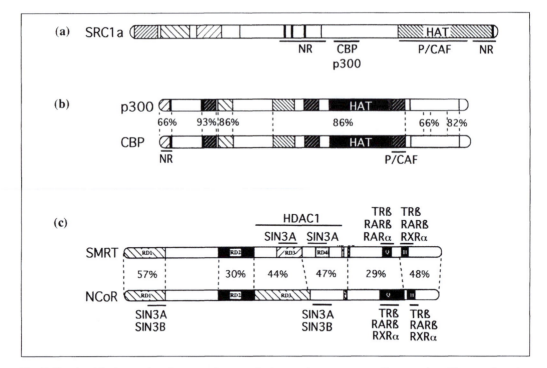

Fig. 5 Structural features of nuclear receptor co-activators and co-repressors. Conserved motifs or regions important for enzymatic activity are shown within boxes. Factors known to interact with these proteins, which were discussed in the text, are labelled at the underlined regions of protein interaction. Homology is represented as identity. (a) The structure of SRC1a is representative of a class of highly homologous co-activators. Regions of interaction with nuclear receptors (NR), CBP/p300, and P/CAF are indicated. HAT indicates the histone acetyltransferase activity domain. (b) p300 compared to CBP. Percent homology between the two proteins and sites of interaction with NR and P/CAF are indicated. (c) SMRT compared to NCoR. Interactions with TR, RAR, RXR, Sin3A, and HDAC family members are noted, as well as specific protein regions: Q, glutamine rich; RD, repression domain; H, α-helical.

The facts that: (1) core histone acetylation facilitates the access of transcription factors to DNA in a nucleosome; (2) transcriptional repressors recruit histone deacetylases; and (3) transcriptional co-activators are histone acetyltransferases, led to a model for transcriptional regulation in which the recruitment of repressors could direct the local stabilization of repressive histone–DNA interactions and the recruitment of activators could destabilize these interactions (Fig. 4). Repressive nucleosomes might prevent either the association or function of the basal transcriptional machinery on a particular promoter. The recruitment of histone deacetylase by chromatin-bound repressors eliminates basal levels of histone acetylation and transcription by impeding the recruitment and/or function of the basal transcriptional machinery. Targeted acetylation provides a means of allowing the basal machinery to displace nucleosomes and assemble a functional transcription complex.

Thus, we can propose three steps in the regulation of transcription by thyroid hormone receptor:

(1) TR binds to chromatin at a TRE on the surface of a positioned nucleosome and facilitates the assembly of a repressive chromatin structure;

(2) in response to hormone, the receptor recruits molecular machines or enzymes that disrupt local chromatin structure; and

(3) the hormone-bound receptor and associated co-activators facilitate the recruitment and activity of the basal transcriptional machinery to further activate transcription.

Additional interesting possibilities include the regulated association and activity of histone acetyltransferases and deacetylases within a common complex. In this way transcriptional activity could be continually modulated through variation in chromatin conformation in response to the availability of TH (Fig. 4).

3.3 Transcriptional control during metamorphosis

We can now combine our information concerning the expression of TR–RXR, the synthesis of TH, and the molecular mechanisms by which TR–RXR regulates transcription to explain the global regulation of metamorphosis by TR–RXR. Early in *Xenopus* development prior to the activation of TR and RXR genes, there is little TR and RXR protein present (42). Genes in the TR–RXR responsive network will be transcribed at levels dependent on the availability of other specific transcription factors that associate with their regulatory DNA, transcriptional co-activators/co-repressors, and the basal transcriptional machinery. As TRα and the RXRs begin to be upregulated after tadpole hatching (Fig. 3, stage 36), the transcription of genes containing TREs in their regulatory DNA will be repressed until the thyroid gland becomes functional and TH is synthesized at the beginning of metamorphosis. The unliganded TR–RXR bound to TREs represses gene expression of proteins that may trigger metamorphosis, and thus the unliganded receptor limits inappropriate expression of adult-specific genes in the tadpole. This represents an important differentiation control step during larval life.

Once TH is synthesized, the addition of TH to TR–RXR will lead to increased expression of the 'immediate early response' TRβ genes. The subsequent increase in the abundance of TR–RXR, coupled to the association of TH with existing TR–RXR, will activate the genes necessary for tissue remodelling and lead to adult-specific patterns of gene expression. It is important to note that the consequences of gene activation in response to TR–RXR will be as diverse as directing the programmed death of certain cell types and the proliferation and differentiation of others. In this view, TR–RXR, in the presence of TH, is a master control regulator for the activation of genes encoding other transcriptional regulators, that will, in turn, help activate the expression of the structural proteins and enzymes necessary to eventually transform the tadpole into a frog. The eventual outcome of the execution of this developmental programme is that many adult genes are activated and can maintain their expression patterns even in the presence of reduced levels of TR–RXR and TH in the adult. These genes may make use of transcriptional regulators that are up-regulated by hormone-bound

TR–RXR, but that can then sustain their own expression through autoregulatory or other mechanisms that are independent of, or require only low levels of, TR–RXR and TH.

4. Final comments

4.1 Mechanisms of gene regulation in chromatin

Regulation of gene expression is continuously dependent on chromatin structure and chromosomal environment. Chromatin structure is dynamic; specific, highly regulated changes in structure lead to repression or activation of gene expression during differentiation and development. We have outlined the regulatory mechanisms involved in (1) progressive, stepwise activation and repression of developmentally restricted genes localized within a chromosomal locus, and (2) the organism-wide effects on gene expression and morphology triggered at a critical point in development by hormonal activation of a master regulatory switch.

Analysis of chromatin structure and its effects on gene regulation has progressed from definition of chromatin structural changes at specific stages of development *in vivo* to *in vitro* recapitulation of regulatory structural patterns and transcription function. These efforts have defined an interplay of ubiquitous, tissue- and developmental stage-specific proteins that act in regulation of β-type globin gene expression. Certain of these factors are essential in chromatin activation/repression and possible targeting of chromatin remodelling machinery to promoters, enhancers, and the LCR. A more complete understanding of how promoters and enhancers communicate, as well as the role of the LCR in locus regulation, relies on continued *in vitro* functional analyses and comparison with *in vivo* landmarks. Further exploitation of globin model systems, both *in vivo* and *in vitro*, will likely shed light on basic mechanisms of chromatin structure perturbation and transcription activation or repression during development.

Amphibian metamorphosis is an exquisite system for the study of the hormonal control of gene expression. Although there is still much to understand concerning the mechanisms by which TR–RXR and TH regulate this process, it is possible to define responsive genes and to demonstrate targeted functional consequences from the activity of the receptor in the developing tadpole. Transcriptional silencing and activation mediated by TR–RXR in the absence or presence of TH have been shown to have biological consequences. Because TR–RXR is expressed in all tissues of the animal and can associate with genes in their natural chromosomal context, there are many future opportunities for exploring the general and tissue-specific functions of this master regulator for metamorphosis.

4.2 Discussion

Key mechanistic questions that remain unanswered include the determinants of chromosomal association and activity. A few of the many questions that could be addressed are listed below: What is the initial signal 'marking' tissue-specific genes for expression in the appropriate cell? How can we determine if lack of tissue-specific expression in an inappropriate cell is the result of active repression of chromatin structure and function, or simply the absence of specific transcription activators? How do proteins that target chromatin remodelling complexes first bind to chromatin, and why do proteins apparently differ in their ability to bind chromatin? If a *trans-acting* factor's DNA binding site is accessible on the surface of a nucleosome, how is that nucleosome positioned in this manner? How can a unifying theory of LCR function and gene locus regulation be developed when transgenic expression of smaller pieces of the LCR versus larger pieces of the LCR and genomic knockout data yield differing results?

Is it possible for TR–RXR to access all TREs in all genes or is the genome compartmentalized into accessible and inaccessible domains? How do the TRs function in chromosomes to silence transcription; which of the different co-repressors do they contact and how do these enigmatic proteins prevent transcription? Since several co-repressors have histone deacetylase activities, which are known to be important for developmental transitions, will inhibition of deacetylase promote premature activation of TH responsive genes? Once TH accumulates, which co-activators are recruited to TR–RXR and what are the consequences for association with histone acetylase? Are these tissue-specific or global transitions? What turns off TR gene expression after metamorphic climax and restricts the necessity for TR–RXR in the adult animal? These are important questions on the interface between transcriptional control and organismal biology.

References

1. Thompson, E. M., Legouy, E., and Renauld, J-P. (1998) Mouse embryos do not wait for the MBT: chromatin and RNA polymerase remodeling in genome activation at the onset of development. *Dev. Genet.,* **22**, 31.
2. Vermaak, D., and Wolffe, A. P. (1998) Chromatin and chromosomal controls in development. *Dev. Genet.,* **22**, 1.
3. Bouvet, P., Dimitrov, S., and Wolffe, A. P. (1994) Specific regulation of *Xenopus* chromosomal 5S rRNA gene transcription *in vivo* by histone H1. *Genes Develop.,* **8**, 1147.
4. Steinbach, O. C., Wolffe, A. P., and Rupp, R. A. W. (1997) Accummulation of somatic linker histones causes loss of mesodermal competence in *Xenopus. Nature,* **389**, 406.
5. Orkin, S. K. (1995) Regulation of globin gene expression in erythroid cells. *Eur. J. Biochem.,* **231**, 271.
6. Emerson, B. M. (1993) In *Gene expression: General and cell-type-specific,* (ed. M. Karin), Vol. 1, pp. 116–161. Birkhauser, Boston.
7. Hebbes, T. R., Clayton, A. L., Thorne, A. W., and Crane-Robinson, C. (1994) Core histone hyperacetylation co-maps with generalized DNase I sensitivity in the chicken beta-globin chromosomal domain. *EMBO J.,* **13**, 1823.
8. Hebbes, T. R., Thorne, A. W., and Crane-Robinson, C. (1988) A direct link between core histone acetylation and transcriptionally active chromatin. *EMBO J.,* **7**, 1395.
9. McGhee, J. D., Wood, W. I., Dolan, M., Engel, J. D., and Felsenfeld, G. (1981) A 200 base pair

region at the 5′ end of the chicken adult β-globin gene is accessible to nuclease digestion. *Cell, 27,* 45.

10. Forrester, W., Takagawa, S., Papayannopoulou, T., Stamatoyannopoulos, G., and Groudine, M. (1987) Evidence for a locus activation region: the formation of developmentally stable hypersensitive sites in globin-expressing hybrids. *Nucleic Acids Res., 15,* 10159.

11. Martin, D. I. K., Fiering, S., and Groudine, M. (1996) Regulation of β-globin gene expression: straightening out the locus. *Curr. Opin. Genet. Develop., 6,* 488.

12. Grosveld, F., Van Assendelft, G. B., Greaves, D. R., and Kollias, G. (1987) Position independent, high level expression of the human β-globin gene in transgenic mice. *Cell, 51,* 75.

13. Gribnau, J., De Boer, E., Trimborn, T., Wijgerde, M., Milot, E., Grosveld, F., and Fraser, P. (1998) Chromatin interaction mechanism of transcriptional control *in vivo. EMBO J., 17,* 6020.

14. Epner, E., Reik, A., Cimbora, D., Telling, A., Bender, M. A., Fiering, S., Enver, T., Martin, D. I. K., Kennedy, M., Keller, G., *et al.* (1998) The β-globin LCR is not necessary for an open chromatin structure or developmentally regulated transcription of the native mouse β-globin locus. *Mol. Cell, 2,* 447.

15. Engel, J.D. and Tanimoto, K. (2000) Looping, linking and chromatin activity: New insights into β-globin locus regulation. *Cell, 100,* 499.

16. Bender, M.A., Bulger, M., Close, J., and Groudine, M. (2000) β-globin gene switching and DNase I sensitivity of the endogenous β-globin locus in mice do not require the locus control region. *Mol. Cell, 5,* 387.

17. Gribnau, J., Diderich, K., Pruzina, S., Calzolari R., and Fraser, P. (2000) Intergenic transcription and developmental remodeling of chromatin subdomains in the human β-globin locus. *Mol. Cell, 5,* 377.

18. Stadler, J., Larsen, A., Engel, J. D., Dolan, M., Groudine, M., and Weintraub, H. (1980) Tissue-specific DNA cleavages in the globin chromatin domain introduced by DNase I. *Cell, 20,* 451.

19. McGhee, J.D., Wood, W.I., Dolan, M., Engel, J.D., and Felsenfeld, G. (1981) A 200 base pair region at the 5′ end of the chicken adult β-globin gene is accessible to nuclease digestion. *Cell, 27,* 45.

20. Armstrong, J. A., Bieker, J. J., and Emerson, B. M. (1998) A SWI/SNF-related chromatin remodeling complex, E-RC1, is required for tissue-specific transcriptional regulation by EKLF *in vitro. Cell, 95,* 93.

21. Bieker, J. J. and Southwood, C. M. (1995) The erythroid Kruppel-like factor transactivation domain is a critical component for cells specific inducibility of the β-globin promoter. *Mol. Cell. Biol., 15,* 852.

22. Emerson, B. M., Nickol, J. M., and Fong, T. C. (1989) Erythroid-specific activation and derepression of the chick β-globin promoter *in vitro. Cell, 57,* 1189.

23. Raich, N., Clegg, C. H., Grofti, J., Romeo, P.-H., and Stamatoyannopoulos, G. (1995) GATA1 and YY1 are developmental repressors of the human ε-globin gene. *EMBO J., 14,* 801.

24. Foley, K. P. and Engel, J. D. (1992) Individual stage selector element mutations lead to reciprocal changes in β- vs. ε-globin gene transcription: genetic confirmation of promoter competition during globin gene switching. *Genes Develop., 6,* 730.

25. Almouzni, G. and Wolffe, A. P. (1993) Nuclear assembly, structure, and function: the use of *Xenopus in vitro* systems. *Exp. Cell Res., 205,* 1.

26. Becker, P. B. and Wu, C. (1992) Cell-free system for assembly of transcriptionally repressed chromatin from *Drosophila* embryos. *Mol. Cell. Biol., 12,* 2241.

27. Emerson, B. M. and Felsenfeld, G. (1984) Specific factor conferring nuclease hypersensitivity at the 5′ end of the chicken adult β-globin gene. *Proc. Natl Acad. Sci., USA,* **81,** 95.

28. Emerson, B. M., Lewis, C. D., and Felsenfeld, G. (1985) Interaction of specific nuclear factors with the nuclease-hypersensitive region of the chicken adult β-globin gene: nature of the binding domain. *Cell,* **41,** 21.

29. Gallarda, J. L., Foley, K. P., Yang, Z., and Engel, J. D. (1989) The β-globin stage selector element factor is erythroid-specific promoter/enhancer binding protein NF-E4. *Genes Develop.,* **3,** 1845.

30. Fong, T. C. and Emerson, B. M. (1992) The erythroid-specific protein cGATA-1 mediates distal enhancer activity through a specialized β-globin TATA box. *Genes Develop.,* **6,** 521.

31. Barton, M. C., Madani, N., and Emerson, B. M. (1993) The erythroid protein, cGATA-1, functions with a stage-specific factor to activate transcription of chromatin-assembled β-globin genes. *Genes Develop.,* **7,** 1796.

32. Boyes, J. and Felsenfeld, G. (1996) Tissue-specific factors additively increase the probability of the all-or-none formation of a hypersensitive site. *EMBO J.,* **15,** 2496.

33. Wolffe, A. P. and Brown, D. D. (1986) DNA replication *in vitro* erases a *Xenopus* 5S RNA gene transcription complex. *Cell,* **47,** 217.

34. Forrester, W. C., Epner, E., Driscoll, M. C., Enver, T., Brice, M., Papayannopoulou, T., and Groudine, M. (1990) A deletion of the human β-globin locus activation region causes a major alteration in chromatin structure and replication across the entire β-globin locus. *Genes Develop.,* **4,** 1637.

35. Newport, J. (1987) Nuclear reconstitution *in vitro*: stages of assembly around protein-free DNA. *Cell,* **48,** 205.

36. Newmeyer, D. D., Finlay, D. R., and Forbes, D. J. (1986) *In vitro* transport of a fluorescent nuclear protein and exclusion of non-nuclear proteins. *J. Cell Biol.,* **103,** 2091.

37. Blow, J. J. and Laskey, R. A. (1986) Initiation of DNA replication in nuclei and purified DNA by a cell-free extract of *Xenopus* eggs. *Cell,* **47,** 577.

38. Barton, M. C. and Emerson, B. M. (1994) Regulated expression of the β-globin gene locus in synthetic nuclei. *Genes Develop.,* **8,** 2453.

39. Schmid, A., Fascher, K.-D., and Horz, W. (1992) Nucleosome disruption at the yeast PHO5 promoter on PHO5 induction occurs in the absence of DNA replication. *Cell,* **71,** 853.

40. Wu, C. (1997) Chromatin remodeling and the control of gene expression. *J. Biol. Chem.,* **272,** 28171.

41. Armstrong, J. A. and Emerson, B. M. (1996) NF-E2 disrupts chromatin structure at human β-globin locus control region hypersensitive site 2 *in vitro*. *Mol. Cell. Biol.,* **16,** 5634.

42. Shi, Y.-B. (2000) *Amphibian metamorphosis.* John Wiley, New York.

43. Baniahmad, A., Leng, X., Burris, T. P., Tsai, S. Y., Tsai, M. J., and O'Malley, B. W. (1995) The T4 activation domain of the thyroid hormone receptor is required for release of a putative corepressor (s) necessary for transcriptional silencing. *Mol. Cell. Biol.,* **15,** 76.

44. Puzianowska-Kuznika, M., Damjanovski, S., and Shi, Y.-B. (1997) Both thyroid hormone and 9-cis retinoic acid receptors are required to efficiently mediate the effects of thyroid hormone on embryonic development and specific gene regulation in *Xenopus laevis*. *Mol. Cell. Biol.,* **17,** 4738.

45. Wong, J., Shi, Y.-B., and Wolffe, A. P. (1995) A role for nucleosome assembly in both silencing and activation of the *Xenopus* TRβA gene by the thyroid hormone receptor. *Genes Develop.,* **9,** 2696.

46. Almouzni, G. and Wolffe, A.P. (1993) Replication coupled chromatin assembly is required for the repression of basal transcription *in vivo*. *Genes Develop.,* **7,** 2033.

47. Wong, J., Li, Q., Levi, B.-Z., Shi, Y.-B., and Wolffe, A. P. (1997) Structural and functional features of a specific nucleosome containing a recognition element for the thyroid hormone receptor. *EMBO J.*, **16**, 130.

48. Knoepfler, P. S. and Eisenman, R. N. (1999) Sin meets NuRD and other tales of repression. *Cell*, **99**, 447.

49. Horlein, A. J., Naar, A. M., Heinzel, T., Torchia, J., Gloss, B., Kurokawa, A., *et al.* (1995) Ligand-independent repression by the thyroid hormone receptor mediated by a nuclear receptor corepressor. *Nature*, **377**, 397.

50. Wong, J., Patterton, D., Imhof, A., Guschin, D., Shi, Y.-B., and Wolffe, A. P. (1998) Distinct requirements for chromatin assembly in transcriptional repression by thyroid hormone receptor and histone deacetylase. *EMBO J.*, **17**, 520.

51. Wong, J., Shi, Y.-B., and Wolffe, A. P. (1997) Determinants of chromatin disruption and transcriptional regulation instigated by the thyroid hormone receptor: Hormone regulated chromatin disruption is not sufficient for transcriptional activation. *EMBO J.*, **16**, 3158.

52. Li, Q., Imhof, A., Collingwood, T. N., Urnov, F. D., and Wolffe, A. P. (1999) p300 stimulates transcription instigated by ligand-bound thyroid hormone receptor at a step subsequent to chromatin disruption. *EMBO J.*, **18**, 5634.

10 | Chromatin contributions to epigenetic transcriptional states in yeast

LISA FREEMAN-COOK, ROHINTON KAMAKAKA, and
LORRAINE PILLUS

1. Introduction

Each cell in a single yeast colony is genetically identical, yet the individual cells may vary in their transcriptional profiles, particularly at loci that are known to be subject to silencing. Such variable or epigenetic transcriptional states are correlated with local differences in chromatin structure. Powerful genetic screens have identified many components of silenced chromatin and analysis of their function reveals links with DNA replication and protein modification. To understand the mechanisms of silencing, it will be necessary to understand how silenced chromatin is assembled and propagated through mitotic cell lineages and how boundaries between active and silent chromatin are established and maintained. Significant progress has been made in each of these areas, but many unanswered questions remain. Because of the generality of silencing as a form of genomic control, it is likely that future studies will have an impact on understanding not only transcription, but other key processes in growth regulation, development, DNA damage repair, and ageing.

2. Position effect control and heterochromatin

Transcriptional silencing is a gene non-specific mechanism of regulating large regions of chromatin that functions, at least in part, through alterations in chromatin structure. Silenced states may be stable and heritable through clonal cell lineages, yet the states are not permanent, in that switches between 'on' and 'off' states are readily observed by monitoring reporter genes. Phenomena of regionally controlled gene expression have been studied extensively in *Drosophila melanogaster*, and in the yeasts *Saccharomyces cerevisiae* and *Schizosaccharomyces pombe*. Silenced loci have many of the properties ascribed to cytogenetically defined heterochromatin. For example, the more compact chromatin at these loci is visible microscopically in *D. melanogaster* as regions that remain condensed throughout the cell cycle, and is inferred in *S. cerevisiae* and *S. pombe* by the inaccessibility of silenced loci to DNA modifying enzymes (see below). When *D. melanogaster* chromosomes become rearranged such that a normally euchromatic gene is juxtaposed with heterochromatin, the gene is transcriptionally repressed or silenced in some cells yet remains active in others. This phenomenon, termed *position effect variegation* (PEV), was first reported by Muller almost 70 years ago. PEV is gene non-specific and can act over great distances. In fact, the transcription levels of genes over one megabase away from the heterochromatin can be affected. Several pieces of evidence suggest that this repression is a result of altered chromatin structure. First, the variegating locus becomes visibly condensed in polytene chromosomes. Secondly, decreasing the copy number of the histone genes suppresses PEV. Further, feeding flies *n*-butyrate or carnitine, inhibitors of histone deacetylation, also suppresses PEV (see Chapter 11 for more details).

Position effects on transcription in mammals are also observed. These effects have been documented in transgenic animal studies and are implicated in human diseases with strong epigenetic or imprinting components, including several paediatric tumours and developmental syndromes. Epigenetic regulation in mammals is discussed in detail in Chapter 12.

In addition to the effects of a gene's particular chromosomal position and chromatin structure on its expression, the absolute position of that chromosomal region within the nucleus may also be crucial for regulation. Heterochromatin is often observed at the periphery of the nucleus, whereas euchromatin is localized in more interior nuclear space. These issues are considered in Chapter 14. Clearly, understanding how the nuclear membrane may participate in the assembly or condensation of chromatin domains is a key question. Recent genetic experiments implicating direct membrane and nuclear pore association (1, 2) with silencing (see below) provide provocative new leads.

Silencing has been studied in both budding and fission yeasts. Strong arguments can be made for such comparative analyses (3), and the case of chromatin structure and function is no exception. Indeed, in some cases the identification of silencing genes in one organism has propelled studies leading to discovery in the other, and functional divergences are arguably as important as similarities.

2.1 Overview of silencing in *S. cerevisiae*

In *S. cerevisiae*, three loci are known to be silenced: the silent mating-type loci, the telomeres, and the ribosomal DNA (rDNA) repeats (4–6). Chromosome III contains the *MAT* locus and the two silent mating-type loci *HMR* and *HML* (also referred to as the *HM* loci). *S. cerevisiae* haploid cells exist as one of two mating types, *MAT**a*** or *MAT*α, and mating-type information is expressed from the centromere-proximal *MAT* locus. Additional copies of the mating information are present at the two silent mating-type loci, located on opposite arms of the chromosome, but at these sites the information is transcriptionally inert (see Fig. 1). Mutation of the *cis-* or *trans*-acting factors required for this transcriptional silencing results in derepression and con-

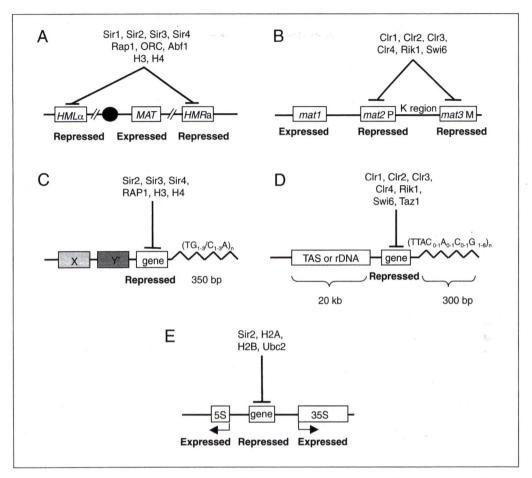

Fig. 1 *S. cerevisiae* and *S. pombe* silenced loci. (A) The *S. cerevisiae* chromosome III, containing the *MAT* locus and the two silent mating loci, *HML* and *HMR*. Many proteins are required to silence the mating information at the silent loci, including the four Sir proteins, Rap1p, the six subunit origin recognition complex ORC, Abf1p, and the histones H3 and H4. (B) A portion of *S. pombe* chromosome II containing the mating locus *mat1*, and the silent mating loci *mat2* and *mat3*. The transcriptional silencing of *mat2* and *mat3* requires many proteins, including Clr1p, Clr2p, Clr3p, Clr4p, Rik1p, and Swi6p. (C) A representative *S. cerevisiae* telomere. Each telomere contains approximately 350 bp of telomeric repeats. Additionally, most telomeres contain X and Y′ middle repetitive elements, although many silencing studies were done using reporter genes at telomeres without these elements. The repression of genes placed adjacent to the telomeres requires several of the proteins that are important in mating loci silencing, including Sir2p, Sir3p, Sir4p, Rap1p, and the histones H3 and H4. (D) An *S. pombe* telomere. Each of the six telomeres contains approximately 300 bp of repeats. The four telomeres of chromosomes I and II also contain 20 kb of telomeric-associated sequence (TAS). On chromosome III, the repeats at both telomeres directly abut more than 1 Mb of rDNA, split between the two telomeres. The repression of genes placed adjacent to telomeres requires several genes that are also required at the silent mating loci, including $clr1^+$, $clr2^+$, $clr3^+$, $clr4^+$, $rik1^+$, and $swi6^+$, as well as genes that do not have a defined role at the mating loci, including $taz1^+$. (E) A single *S. cerevisiae* rDNA repeat (of the 100 or so copies on chromosome XII), with arrows indicating the direction of transcription of the 5S and 35S rRNA genes. Reporter genes integrated into this repeat are transcriptionally silenced, especially when integrated into the normally non-transcribed region between the 5S and 35S genes. This transcriptional repression requires several proteins, including Sir2p and the histones H2A and H2B, as well as the ubiquitin conjugating enzyme Ubc2p.

sequent expression of both **a** and α information. When **a** and α information are expressed simultaneously, haploid cells acquire many properties of non-mating diploids.

Similar transcriptional repression is seen when genes are placed adjacent to telomeres (5), and, recently, silencing has also been discovered for reporter genes integrated within the rDNA repeats (6–8). Silencing at telomeres and within the rDNA meets many of the classic definitions of PEV and most notably, like *HM* silencing, is epigenetically controlled and sensitive to the gene dosage of its regulators.

2.2 Similarities and differences in *S. pombe*

S. pombe diverged from the *S. cerevisiae* lineage over 1 billion years ago and the similarities and differences in silencing between the two organisms are instructive. In *S. pombe*, at least three loci are silenced (Fig. 1 and 2). Genes placed within the silent mating-type loci (9, 10) or adjacent to telomeres (11) are silenced in *S. pombe* as in *S. cerevisiae*. The general organization of the mating-type region is similar in both organisms. *S. pombe* exists in the haploid state as one of two mating types, plus (P/h$^+$) or minus (M/h$^-$). The P or M information is present at and transcribed from the *mat1* locus. Identical P and M information is found at the silent mating loci *mat2* and *mat3*, respectively, located 15 and 30 kb to the right of *mat1* on chromosome II. At the two silent loci, the P and M genes are transcriptionally repressed even though they contain all of the sequences required for expression (12). This transcriptional repression is not gene-specific. Reporter genes inserted within the silent mating-type region are also silenced. Mutations in *cis*- or *trans*-acting factors that alleviate silencing cause expression of both M and P information (12).

In addition to the telomeres and silent mating loci, *S. pombe* reporter genes are also silenced when placed within the centromeres (13, 14). This feature is a key distinction and likely reflects the fact that *S. pombe* centromeres, like those in multicellular organisms, are larger, more complex, and generally heterochromatic, when compared to the small 'point centromeres' of *S. cerevisiae*. It is to be expected that rDNA silencing also exists in *S. pombe*, although direct evidence for this has not yet been published.

2.3 Silencing in other yeasts

Transcriptional silencing has been reported in *Kluyveromyces lactis*, a budding yeast that is closely related to *S. cerevisiae*. Functional homologues of two *S. cerevisiae* silencing genes, *SIR2* and *SIR4* (see below), have been identified, and mutations in both result in partial derepression of the *K. lactis* silent mating-type genes. Although telomeric and rDNA silencing have not been examined, the *sir4* mutant has altered telomere length, and the *sir2* mutant has increased rDNA recombination, both phenotypes associated with a loss of silencing in *S. cerevisiae* (15, 16).

SIR2 homologues have also been identified in the pathogenic yeast *Candida albicans*. One of these genes partially rescues mating defects of an *S. cerevisiae sir2* null mutant, but silencing has not been examined directly in *C. albicans*. However, this *SIR2* homologue appears to control the epigenetic switch between colony morphologies that is correlated with pathogenicity, and the mutant has reduced chromosome stability (17). Mating-type genes similar to those found in *S. cerevisiae* have also been reported in *C. albicans*, an intriguing discovery since no sexual cycle has been discovered yet for this organism (18). There do not appear to be silenced copies of these genes, nor is it known whether, or how, they contribute to cell-type specification. Further analysis of these mating-type genes and transcriptional silencing will be important for understanding this clinically significant yeast.

3. Silencing sequences and proteins

3.1 Silencers and silenced sequences

As noted above, transcriptional silencing of mating-type loci requires both *cis*- and *trans*-acting factors in both *S. cerevisiae* and *S. pombe*. In *S. cerevisiae*, repression requires the *cis*-acting E and I silencers that flank the *HM* silent mating-type loci. These *cis*- 'silencer' sequences contain combinations of binding sites for the *a*utonomously *r*eplicating *s*equence (ARS) *b*inding *f*actor (Abf1p), the *o*rigin *r*ecognition *c*omplex (ORC), and the *r*epressor-*a*ctivator *p*rotein (Rap1p), which are discussed below (4). The importance of these three binding sites has been confirmed through mutational analysis and through the construction of a functional 'synthetic silencer' that contains the three binding sites (19–21).

Similar to mating-type loci silencing in *S. cerevisiae*, the transcriptional repression of *mat2* and *mat3* in *S. pombe* requires both *cis*- and *trans*-acting factors. For example, four *cis*-acting silencers surround *mat2*, with two to the right and two to the left. Both of the left silencers and one of the two right silencers are required for transcriptional repression of the *mat2*P genes (22). The 15 kb region between *mat2* and *mat3*, known as the K region, is a recombination cold spot. Deletions within the K region relieve the recombination block and affect efficiency of mating-type switching (23).

In *S. cerevisiae*, centromeres can be reduced to a 125 bp element that retains complete mitotic function. In contrast, the centromeres of *S. pombe* are approximately 35 kb, 60 kb, and 110 kb for chromosomes I, II, and III respectively, of which at least 12–15 kb is required for partial function (12). Although *S. pombe* centromeres are much larger than those of *S. cerevisiae*, they are still 20–100 times smaller than *D. melanogaster* and mammalian centromeres.

S. pombe centromeres consist of repeated sequences and a complex structure, features that are also analogous to centromeres of multicellular eukaryotes, including humans. The central core domains of *S. pombe* centromeres do not have a typical nucleosomal array and, like the centromeres of *S. cerevisiae*, they are organized into a special chromatin structure. Further, like human centromeres, *S. pombe* centromeric chromatin is underacetylated (12, 24). Figure 2 shows a diagram of *S. pombe* centromeres 1, 2, and 3. All three contain an approximately 15 kb central region composed of the *cent*ral (*cnt*) domain and the *in*most *repeats* (*imr*). *cnt1* and *cnt3* contain the conserved TM domain, and *cnt2* is unique. The *imr* inverted repeats are not conserved between chromosomes, but interestingly, the left and right repeat are absolutely

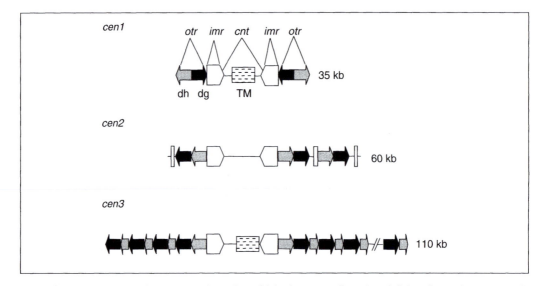

Fig. 2 *S. pombe* centromeric structure. A number of labs have contributed to defining *S. pombe* centromeric structure. *S. pombe* centromeres consist of a central domain (*cnt*) flanked by two sets of inverted repeats, the *imr* repeats (open arrows) and the *otr* repeats. The central domain is not conserved between the three centromeres except for the TM domain found at *cnt1* and *cnt3* (dashed box). The *otr* repeats are conserved between centromeres and contain two types of repetitive sequences, the dg and dh repeats (black and grey arrows, respectively). Portions of the *imr* repeat of *cen2* are repeated in *otr2* (open rectangles). The number of repeats shown within the *otr* domains is representative; the actual number of repeats varies from strain to strain.

identical within the same chromosome (25). Each of the three chromosomes also contains various classes of repeats forming the large inverted repeats surrounding the central region, the *outer repetitive domains* (*otr*). The sequence of the *otr* repeats is conserved between the three chromosomes, and is also present in the K region between the *mat2* and *mat3* silenced loci, yet the number of repeats seen at each chromosome varies. The *otr* repeats are similar to repeats found at mammalian centromeres (12, 25).

Reporter genes inserted at various positions within *cnt1*, *cnt3*, *imr1*, and *otr1* are transcriptionally silenced. Similar to PEV and telomeric silencing in *S. cerevisiae* and *S. pombe*, these genes have a mosaic expression pattern; in some cells the reporter gene is expressed, yet in other cells the reporter is silenced (12, 13).

Transcriptional silencing is observed at the telomeres of both yeasts. *S. cerevisiae* telomeres, like those of many other organisms, are composed of a series of short tandem repeats. There are, on average, $300 + 75\,\mathrm{bp}$ of the telomeric consensus sequence $C_{1-3}A/TG_{1-3}$ at each *S. cerevisiae* telomere (5). *S. pombe* telomeres are also composed of G-rich repeated sequences, with about 300 bp of the consensus telomeric repeat ($TTAC_{0-1}A_{0-1}C_{0-1}G_{1-8}$) present at each of the six telomeres (12). The telomeric repeats are organized into a specialized chromatin structure, like other silenced regions in a variety of organisms. The chromatin of the telomeric repeats is micrococcal nuclease resistant and has a non-nucleosomal structure in both yeasts.

Further, as is true for *D. melanogaster*, *S. cerevisiae* and *S. pombe* telomeres cluster at the nuclear periphery (5, 12).

Like the silent mating-type loci, telomeric DNA in *S. cerevisiae* contains binding sites for Rap1p. *S. pombe* telomeres are bound by Taz1p, which does not share sequence similarity to Rap1p except that both contain motifs that are similar to the proto-oncogene Myb (5, 26). Both Rap1p and Taz1p are required for telomeric silencing (see below).

In *S. cerevisiae*, homologous recombination between repeated sequences can occur with high frequency. However, the rDNA repeats are ordinarily quite stable. Only about 50% of the repeats are estimated to be transcribed. It was originally observed that the *SIR2* silencing gene was required to suppress rDNA recombination, and it was later found that reporter constructs integrated into the rDNA were silenced in a *SIR2*-dependent manner (6–8). Although rDNA silencing in *S. pombe* has not yet been examined, at least two *S. pombe* silencing proteins localize to the nucleolus (see below).

3.2 Silencer-bound proteins

The idea that transcriptional silencing is due to a special, local chromatin structure is given direct support by the fact that the histone proteins themselves are required for silencing. Both histone H3 and H4 can be separated into two domains: a globular hydrophobic domain that participates in the formation of the core nucleosome, and a hydophilic N-terminal tail that is relatively unstructured (see Chapter 1). Mutations or deletions in the N-terminal tails of histones H3 and H4 in *S. cerevisiae* disrupt silencing (see Chapter 3). These histone tails contain lysine residues that can be acetylated *in vivo*, but in silenced regions, the tails are generally hypoacetylated (see Chapter 8). Such hypoacetylation is likely to be important for the silenced state, as mutations that disrupt silencing are correlated with alterations of the normal acetylation pattern of silenced regions (see below and Chapter 7). Several non-histone chromosomal proteins play critical roles in silencing. These include Rap1p, Abf1p, Taz1p, and ORC, which will be discussed in detail below.

Rap1p is an essential DNA-binding protein with a dual role in the cell: it is required for both gene repression and activation (27). Binding sites for Rap1p lie within the *cis*-acting silencing sequences at the silent mating-type loci and the telomeric repeats. Indeed, several lines of evidence indicate that Rap1p binds to DNA at the silent mating loci and telomeres both *in vitro* and *in vivo*. Mutational analysis of *RAP1* yielded mutants that are specifically silencing-defective. Additionally, overexpressing a C-terminal fragment of Rap1p that lacks the full DNA-binding domain causes a loss of telomeric silencing, perhaps due to titration of other factors required for silencing (27).

Abf1p is an essential DNA-binding transcription factor that binds to the silent mating-type loci within the *cis*-acting silencers. When the Abf1p binding site is deleted from a synthetic silencer, a loss of transcriptional silencing occurs (21). Further, *abf1* mutants disrupt silencing at the mating-type loci (28). Recently, *ABF1* was found to play a role in the nucleotide excision repair process (29). No roles for Abf1p have been reported in telomeric or rDNA silencing. Interestingly, Abf1p and Rap1p share blocks of sequence similarity throughout their structure (4).

Taz1p (*telomere associated in Schizosaccharomyces*) is an *S. pombe* telomere-binding protein with sequence similarity to the DNA-binding domain of the Myb proto-oncogene. Interestingly, *S. cerevisiae* Rap1p also shares structural similarity with the Myb motif but has limited overall sequence similarity. Mutations in Taz1p cause defects in telomeric silencing as well as telomere length control, and Taz1p is also required for normal meiosis (26).

ORC is a six subunit protein complex that binds to chromosomal origins of replication and is essential for DNA replication initiation. In addition, ORC is also required for silencing. ORC binds to the silencers at the mating-type loci, and mutations in several of the components have defects in silencing at the silent mating loci and telomeres. Another link between DNA replication and silencing is the requirement for passage through S phase to re-establish silencing at the mating loci. However, this S-phase requirement cannot be simply explained by a need for the silencer to serve as an ORC-dependent origin of replication, as the two functions of ORC are genetically separable. Alleles have been identified that are replication competent but silencing defective, and vice versa. Interestingly, tethering Sir1p to the silencer bypasses the requirement for ORC in silencing, but passage through S phase is still required (summarized in 30, 31).

The six subunits of ORC have recently been identified and cloned from *S. pombe* (32). Although the *S. pombe* silencers contain ARS elements, the role these proteins play in transcriptional silencing remains to be determined.

3.3 Silencing complexes: the Sir proteins

The proteins described above directly bind DNA at the silenced loci. However, the roles of these proteins appear not to be limited to silencing, but may also include activation and replication functions. The specification of silencing functions may thus be determined through additional protein–protein interactions. Indeed, silenced chromatin also includes proteins that establish and maintain elements of chromatin structure. The most notable of these are the Sir proteins, which were originally identified by their mutant phenotypes.

SIR1 encodes one of four *silent information regulator* proteins found in *S. cerevisiae* (reviewed in 33), and *sir1* null mutants first revealed the epigenetic nature of *HM* silencing (34). In *sir1* null mutant strains, two populations of cells exist: 80% of the cells are derepressed at the silent mating loci and are incompetent for mating, whereas the other 20% have intact silencing and mating abilities. The two populations are relatively stable and the transcriptional states are heritable, but changes between the two states occur at detectable frequencies. Sir1p is thought to be important for the establishment of silencing at the mating-type loci, although any role at telomeres may be more modest (35–37). Sir1p does not bind DNA directly at the silent mating loci, but instead interacts with ORC by binding to Orc1p (38, 39). The mechanistic significance of this interaction is not fully understood, but again points to connections between DNA replication and the establishment of heritable chromatin states (see below).

SIR2 is unique among the *SIR* genes as it is required for silencing at all three loci in *S. cerevisiae*. A *sir2Δ* mutant has pleiotropic phenotypes, including increased rDNA recombination, and defects in the meiotic pachytene checkpoint arrest. Biochemical and microscopic assays reveal that Sir2p localizes to mating-type loci, telomeric foci, and the nucleolus, all sites of its silencing activity (6–8).

A suggestion for the function of *SIR2* was made several years ago, when it was found that increasing Sir2p levels correlated with global decreases in cellular histone deacetylation (40). The connection between acetylation states and silenced chromatin is detailed below and in Chapter 8. Recent mechanistic insights into the role of Sir2 come from reports of its catalytic activity (see below).

Sir2 appears to be targeted to chromatin at the telomeres and mating loci through its interactions with Sir3p and Sir4p, which in turn interact with the histones H3 and H4 (5, 7). In the nucleolus, Sir2p may interact with the rDNA by association with Net1 protein (*n*ucleolar silencing *e*stablishing factor and *t*elophase regulator). Indeed, Net1p, Sir2p, Cdc14p, and Nan1p (*N*et1 *a*ssociated *n*ucleolar protein) are members of the nucleolar complex RENT (*r*egulator of *n*ucleolar silencing and *t*elophase exit) that may function in cell cycle regulation. Interestingly, *SIR2* may function in a different complex to contribute to meiotic regulation (reviewed in 8).

SIR2 is the founding member of a family of related genes named *h*omologues of *SIR two* (HSTs) in *S. cerevisiae* (41); additional homologues have been found in more than 20 other organisms, including human, rat, mouse, *Caenorhabditis elegans, Arabidopsis thaliana, C. albicans, S. pombe,* and *Leishmania*. Eubacterial and archaebacterial homologues also exist, raising the likelihood that members of this family will prove to be among the most ancient and far-ranging silencing proteins. In addition to the molecular similarities, three of the four *S. cerevisiae* HSTs can function in silencing, as can the *K. lactis , C. albicans,* and *S. pombe* homologues (15, 41, 42). Indeed, a chimeric protein containing the core domain of a human *SIR2* homologue can restore partial silencing in *S. cerevisiae* mutants (43). Clearly, understanding the *in vivo* catalytic activity of this class of proteins, the complexes in which they interact, and their genomic targeting and effects are important future goals.

Sir3p is required for silencing at the silent mating loci and telomeres, and is localized to the telomeric clusters in the nucleus (reviewed in 44). Sir3p interacts with Sir4p, Rap1p, and the histones to form a silencing complex at telomeres and the silent mating loci. *SIR3* shares extensive sequence similarity with *ORC1*, again underscoring links between replication and silencing. Sir3p is a phosphoprotein, and its phosphorylation is regulated through the pheromone response MAP kinase cascade, linking silencing with MAP kinase signal transduction. Sir3p may play a direct role in the formation of the silenced chromatin, as it has been found to spread along the silenced chromatin. Overexpression of *SIR3* promotes spreading of silencing to genes farther from the telomere, and Sir3p has been shown under these conditions to bind chromatin at sites increasingly distal from the telomere.

Sir4p, like Sir3p, is required for silencing at the telomeres and silent mating loci, and localizes to telomeric clusters (33, 45). Sir4p interacts with Sir3p, Sir2p, and the histones H3 and H4. In addition, Sir4p interacts with Ubp3p, a deubiquitinating enzyme that has a role in transcriptional silencing (46, 47). In contrast to Sir2p, Sir4p appears to antagonize rDNA silencing (48). Unlike *SIR3*, overexpression of either full-length *SIR4* or a C-terminal fragment disrupts silencing (49, 50).

Several reports link the Sir2, 3, and 4 proteins to ageing in *S. cerevisiae* (reviewed in 6). Mutations in *sir2*, *sir3*, and *sir4* all decrease life span, although apparently through distinguishable mechanisms. The decreased lifespan in *sir* mutants is correlated with instability of the rDNA within the nucleolus, which the Sir proteins, especially Sir2, are required to suppress (51). Whether loss of silencing at multiple loci contributes to replicative ageing may emerge from further studies on distinct silencing proteins.

3.4 *S. pombe* silencing genes

Many genes have been discovered to play a role in silencing in *S. pombe*, predominantly through screens focused on regulation of silencing at the silent mating loci. These genes include the cryptic loci regulators, *clr1+*, *clr2+*, *clr3+*, *clr4+*, *rik1+* (recombination in K region), *swi6+* (switching) and *hst4+*. Mutations in *clr1*, *clr2*, and *clr3* cause derepression of both the endogenous mating-type genes and a reporter gene inserted adjacent to *mat3*. The mutations also cause increased recombination in the normally cold K region between *mat2* and *mat3* (9, 10, 52). In addition to effects at the silent mating-type loci, *clr1*, *clr2*, and *clr3* mutations also partially derepress reporter genes inserted adjacent to a minichromosome telomere or within *cen1* (14). Molecular identification of *clr3+* revealed that it is likely to encode a histone deacetylase (53). This suggests that *S. pombe* silencing genes, perhaps including *SIR2* homologues (below), may act by regulating the acetylation state of histones.

clr4, *rik1*, and *swi6* share the same silencing phenotypes as *clr1*, *clr2*, and *clr3*. All cause derepression of the silent mating loci, and of reporter genes at centromeres and the telomere of a minichromosome, and all increase recombination in the K region (9, 10, 14, 52, 54). However, *clr4+*, *rik1+*, and *swi6+* may play a more direct role in centromeric structure. Mutations in these three genes result in more dramatic derepression of centromeric reporter genes than the derepression seen in *clr1*, *clr2*, and *clr3* mutants (14). The mutants also have elevated chromosome loss rates, perhaps explained by the prevalence of lagging chromosomes in anaphase cells. In addition, *clr4*, *rik1*, and *swi6* mutants show genetic interactions with the β-tubulin gene *nda3+*, supporting a direct role for these genes in centromeric function. The mutants are sensitive to the microtubule destabilizing drugs thiabendazole (TBZ) and methylbenzimidazole-2yl carbamate (MBC) (14, 55, 56). These data support the hypothesis that *clr4+*, *rik1+*, and *swi6+* interact directly or indirectly with microtubules at the kinetochore.

Clr4p contains both a chromo (*chr*omatin *o*rganization *mo*difier) domain and a SET (*s*uppressor of variegation, *e*nhancer of zeste, *t*rithorax) domain, motifs originally identified in *D. melanogaster* proteins that regulate chromatin structure (see Chapter

11). Clr4p shares high sequence similarity with Su(var)3–9p, a *D. melanogaster* protein that affects PEV (57). *SET1*, an *S. cerevisiae* gene that encodes a SET domain protein, is also important for transcriptional control, and its mutation leads to silencing defects (58). Other chromo or SET domain containing proteins have been identified in *C. elegans*, mouse, and humans, suggesting that the silencing function of these domains may be universally conserved (57, 59).

swi6$^+$ is a homologue of the *D. melanogaster* heterochromatin protein 1 (HP1), which localizes to centromeric heterochromatin and is important for PEV (60, 61). Also like *swi6*, mutations in HP1 result in abnormal chromosome segregation (62). Both proteins contain chromo domains and the phenotypes of *swi6* mutants are consistent with the hypothesis that Swi6p plays an integral role at fission yeast centromeres, telomeres, and the silent mating loci.

S. pombe hst4$^+$ is most closely related to the *S. cerevisiae* genes *HST3* and *HST4*, members of the *SIR2* gene family. Like its *S. cerevisiae* homologues, it likely plays a role in chromatin structure; *hst4*$^+$ is required for transcriptional silencing at centromeres, and at the telomere of a minichromosome. *hst4*$^+$ also has an apparent role in the chromosomal transmission function of centromeres, as mutants have elevated chromosome loss rates and are sensitive to TBZ. Hst4p localizes to the nucleolus, like overexpressed Clr4p, again raising the possibility of rDNA silencing in *S. pombe* (42). In addition to *hst4*$^+$, there are at least two additional *SIR2* homologues in *S. pombe* (L. Freeman-Cook and L. Pillus, unpublished data; M. Derbyshire and J. Strathern, unpublished data).

3.5 Other regulators of silencing

Tables 1 and 2 list *S. cerevisiae* and *S. pombe* genes reported to have some role in transcriptional silencing, often inferred from analysis of mutant phenotypes. Some of the genes in this compilation may therefore play only indirect roles. However, at least some of the genes can be grouped based on their proposed biological activity. Many may affect cell-cycle progression, for example *S. cerevisiae SWI6* and the cyclin genes. Other genes may act by modifying histones or other cellular proteins by acetylation (e.g. *S. cerevisiae NAT1*, *SAS2*, *SAS3*, and *GCN5*), deacetylation (e.g. *S. cerevisiae RPD3* and *S. pombe clr3*$^+$), or ubiquitination (e.g. *S. cerevisiae UBC2* and *UBP3*). Still others may be more directly involved with protein degradation (e.g. *S. pombe mts2*$^+$ and *pad1*$^+$), and some indicate a link between silencing and DNA repair (e.g. *S. cerevisiae HDF1* and *HDF2*, and *S. pombe rhp6*$^+$). Roles for many of the other genes remain speculative.

4. Building and regulating silenced chromatin

4.1 The structure of silenced chromatin

Enzymatic probes have been used in numerous studies of the chromatin structure of silenced domains. Many of these have compared the differences in accessibility for

Table 1 *S. cerevisiae* silencing genes[a]

Chromatin structure	Cell-cycle progression	(De)acetylation	Other (cont.)
ASF1	CDC7	ARD1	NPL3
ASF2	CDC44	GCN5	NPT1
CAC2	CDC45	HDA1	ORC
CAC3	CIN8	NAT1	PAS1
HHT1–2	CLB2	RPD3	PCH2
HHF1–2	CLB5	SAS2	SAN1–3
HTA1	CLN3	SAS3	SAS4,5,10
HTB1	MBP1	SIN3	SDS3
RAP1	PCNA (POL30)	SIR2 (and HSTs 1–4)	SET1
RIF1	POL2		SIF2
RIF2	RNR1	**Other**	SIR1
RLF2 (CAC1)	SWI4	ACT3	SPT3
SIR3	SWI6	DIS1	SPT6
SIR4		DOT 5,6	SUM1–3
	DNA repair	FKH1,2	TLC1
	ABF1	GAL11	UME6
Ubiquitination	HDF1	HIR1–3	WTM1–3
UBC2/RAD6	HDF2	IRS	YCL54
UBP3	MEC3	LRS	ZDS1–2
UBP10 (DOT4)	SGS1	NET1	

[a] The genes listed here have been reported to function in silencing through a variety of different studies, although in many cases the mechanism is poorly understood. These include overexpression studies, mutant phenotype analysis, and biochemical and cell biological assays. For detailed references and description of these functions, consult the Saccharomyces Genome Database (http://genome-www.stanford.edu/Saccharomyces/) and links to YPD@PROTEOME (http://www.proteome.com/database/YPD/reports).

Table 2 *S. pombe* silencing genes[a]

Chromatin structure	DNA repair	Other	Other (cont.)
swi6	rhp6	cep3	mrf
taz1		clr1	rat1
	(De)acetylation	clr2	rik1
	clr3	clr4	
Protein degradation	clr6	csp1–13	
mts2	hda1	esp1–3	
pad1	hst4	lot2	

[a] For detailed references and description of these functions, consult the Sanger Centre database (http://webace.sanger.ac.uk).

enzymes to the same DNA sequence when it is packaged into repressed or active chromatin.

Analysis using sequence-specific nucleases such as restriction enzymes or the yeast HO endonuclease (63), and methyltransferases (64), have shown that silenced regions are more resistant to modification by these enzymes compared to active loci. In *S. cerevisiae*, these techniques have also allowed the delineation of the entire silenced domains and demonstrate, for example, that the *HMR* domain is not limited to the

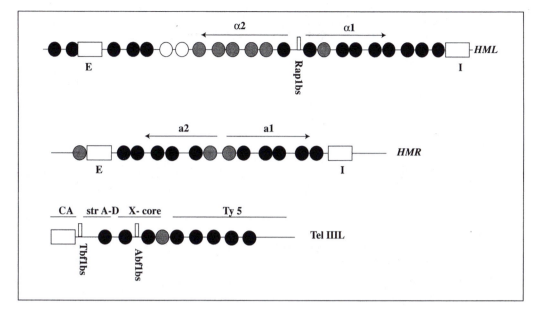

Fig. 3 Schematic representation of the chromatin structures at various silenced loci in the yeast *Saccharomyces cerevisiae*. Dark circles indicate precisely positioned nucleosomes, while lightly shaded circles represent loosely positioned nucleosomes and clear circles indicate undefined nucleosomes. Arrows indicate the genes at the silenced loci, while the boxes identify the silencers and binding sites for various transcription factors.

region between the two silencers E and I but extends several hundred base pairs beyond the silencers (63, 65).

Studies have also been performed with nucleases that do not possess any sequence specificity, such as micrococcal nuclease. This enzyme preferentially digests linker DNA between nucleosomes, allowing the mapping of nucleosomes over the entire repressed chromatin domain. The results of this analysis at *HML* and *HMR* suggest significant differences in the chromatin structure at these loci. The *HMR* locus is characterized by a 1.9 kb domain where nucleosomes are positioned as closely packed dimers with linkers of less than 5 bp, separated by linkers of *c*. 20 bp across the entire domain (Fig. 3) (66). In contrast to this positioned pattern, mutations in *sir3* and the histone H4 gene were shown to lead to randomly positioned nucleosomes. The two silencers, E and I, are nucleosome-free sites of DNAase I hypersensitivity. Although studies at the *HMR* locus suggest that the di-nucleosome packaging is not a consequence of silencing per se, (since this pattern was not seen on a silenced reporter *TRP1* gene), close packing of nucleosomes might be a consequence of silencing (67).

The *HML* locus has some of the characteristics of *HMR*. Precisely positioned nucleosomes abut the two silencers and extend over the α1 and α2 genes. Positioned nucleosomes are present at the transcription start sites of the two genes, although the promoters of the genes are nucleosome-free and in a more accessible configuration. Mutant Sir proteins disrupt the positioned nucleosomes near *HML*-I but not around *HML*-E (68).

The observation that the promoter of silenced genes is in an open conformation may at first be surprising, but recent studies of the *HSP82* gene shed some light (69). Data suggest that the HSF factor and the TBP are still bound to the promoter in the silenced *HSP82* gene, suggesting that silencing mediated by the Sir proteins acts primarily downstream of activator and TBP binding.

The chromatin structure of telomeres is distinct from that observed at the silent *HM* loci. The terminal C_{1-3}-A repeats are devoid of nucleosomes and are bound by non-histone proteins to form a structure referred to as the telosome. Rap1p is believed to be one of the components of the telosome (70). A DNAase I hypersensitive site is located adjacent to the repeats. Nucleosomes are positioned across the X and Y' repeat regions. On the chromosome III left arm, the X repeat is packaged into irregularly spaced nucleosomes whereas an adjacent silenced Ty5 LTR is packaged into a regular array of positioned nucleosomes (71). Similar to the results obtained at *HML* and *HMR*, in *sir* mutants this array is disrupted, and the formation of the array requires the histone H4 tail (72).

See Fig. 3 for a graphical representation of chromatin structures at the silenced loci in *S. cerevisiae*. Although comparable studies have been somewhat more limited in *S. pombe*, the same general image exists of altered chromatin structure in silenced domains.

4.2 Silencing domains and boundary elements

It appears that formation of the chromatin structure characteristic of silenced domains initiates from the silencer and spreads along the DNA fibre. Data indicate that this structure is restricted to specific regions of the chromosome by boundary elements, and these boundary elements ensure that temporally and spatially regulated factors act only on the appropriate target domain. Analyses of several silenced chromatin domains in yeast demonstrate that different elements act as boundaries at different loci.

A deletion analysis of the right *HMR* boundary indicated that a tRNA gene located downstream of the I silencer acted as a boundary to prevent the spread of silenced chromatin. Deletion of this element resulted in silencing of an adjacent gene, yet ectopic insertion of this element between the silencer and a reporter gene insulated the reporter gene from silencing (65).

At *HML*, the results are different; it appears as though here the I silencer defines the boundary by initiating silencing in only one direction. In this instance, the unidirectional propagation of the silenced domain from the I silencer prevents the spread of silenced chromatin into flanking regions of the chromosome (73).

Boundary elements have also been mapped at native yeast telomeres. The middle-repetitive Y' elements are flanked by the STAR regions that consist of small elements called STAR-D, -C, -B, and -A. The junction between the X and Y' element usually has variable numbers of the STAR repeats. Detailed analysis of the sequences abutting the telomeric repeats and the X and Y' elements demonstrate that the STAR sequences act as a barrier to the spread of silenced chromatin from the telomeric ends. The Y' element contains genes that are expressed in meiosis, and the insulation activity of the STAR elements could protect these genes from silencing during meiosis. The STAR sequences can protect a reporter gene from the repressing effects of silencers when interposed between the gene and the silencer. Multimerization of the STAR submotifs recapitulates the anti-silencing activity of the intact element (36, 37).

Studies on the *TEF1* and *TEF2* genes also identified elements in the promoters of these genes that act as boundaries when interposed between a silencer and a reporter gene (73). Together, these studies demonstrate that an array of binding sites for transcription factors can block the spread of silencing, thus raising the possibility that in yeast, strong promoters act as boundaries by actively disrupting silenced chromatin.

4.3 Assembling silenced chromatin

Silencing of the *HMR* domain is believed to occur through interactions between proteins bound to the silencer and the proteins associated with nucleosomes. The role of the silencers includes recruiting the Sir proteins to the silent loci. The recruitment and subsequent binding of the proteins to nucleosomes generates a chromatin domain that is inaccessible to enzymatic or chemical probes and is transcriptionally inactive. Though initial results suggested that once a domain was silenced the silencers were not necessary for maintaining the repressed state, recent results suggest that the DNA silencers, proto-silencers, and associated proteins present are required to maintain the locus in a stably silenced state throughout the cell cycle (74, 75). Uncoupling of the silenced domain from the DNA silencers resulted in the rapid derepression of the silenced genes. In addition, the proteins that bind the silencers, such as ORC (76), and the proteins that bind the nucleosomes in the silenced domain, such as Sir3p (77), are necessary to maintain the repressed state. The use of conditional alleles of these proteins revealed that disruption of either *sir3* or ORC leads to an immediate derepression of the silent genes.

The establishment of the repressed state is defined as the formation of silenced chromatin, in contrast to inheritance of the repressed state, which is defined as stable duplication of the state following replication. When a silent domain is experimentally derepressed, the silenced state can be established only following passage of the cells through the S phase of the cell cycle (76, 77). The exact nature of the S-phase requirement is not clear, but is probably dependent on replication-coupled chromatin assembly.

Several non-exclusive models can account for the replication-coupled establishment of silencing. One of the S-phase requirements could be the binding of the silencer proteins to the silencer. Replication can disrupt chromatin, allowing a window of opportunity for binding of sequence-specific factors even in the absence of chromatin remodelling machines. It is possible that regions of the nucleus that contain high concentrations of Sir proteins are deficient in chromatin remodelling machines, and therefore binding of silencer proteins to the silencer can only occur during replication.

A second possibility for requiring passage through S phase could be the need for the proper assembly of nucleosomes with specific modifications in the histones. It should be noted that mutations in the histone-tail domains and Sir proteins both affect the spacing of nucleosomes. Deposition of the Sir proteins might be coupled to histone deposition during replication. The demonstration that mutants in chromatin assembly factors affect silencing is of particular interest in this regard. The observation that Rap1p is mislocalized in *cac1* (chromatin *a*ssembly *c*omplex 1) mutants and the demonstration that mouse HP1 is associated with mouse CAF-1 (chromatin *a*ssembly *f*actor) supports this model. It is therefore possible that the formation of close-packed nucleosomes observed at the silenced loci may be mediated by specific interactions between the Sir proteins and histone tails during chromatin maturation immediately following histone deposition. Unfortunately not much is known about the modification states of the histones or nucleosome positioning at silenced loci in *cac1* mutants. The recent characterization of Asf1p, previously known to influence silencing, as such a coupling factor may lead to a greater understanding of the S-phase requirement (78, 79).

Efficient establishment requires Sir1p and an intact silencer, since mutations in the silencer-bound proteins affect efficient establishment. Mutations in the Rap1 protein as well as mutations in *cis*-acting elements at the silencers are like the *sir1* mutants described above, leading to two populations of cells—repressed and derepressed (34, 80, 81). This all-or-nothing mechanism of silencing suggests that the silencers may act as binary switches, the set-points of which are then maintained until being disrupted stochastically or perturbed directly.

Since chromatin structures are at least transiently disrupted during replication and need to be duplicated, inheritance of silencing can be visualized as the re-establishment of the repressed state in every cell cycle. Mutations in Sir1p and the silencer affect inheritance. In *sir1* mutant cells, around 90% of repressed mother cells give rise to repressed daughter cells (34). Unfortunately the mechanism by which the silent state is stably inherited is not understood. However, the silencers are critical for the inheritance of the repressed state, since it has been shown that deletion of the silencer from a silenced chromatin domain in cells progressing through the cell cycle leads to complete derepression of the silent state within a single generation (74). These results suggest that in the absence of the silencer, the repressed state cannot be inherited. Cells with a defective silencer (such as *sir1* mutants or *rap1* mutants) can still faithfully propagate the repressed state to their progeny, but the inheritance in every cell cycle is more prone to failure.

There appear to be two subcompartments in the nucleus that contain increased concentrations of silencing factors. Telomeres are clustered at a limited number of foci in nuclei; these foci contain Rap1p and the Sir proteins Sir2p, Sir3p, and Sir4p. The nucleolus is the second subcompartment and appears enriched for Sir2p, but not the other Sir proteins. Many of these silenced compartments appear to be near the nuclear periphery, a location common for heterochromatin in many organisms (see Chapter 14).

The observed correlation between repression and telomeric foci has led to models suggesting that the probability of forming the silenced state may be increased by the interactions of telomeres with each other and with the nuclear envelope. The Sir-dependent repression mediated by the *HM* silencers is also more efficient when silencers are located near telomeres (82). The unequal distribution of Sir proteins in the nucleus is most likely responsible for the inability of silencers to repress efficiently at internal loci. This model is consistent with the observation that overexpression of *SIR3* and *SIR4* improved repression at internal sites (83, 84). Artificial tethering of Sir3p and Sir4p near a reporter also repressed the reporter at an internal site in Rap1p mutants that cannot silence telomeres (85). These mutants presumably cannot recruit Sir3p and Sir4p to telomeres, and the mobilization of the Sir proteins from the telomeres presumably allows a critical concentration to be achieved at internal loci, resulting in silencing at these loci. It is therefore believed that the local concentration of silencing factors in the nucleus is a function of the distance from the telomeric ends (86) but is not due to a continuous extension of silencing from the telomeres (82, 84). In addition, results from several laboratories support the idea that silencing at various loci is the end-point of a competition for recruitment of silencing factors between the *HM* loci and telomeres, and between telomeres and the nucleolus (87, 88).

The recruitment of the Sir proteins to the various subcompartments is dependent on specific protein–protein interactions. Rap1p interacts with both Sir3p and Sir4p; deletion of the interacting domain in Rap1p abolishes silencing at telomeres and *HMR* and causes a redistribution of these proteins throughout the nucleus. The formation of telomeric foci appears to be dependent on other silencing factors, such as the histone tails and Hdf1p (*high affinity DNA binding factor*) since mutations in these leads to loss of localization (89). Mutations in nuclear pore complex proteins may disrupt the clustering of telomeres in the nucleus and causes loss of telomeric silencing (2). These effects are probably mediated through interaction between Hdf1p and the nuclear pore. Similarly, the localization of Sir2p in the nucleolus is dependent on Net1p and possibly on other factors that have binding sites in the rDNA repeats (8).

An interesting corollary to the observation that silenced loci are localized to the nuclear periphery is the demonstration that anchoring the locus to the nuclear periphery (1) can silence a defective *HMR* locus. These results suggest that the probability of establishing the silent state is increased when the domain is sequestered in a compartment of the nucleus rich in silencing factors.

4.4 Regulating silenced chromatin by protein modification

As noted above and detailed in Chapters 1, 7, and 8, a link has been found between the transcriptional state of a gene and the modification of the amino-terminal tails of the histones H3 and H4. Nucleosomes in silenced regions appear to possess hypo-acetylated histones. Mutations in histones H3 and H4 that alter the acetylation of lysine residues cause silencing defects. Consonant with these observations, several different acetylase and deacetylase enzymatic activities have been implicated in silencing. Both N-terminal and lysine-specific acetyltransferases affect the silenced state. Deletion of *NAT1* and *ARD1*, which encode a N-terminal acetyltransferase, cause silencing defects primarily at *HML* (90). Similarly, mutations in *SAS2* and *SAS3* have locus-specific silencing defects. These proteins have limited but significant homology to a protein superfamily of acetyltransferases. Surprisingly, mutations in *SAS2* or *SAS3* restore silencing at *HMR* where the silencer is defective, and *SAS2*, but not *SAS3*, is absolutely required for telomeric silencing (91, 92). Likewise, *SAS2*, but not *SAS3*, has a role in *HML* silencing. These results imply that Sas2p contributes to silencing at *HML* and telomeres but is antagonistic to silencing at *HMR*. The opposite phenotypes may reflect differences in the chromatin structure present at *HML* and *HMR*, or differences in the silencers. It is also possible that Sas2p may have two or more targets that have separate roles at *HML* and *HMR*. One of these targets may be ORC or an accessory protein of ORC since *sas2* mutants suppress the temperature sensitivity of *orc2-1* mutant strains (92).

The histone deacetylases *SIN3/RPD3* have also been isolated in genetic screens for mutants that affect silencing in both yeast and *Drosophila*. Deletions in these genes in yeast increase silencing at rDNA and telomeres and restore silencing at a defective *HMR* silencer. It is not clear if these enzymes act directly or indirectly to affect silencing.

Sir2p and its homologues have recently been shown to possess at least two enzymatic activities, further suggesting that silenced chromatin is highly modified. Sir2p is able to utilize NAD to mono-ADP-ribosylate proteins, including itself and histones, under some conditions *in vitro* (93). The significance of this potential ADP-ribosylation is not yet clear, since other data demonstrate that Sir2p and its homologues are NAD-dependent deacetylases (94–96). These results are consistent with previous data demonstrating that overexpression of *SIR2* led to severe growth defects and was accompanied by a general decrease in histone acetylation in the cell (40). It will be critical to understand the nature of these activities and to identify their relevant *in vivo* targets.

In addition to acetylation/deacetylation, ubiquitination might also have a role in silencing. This hypothesis is based on the recent demonstration that Rad6p, an E2 ubiquitin-conjugating enzyme involved in DNA repair and protein degradation, is also required for silencing at telomeres and the *HM* loci (97). It is the ubiquitin conjugating activity of this enzyme, but not the protein degradation activity, that is essential for silencing. In addition, Rad6p ubiquitinates histones H2A and H2B *in vitro* and H2B *in vivo* (98), and it is therefore possible that Rad6p may be a modifier of chromatin structures at the silenced loci. Consistent with a role for ubiquitination in silencing is the observation that Ubp3p (a ubiquitin hydrolase) interacts with Sir4p (47); deletion of this hydrolase increases silencing while a second ubiquitin hydrolase, Dot4p, disrupts silencing when overexpressed (99).

5. Final comments

5.1 Silencing mechanisms and biological regulation

Progress in recent years has emphasized the significance, in budding and fission yeasts, of chromatin structure to gene regulation and chromosomal structure and function. Powerful genetic screens and *tour de force* biochemical analyses have identified structural and catalytic components that are necessary for the establishment, maintenance, and inheritance of both active and inactive transcriptional states. It is likely, though, that chromatin contributions to biological regulation will extend beyond transcription.

Evidence for the generality of chromatin control of nuclear transactions comes from clear links to DNA replication and repair. The fact that yeast silencing mutants are also defective in these processes strengthens these connections. Likewise, evidence is accumulating that genes originally defined for their roles in silencing may contribute to cell-cycle control, limits on replication competence or ageing, and meiosis (6, 8). The discovery of human homologues of silencing factors, including some that may be mutated in cancer, underscores the value of studying basic mechanisms of silencing.

5.2 Discussion

Several interesting points were raised in discussion with our fellow authors. We noted significant progress identifying the genes and mechanisms involved in silencing in both *S. cerevisiae* and *S. pombe*. However, it will be important to determine whether similar mechanisms are responsible for the epigenetic silencing seen in multicellular eukaryotes, including humans. At least some genes and/or mechanisms do appear to be broadly conserved. For example, homologues of the *S. cerevisiae SIR2* gene are found in organisms ranging from bacteria to humans (41), and the recently defined enzymatic activities of *SIR2* may play a role in silencing in these diverse organisms. Homologues of the *Drosophila* HP1 protein are also found in many organisms, including *S. pombe*, but a homologue does not exist in *S. cerevisiae*.

One major difference between the yeasts and multicellular organisms is the presence of transcriptionally relevant DNA methylation in mammals and plants. Silencing in both plants and mammals is often accompanied by DNA methylation, whereas this modification is not present in either yeast (nor in flies). Although it is clear that methylation is important to the development of these organisms (see Chapter 12), perhaps the function of this DNA modification is supplied in another fashion in the yeasts. For example, perhaps the silenced state is established similarly in all organisms, but the maintenance of the altered chromatin structure requires methylation in plants and mammals but requires only DNA–protein interactions in the yeasts. Further exploration of parallels and differences in chromatin-based gene regulation is likely to sharpen understanding of the most significant elements for its control.

Acknowledgements

We thank our colleagues for their critical and constructive review of this chapter. LFC was partially supported by a predoctoral fellowship from the Howard Hughes Medical Institute. Research in the laboratories of RK and LP is supported by the National Institutes of Health.

References

1. Andrulis, E. D., Neiman, A. M., Zappulla, D. C., and Sternglanz, R. (1998) Perinuclear localization of chromatin facilitates transcriptional silencing. *Nature*, **394**, 592.
2. Galy, V., Olivo-Marin, J. C., Scherthan, H., Doye, V., Rascalou, N., and Nehrbass, U. (2000) Nuclear pore complexes in the organization of silent telomeric chromatin. *Nature*, **403**, 108.
3. Forsburg, S. L. (1999) The best yeast? *Trends Genet.*, **15**, 340.
4. Loo, S. and Rine, J. (1995) Silencing and heritable domains of gene expression. *Annu. Rev. Cell. Dev. Biol.*, **11**, 519.
5. Lowell, J. E. and Pillus, L. (1998) Telomere tales: chromatin, telomerase and telomere function in *Saccharomyces cerevisiae*. *Cell. Mol. Life Sci.*, **54**, 32.
6. Guarente, L. (1999) Diverse and dynamic functions of the Sir silencing complex. *Nature Genet.*, **23**, 281.
7. Sherman, J. and Pillus, L. (1997) An uncertain silence. *Trends Genet.*, **13**, 308.
8. Garcia, S. N. and Pillus, L. (1999) Net results of nucleolar dynamics. *Cell*, **97**, 825.
9. Thon, G. and Klar, A. J. (1992) The *clr1* locus regulates the expression of the cryptic mating-type loci of fission yeast. *Genetics*, **131**, 287.
10. Thon, G., Cohen, A., and Klar, A. J. (1994) Three additional linkage groups that repress transcription and meiotic recombination in the mating-type region of *Schizosaccharomyces pombe*. *Genetics*, **138**, 29.
11. Nimmo, E. R., Cranston, G., and Allshire, R. C. (1994) Telomere-associated chromosome breakage in fission yeast results in variegated expression of adjacent genes. *EMBO J.*, **13**, 3801.
12. Allshire, R. C. (1995) Elements of chromosome structure and function in fission yeast. *Semin. Cell Biol.*, **6**, 55.
13. Allshire, R. C., Javerzat, J. P., Redhead, N. J., and Cranston, G. (1994) Position effect variegation at fission yeast centromeres. *Cell*, **76**, 157.
14. Allshire, R. C., Nimmo, E. R., Ekwall, K., Javerzat, J. P., and Cranston, G. (1995) Mutations derepressing silent centromeric domains in fission yeast disrupt chromosome segregation. *Genes Develop.*, **9**, 218.
15. Chen, X. J. and Clark-Walker, G. D. (1994) *sir2* mutants of *Kluyveromyces lactis* are hypersensitive to DNA- targeting drugs. *Mol. Cell. Biol.*, **14**, 4501.
16. Astrom, S. U. and Rine, J. (1998) Theme and variation among silencing proteins in *Saccharomyces cerevisiae* and *Kluyveromyces lactis*. *Genetics*, **148**, 1021.
17. Perez-Martin, J., Uria, J. A., and Johnson, A. D. (1999) Phenotypic switching in *Candida albicans* is controlled by a SIR2 gene. *EMBO J.*, **18**, 2580.
18. Hull, C. M. and Johnson, A. D. (1999) Identification of a mating type-like locus in the asexual pathogenic yeast *Candida albicans*. *Science*, **285**, 1271.
19. Brand, A. H., Micklem, G., and Nasmyth, K. (1987) A yeast silencer contains sequences that can promote autonomous plasmid replication and transcriptional activation. *Cell*, **51**, 709.
20. Kimmerly, W., Buchman, A., Kornberg, R., and Rine, J. (1988) Roles of two DNA-binding factors in replication, segregation and transcriptional repression mediated by a yeast silencer. *EMBO J.*, **7**, 2241.
21. McNally, F. J. and Rine, J. (1991) A synthetic silencer mediates SIR-dependent functions in *Saccharomyces cerevisiae*. *Mol. Cell. Biol.*, **11**, 5648.
22. Ekwall, K., Nielsen, O., and Ruusala, T. (1991) Repression of a mating type cassette in the fission yeast by four DNA elements. *Yeast*, **7**, 745.

23. Grewal, S. I. and Klar, A. J. (1997) A recombinationally repressed region between mat2 and mat3 loci shares homology to centromeric repeats and regulates directionality of mating-type switching in fission yeast. *Genetics*, **146**, 1221.

24. Karpen, G. H. and Allshire, R. C. (1997) The case for epigenetic effects on centromere identity and function. *Trends Genet.*, **13**, 489.

25. Takahashi, K., Murakami, S., Chikashige, Y., Funabiki, H., Niwa, O., and Yanagida, M. (1992) A low copy number central sequence with strict symmetry and unusual chromatin structure in fission yeast centromere. *Mol. Biol. Cell*, **3**, 819.

26. Cooper, J. P., Nimmo, E. R., Allshire, R. C., and Cech, T. R. (1997) Regulation of telomere length and function by a Myb-domain protein in fission yeast. *Nature*, **385**, 744.

27. Shore, D. (1994) RAP1: a protean regulator in yeast. *Trends Genet.*, **10**, 408.

28. Loo, S., Laurenson, P., Foss, M., Dillin, A., and Rine, J. (1995) Roles of ABF1, NPL3, and YCL54 in silencing in *Saccharomyces cerevisiae*. *Genetics*, **141**, 889.

29. Reed, S. H., Akiyama, M., Stillman, B., and Friedberg, E. C. (1999) Yeast autonomously replicating sequence binding factor is involved in nucleotide excision repair. *Genes Develop.*, **13**, 3052.

30. Dillin, A. and Rine, J. (1997) Separable functions of ORC5 in replication initiation and silencing in *Saccharomyces cerevisiae*. *Genetics*, **147**, 1053.

31. Stone, E. M. and Pillus, L. (1998) Silent chromatin in yeast: an orchestrated medley featuring Sir3p. *Bioessays*, **20**, 30.

32. Moon, K.-Y., Kong, D., Lee, J.-K., Raychaudhuri, S., and Hurwitz, J. (1999) Identification and reconstitution of the origin recognition complex from *Schizosaccharomyces pombe*. *Proc. Natl Acad. Sci., USA*, **96**, 12367.

33. Laurenson, P. and Rine, J. (1992) Silencers, silencing, and heritable transcriptional states. *Microbiol. Rev.*, **56**, 543.

34. Pillus, L. and Rine, J. (1989) Epigenetic inheritance of transcriptional states in *S. cerevisiae*. *Cell*, **59**, 637.

35. Aparicio, O. M., Billington, B. L., and Gottschling, D. E. (1991) Modifiers of position effect are shared between telomeric and silent mating-type loci in *S. cerevisiae*. *Cell*, **66**, 1279.

36. Fourel, G., Revardel, E., Koering, C. E., and Gilson, E. (1999) Cohabitation of insulators and silencing elements in yeast subtelomeric regions. *EMBO J.*, **18**, 2522.

37. Pryde, F. E. and Louis, E. J. (1999) Limitations of silencing at native yeast telomeres. *EMBO J.*, **18**, 2538.

38. Triolo, T. and Sternglanz, R. (1996) Role of interactions between the origin recognition complex and SIR1 in transcriptional silencing. *Nature*, **381**, 251.

39. Gardner, K. A., Rine, J., and Fox, C. A. (1999) A region of the Sir1 protein dedicated to recognition of a silencer and required for interaction with the Orc1 protein in *Saccharomyces cerevisiae*. *Genetics*, **151**, 31.

40. Braunstein, M., Rose, A. B., Holmes, S. G., Allis, C. D., and Broach, J. R. (1993) Transcriptional silencing in yeast is associated with reduced nucleosome acetylation. *Genes Develop.*, **7**, 592.

41. Brachmann, C. B., Sherman, J. M., Devine, S. E., Cameron, E. E., Pillus, L., and Boeke, J. D. (1995) The SIR2 gene family, conserved from bacteria to humans, functions in silencing, cell cycle progression, and chromosome stability. *Genes Develop.*, **9**, 2888.

42. Freeman-Cook, L. L., Sherman, J. M., Brachmann, C. B., Allshire, R. C., Boeke, J. D., and Pillus, L. (1999) The *Schizosaccharomyces pombe* hst4$^+$ gene is a SIR2 homologue with silencing and centromeric functions. *Mol. Biol. Cell*, **10**, 3171.

43. Sherman, J. M., Stone, E. M., Freeman-Cook, L. L., Brachmann, C. B., Boeke, J. D., and

Pillus, L. (1999) The conserved core of a human SIR2 homologue functions in yeast silencing. *Mol. Biol. Cell*, **10**, 3045.

44. Stone, E. M. and Pillus, L. (1998) Silent chromatin in yeast: an orchestrated medley featuring Sir3p. *BioEssays*, **20**, 30.

45. Palladino, F., Laroche, T., Gilson, E., Axelrod, A., Pillus, L., and Gasser, S. M. (1993) SIR3 and SIR4 proteins are required for the positioning and integrity of yeast telomeres. *Cell*, **75**, 543.

46. Hecht, A., Laroche, T., Strahl-Bolsinger, S., Gasser, S. M., and Grunstein, M. (1995) Histone H3 and H4 N-termini interact with SIR3 and SIR4 proteins: a molecular model for the formation of heterochromatin in yeast. *Cell*, **80**, 583.

47. Moazed, D. and Johnson, D. (1996) A deubiquitinating enzyme interacts with SIR4 and regulates silencing in *S. cerevisiae*. *Cell*, **86**, 667.

48. Smith, J. S. and Boeke, J. D. (1997) An unusual form of transcriptional silencing in yeast ribosomal DNA. *Genes Develop.*, **11**, 241.

49. Ivy, J. M., Klar, A. J., and Hicks, J. B. (1986) Cloning and characterization of four SIR genes of *Saccharomyces cerevisiae*. *Mol. Cell. Biol.*, **6**, 688.

50. Marshall, M., Mahoney, D., Rose, A., Hicks, J. B., and Broach, J. R. (1987) Functional domains of SIR4, a gene required for position effect regulation in *Saccharomyces cerevisiae*. *Mol. Cell. Biol.*, **7**, 4441.

51. Kaeberlein, M., McVey, M., and Guarente, L. (1999) The SIR2/3/4 complex and SIR2 alone promote longevity in *Saccharomyces cerevisiae* by two different mechanisms. *Genes Develop.*, **13**, 2570.

52. Ekwall, K. and Ruusala, T. (1994) Mutations in *rik1*, *clr2*, *clr3* and *clr4* genes asymmetrically derepress the silent mating-type loci in fission yeast. *Genetics*, **136**, 53.

53. Grewal, S. I., Bonaduce, M. J., and Klar, A. J. (1998) Histone deacetylase homologs regulate epigenetic inheritance of transcriptional silencing and chromosome segregation in fission yeast. *Genetics*, **150**, 563.

54. Egel, R., Willer, M., and Nielsen, O. (1989) Unblocking of meiotic crossing-over between the silent mating-type cassettes of fission yeast, conditioned by the recessive, pleiotropic mutant *rik1*. *Curr. Genet.*, **15**, 407.

55. Ekwall, K., Javerzat, J. P., Lorentz, A., Schmidt, H., Cranston, G., and Allshire, R. (1995) The chromodomain protein Swi6: a key component at fission yeast centromeres. *Science*, **269**, 1429.

56. Ekwall, K., Nimmo, E. R., Javerzat, J. P., Borgstrom, B., Egel, R., Cranston, G., and Allshire, R. (1996) Mutations in the fission yeast silencing factors *clr4+* and *rik1+* disrupt the localisation of the chromo domain protein Swi6p and impair centromere function. *J. Cell Sci.*, **109**, 2637.

57. Ivanova, A. V., Bonaduce, M. J., Ivanov, S. V., and Klar, A. J. (1998) The chromo and SET domains of the Clr4 protein are essential for silencing in fission yeast. *Nature Genet.*, **19**, 192.

58. Nislow, C., Ray, E., and Pillus, L. (1997) SET1, a yeast member of the trithorax family, functions in transcriptional silencing and diverse cellular processes. *Mol. Biol. Cell*, **8**, 2421.

59. Aagaard, L., Laible, G., Selenko, P., Dorn, R., Schotta, G., Kuhfittig, S., Wolf, A., Lebersorger, A., Singh, P. B., Reuter, G., *et al.* (1999) Functional mammalian homologues of the *Drosophila* PEV-modifier Su(var)3–9 encode centromere-associated proteins which complex with the heterochromatin component M31. *EMBO J.*, **18**, 1923.

60. Lorentz, A., Ostermann, K., Fleck, O., and Schmidt, H. (1994) Switching gene *swi6*, involved in repression of silent mating-type loci in fission yeast, encodes a homologue of chromatin-associated proteins from *Drosophila* and mammals. *Gene*, **143**, 139.

61. Weiler, K. S. and Wakimoto, B. T. (1995) Heterochromatin and gene expression in *Drosophila*. *Annu. Rev. Genet.*, **29**, 577.

62. Kellum, R. and Alberts, B. M. (1995) Heterochromatin protein 1 is required for correct chromosome segregation in *Drosophila* embryos. *J. Cell Sci.*, **108**, 1419.

63. Loo, S. and Rine, J. (1994) Silencers and domains of generalized repression. *Science*, **264**, 1768.

64. Singh, J. and Klar, A. J. (1992) Active genes in budding yeast display enhanced *in vivo* accessibility to foreign DNA methylases: a novel *in vivo* probe for chromatin structure of yeast. *Genes Develop.*, **6**, 186.

65. Donze, D., Adams, C. R., Rine, J., and Kamakaka, R. T. (1999) The boundaries of the silenced HMR domain in *Saccharomyces cerevisiae*. *Genes Develop.*, **13**, 698.

66. Ravindra, A., Weiss, K., and Simpson, R. T. (1999) High-resolution structural analysis of chromatin at specific loci: *Saccharomyces cerevisiae* silent mating-type locus HMRa. *Mol. Cell. Biol.*, **19**, 7944.

67. Reimer, S. K. and Buchman, A. R. (1997) Yeast silencers create domains of nuclease-resistant chromatin in an SIR4-dependent manner. *Chromosoma*, **106**, 136.

68. Weiss, K. and Simpson, R. T. (1998) High-resolution structural analysis of chromatin at specific loci: *Saccharomyces cerevisiae* silent mating type locus HMLalpha. *Mol. Cell. Biol.*, **18**, 5392.

69. Sekinger, E. A. and Gross, D. S. (1999) SIR repression of a yeast heat shock gene: UAS and TATA footprints persist within heterochromatin. *EMBO J.*, **18**, 7041.

70. Wright, J. H., Gottschling, D. E., and Zakian, V. A. (1992) *Saccharomyces* telomeres assume a non-nucleosomal chromatin structure. *Genes Develop.*, **6**, 197.

71. Vega-Palas, M. A., Venditti, S., and Di Mauro, E. (1997) Telomeric transcriptional silencing in a natural context. *Nature Genet.*, **15**, 232.

72. Venditti, S., Vega-Palas, M. A., and Di Mauro, E. (1999) Heterochromatin organization of a natural yeast telomere. Recruitment of Sir3p through interaction with histone H4 N terminus is required for the establishment of repressive structures. *J. Biol. Chem.*, **274**, 1928.

73. Bi, X. and Broach, J. R. (1999) UASrpg can function as a heterochromatin boundary element in yeast. *Genes Develop.*, **13**, 1089.

74. Holmes, S. G. and Broach, J. R. (1996) Silencers are required for inheritance of the repressed state in yeast. *Genes Develop.*, **10**, 1021.

75. Cheng, T. H. and Gartenberg, M. R. (2000) Yeast heterochromatin is a dynamic structure that requires silencers continuously. *Genes Develop.*, **14**, 452.

76. Fox, C. A., Loo, S., Dillin, A., and Rine, J. (1995) The origin recognition complex has essential functions in transcriptional silencing and chromosomal replication. *Genes Develop.*, **9**, 911.

77. Miller, A. M. and Nasmyth, K. A. (1984) Role of DNA replication in the repression of silent mating type loci in yeast. *Nature*, **312**, 247.

78. Le, S., Davis, C., Konopka, J. B., and Sternglanz, R. (1997) Two new S-phase-specific genes from *Saccharomyces cerevisiae*. *Yeast*, **13**, 1029.

79. Tyler, J. K., Adams, C. R., Chen, S. R., Kobayashi, R., Kamakaka, R. T., and Kadonaga, J. T. (1999) The RCAF complex mediates chromatin assembly during DNA replication and repair. *Nature*, **402**, 555.

80. Mahoney, D. J., Marquardt, R., Shei, G. J., Rose, A. B., and Broach, J. R. (1991) Mutations in the HML E silencer of *Saccharomyces cerevisiae* yield metastable inheritance of transcriptional repression. *Genes Develop.*, **5**, 605.

81. Sussel, L., Vannier, D., and Shore, D. (1993) Epigenetic switching of transcriptional states: *cis*- and *trans*-acting factors affecting establishment of silencing at the HMR locus in *Saccharomyces cerevisiae*. *Mol. Cell. Biol.*, **13**, 3919.

82. Thompson, J. S., Johnson, L. M., and Grunstein, M. (1994) Specific repression of the yeast silent mating locus HMR by an adjacent telomere. *Mol. Cell. Biol.*, **14**, 446.

83. Wiley, E. A. and Zakian, V. A. (1995) Extra telomeres, but not internal tracts of telomeric DNA, reduce transcriptional repression at *Saccharomyces* telomeres. *Genetics*, **139**, 67.

84. Maillet, L., Boscheron, C., Gotta, M., Marcand, S., Gilson, E., and Gasser, S. M. (1996) Evidence for silencing compartments within the yeast nucleus: a role for telomere proximity and Sir protein concentration in silencer-mediated repression. *Genes Develop.*, **10**, 1796.

85. Lustig, A. J., Liu, C., Zhang, C., and Hanish, J. P. (1996) Tethered Sir3p nucleates silencing at telomeres and internal loci in *Saccharomyces cerevisiae*. *Mol. Cell. Biol.*, **16**, 2483.

86. Marcand, S., Buck, S. W., Moretti, P., Gilson, E., and Shore, D. (1996) Silencing of genes at nontelomeric sites in yeast is controlled by sequestration of silencing factors at telomeres by Rap1 protein. *Genes Develop.*, **10**, 1297.

87. Buck, S. W. and Shore, D. (1995) Action of a RAP1 carboxy-terminal silencing domain reveals an underlying competition between HMR and telomeres in yeast. *Genes Develop.*, **9**, 370.

88. Smith, J. S., Brachmann, C. B., Pillus, L., and Boeke, J. D. (1998) Distribution of a limited Sir2 protein pool regulates the strength of yeast rDNA silencing and is modulated by Sir4p. *Genetics*, **149**, 1205.

89. Laroche, T., Martin, S. G., Gotta, M., Gorham, H. C., Pryde, F. E., Louis, E. J., and Gasser, S. M. (1998) Mutation of yeast *Ku* genes disrupts the subnuclear organization of telomeres. *Curr. Biol.*, **8**, 653.

90. Mullen, J. R., Kayne, P. S., Moerschell, R. P., Tsunasawa, S., Gribskov, M., Colavito-Shepanski, M., Grunstein, M., Sherman, F., and Sternglanz, R. (1989) Identification and characterization of genes and mutants for an N-terminal acetyltransferase from yeast. *EMBO J.*, **8**, 2067.

91. Reifsnyder, C., Lowell, J., Clarke, A., and Pillus, L. (1996) Yeast SAS silencing genes and human genes associated with AML and HIV-1 Tat interactions are homologous with acetyltransferases. *Nature Genet.*, **14**, 42.

92. Ehrenhofer-Murray, A. E., Rivier, D. H., and Rine, J. (1997) The role of Sas2, an acetyltransferase homologue of *Saccharomyces cerevisiae*, in silencing and ORC function. *Genetics*, **145**, 923.

93. Tanny, J. C., Dowd, G. J., Huang, J., Hilz, M., and Moazed, D. (1999) An enzymatic activity in the yeast Sir2 protein that is essential for gene silencing. *Cell*, **99**, 735.

94. Imai, S., Armstrong, C. M., Kaeberlein, M., and Guarente, L. (2000) Transcriptional silencing and longevity protein Sir2 is an NAD-dependent histone deacetylase. *Nature*, **403**, 795.

95. Landry, J., Sutton, A., Tafrov, S. T., Heller, R. C., Stebbins, J., Pillus, L., and Sternglanz, R. (2000) The silencing protein SIR2 and its homologs are NAD-dependent protein deacetylases. *Proc. Natl Acad. Sci., USA*, **97**, 5807.

96. Smith, J. S., Brachmann, C. B., Celic, I., Kenna, M. A., Muhammad, S., Starai, V. J., Avalos, J., Escalante-Semerena, J. C., Grubmeyer, C., Wolberger, C. *et al.* (2000) A phylogenetically conserved NAD^+-dependent protein deacetylase activity in the Sir2 protein family. *Proc. Natl Acad. Sci., USA*, **97**, 6658.

97. Huang, H., Kahana, A., Gottschling, D. E., Prakash, L., and Liebman, S. W. (1997) The ubiquitin-conjugating enzyme Rad6 (Ubc2) is required for silencing in *Saccharomyces cerevisiae*. *Mol. Cell. Biol.*, **17**, 6693.

98. Robzyk, K., Recht, J., and Osley, M. A. (2000) Rad6-dependent ubiquitination of histone H2B in yeast. *Science*, **287**, 501.
99. Singer, M. S., Kahana, A., Wolf, A. J., Meisinger, L. L., Peterson, S. E., Goggin, C., Mahowald, M., and Gottschling, D. E. (1998) Identification of high-copy disruptors of telomeric silencing in *Saccharomyces cerevisiae*. *Genetics*, **150**, 613.

11 | Epigenetic regulation in *Drosophila*: unravelling the conspiracy of silence

JOEL C. EISSENBERG, SARAH C. R. ELGIN, and RENATO PARO

1. Introduction: chromatin structure and gene silencing

Understanding the mechanisms of epigenetic regulation, by which cell-specific gene silencing is established and stably maintained, is a major unresolved challenge. Gene expression in eukaryotes relies upon the accessibility of the DNA template to RNA polymerase and a variety of accessory proteins. It is clear that the accessibility of regulatory elements can be controlled by the chromatin structure at a given locus, and that remodelling of chromatin structure is, in many cases, a prerequisite for transcriptional activation (Chapters 5, 6). In many instances, however, a gene can remain silent and resistant to remodelling. Here we will discuss two mechanisms responsible for stable gene silencing in *Drosophila*, heterochromatic position effect variegation, and homeotic gene repression by the *Polycomb* group complex.

The examples of epigenetic regulation to be discussed here involve the silencing of large chromosomal regions, arguing for a mechanism beyond single nucleosome-based promoter extinction. A description of the distinct composition and organization of heterochromatin is now beginning to emerge; heterochromatic regions do appear to have a highly regular nucleosome array. While silencing might be accomplished in this way, other types of packaging are likely to be involved as well. In *Drosophila*, the inference that transcriptional silencing is associated with higher-level chromatin compaction has rested primarily on the distinctive cytological appearance of heterochromatin. How (or whether) the features of the nucleosome array contribute to folding into a higher-order structure remains to be seen.

In plants and vertebrates, transcriptional repression is frequently associated with hypermethylation of the DNA of the repressed gene and/or associated regulatory sequences (Chapter 12). Since there is little or no detectable DNA methylation in *Drosophila* (1), the mitotically stable repression of genes is presumably the consequence of DNA–protein interactions alone. The ways in which chromosomal proteins conspire to silence genes—the membership of this conspiracy, and their hierarchy of interactions—are amenable to genetic and biochemical investigation, and are now beginning to be unravelled.

2. Heterochromatic position effect variegation

2.1 Gene silencing associated with chromosomal position

An early recognition of the differential packaging of chromatin in eukaryotic cells was made by Heitz (2), who observed a fraction of the nuclear material that failed to decondense after mitosis, which he termed 'heterochromatin'. Subsequent studies have revealed that heterochromatin is rich in repetitive DNA, relatively gene-poor, late replicating, and minimally transcribed (3, 4). Large blocks of heterochromatin are typically observed flanking the centromeres, and are often found adjacent to telomeres. Genetic and cytological evidence suggests that heterochromatin formation interferes with expression of normally euchromatic genes; for example, all but one X chromosome in mammalian cells appear to be heterochromatic, and are generally transcriptionally inactive (Chapter 12). Genes relocated to heterochromatin through rearrangement or transposition experience an abnormal, position-dependent mosaic silencing, termed 'position effect variegation' (PEV), which can serve as a model for mitotically stable gene repression (5).

PEV was initially observed as a result of chromosomal rearrangements (inversions, translocations), generally a product of X-irradiation, which place a euchromatic gene close to a breakpoint in heterochromatin. This position-dependent inactivation suggested a *cis*-acting silencing activity of heterochromatin acting across the rearrangement breakpoint. The acquisition of heterochromatic structure by normally euchromatic material at such breakpoints is visibly manifest in the giant polytene chromosomes of *Drosophila* larval salivary glands. In these nuclei, the euchromatic DNA is amplified a thousandfold; the chromatin strands remain tightly synapsed to give a highly ordered, stereotypical, banded appearance. In contrast, the pericentric heterochromatin in these nuclei is under-represented, densely staining, and disordered in appearance; the heterochromatin surrounding the centromeres of all of the chromosomes aggregates to form a single, cytologically distinct, compact focus (the 'chromocentre') from which the banded euchromatic arms appear to radiate (Fig. 1a). In strains of flies carrying variegating chromosome rearrangements, the regular banded appearance of the euchromatin in the vicinity of the variegating breakpoint is lost in a subset of nuclei, and the region takes on the cytological appearance of heterochromatin (6–8). Thus, the mosaic nature of the silencing shown by variegating loci has a cytological correlate in differential chromatin organization. The decision as to whether the rearranged gene is to be on or off appears to be random, and results in patches of cells in which the gene is expressed and other patches in which the gene is inactive. Generally an adult phenotype is scored, with clones of 'on' cells ranging from one to several hundred cells in size. A similar phenotype is observed in cases where a transgene is inserted into the pericentric heterochromatin,

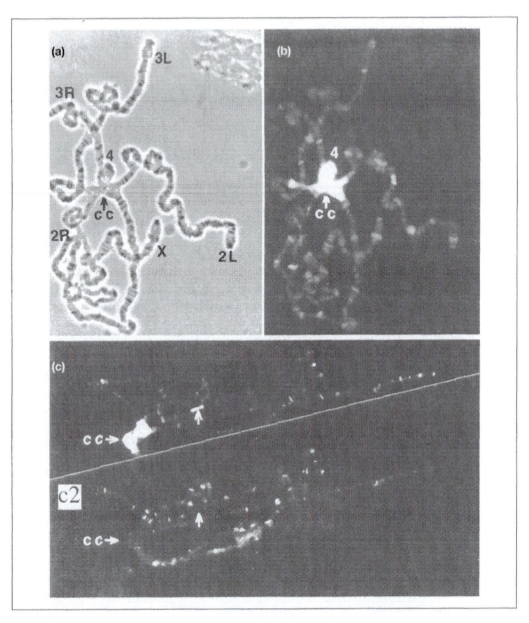

Fig. 1 The polytene chromosomes of *Drosophila melanogaster.* (a) phase contrast; (b) immunofluorescence staining to localize heterochromatin protein 1 (HP1). The chromocentre (cc) and fourth chromosome are intensely stained. (c) Double-label immunofluorescence staining of salivary gland polytene chromosomes, showing the different distributions of the POLYCOMB and HP1 proteins: (upper panel) staining with monoclonal mouse anti-HP1 antibodies and fluorescein-conjugated anti-mouse antibody; (lower panel) staining with polyclonal rabbit anti-PC antiserum and rhodamine-conjugated anti-rabbit antibody. The arrow points to region 31, a euchromatic site consistently associated with HP1. Panels (a) and (b) provided by C. Craig; panels (c) provided by R. F. Clark.

at a subset of sites within the banded region of the small fourth chromosome, or in telomeres (9).

The silencing activity of heterochromatin is dynamic during *Drosophila* development. Using a ubiquitously inducible *lacZ* reporter transgene, Lu *et al.* (10) found that heterochromatin-induced silencing was undetectable during embryogenesis until the onset of gastrulation, about an hour after heterochromatin can first be observed by cytological techniques. Many (if not most) of the chromosomal proteins are maternally provided to the oocyte; what controls the delay in heterochromatin formation and the acquisition of silencing is unknown. Silencing is relatively stable in mitotically active cells during larval development, but is dramatically relaxed with the onset of metamorphosis, leading to expression of the gene in a subset of adult cells. Thus, the silencing activity of heterochromatin appears to be regulated during development, as is the silencing activity of the *Polycomb* group proteins (see below). In several instances, the DNA sequences of loci subject to variegation appear to be under-represented in polytene tissue (11, 12). Such under-representation is a consequence of a failure to complete replication of heterochromatic DNA during polytenization (11). However, copy-number effects cannot account for most known examples of variegation (6, 13, 14).

In *Drosophila*, virtually every locus that has been examined in an appropriate rearrangement has been found to variegate, and rearrangements involving the pericentric heterochromatin of any chromosome can lead to PEV. In the cases studied, variegation has been associated with reduced transcript accumulation for the affected gene (6, 13). The molecular basis for gene inactivation by heterochromatin must involve a fundamental aspect of gene activity common to genes which otherwise differ in their temporal and spatial regulation. One can imagine that the densely compacted quality of heterochromatin might be sufficient to occlude sites of DNA–protein interaction necessary for gene activation, e.g. preventing access for remodelling complexes and/or for the assembly of the transcription initiation complex in promoter regions. Alternatively, association with heterochromatin, which is generally observed clumped in a single mass at the nuclear periphery, may place the gene in a 'compartment' inaccessible to transcription factors. In a study of a variegating transgene present in the small fourth chromosome, an X-ray induced reciprocal translocation between the fourth chromosome banded region and the end of the second chromosome, which results in a change in nuclear location of the transgene, resulted in a loss of silencing (15). This and other observations suggest that proximity to the pericentric heterochromatin is a factor in determining the effectiveness of the silencing mechanism, as suggested by a 'compartment' model (4).

Although heterochromatin appears to be an inhospitable environment for the expression of euchromatic genes, there are a number of loci that behave as classical Mendelian genes which map to heterochromatic regions of the genome (3). Interestingly, several of these genes have been found to variegate in rearrangements placing them next to a *euchromatic* breakpoint, suggesting that certain genes may not only tolerate a heterochromatic environment, but have come to depend on this environment for proper expression (16). The degree of variegation experienced by a heterochromatic locus depends upon how far out into the euchromatic chromosome arm the gene is displaced, with more distal breakpoints giving more extreme variegation (17, 18). This suggests that proximity to the pericentric heterochromatin again has an effect; stable pairing of the variegating region with the pericentric heterochromatin has been suggested to play a role.

The importance of heterochromatin association and chromosome pairing is further underscored by the phenomenon of *trans*-inactivation, illustrated by dominant *brown* (*bw*) variegation. Inactivation of the gene by mutation leads to flies that have brown, rather than red, eyes. Such mutations are generally recessive, but the *brown*^{Dominant} (*bw*^D) allele is an exception. This allele induces a mosaic inactivation of a wild-type *brown* allele located at its normal position on the homologous chromosome, but has no effect on the expression of *bw*⁺ alleles placed at other euchromatic sites (19, 20). *bw*^D is associated with insertion of a *c.* 1.5 Mb fragment of largely uninterrupted AAGAG repeats within the *brown* gene (21). *Trans*-dominant inactivation appears to depend on specific sequences near the *brown* gene that may normally serve to promote pairing. *bw*^D variegation is suppressed by chromosome rearrangements that move the *bw*^D allele further away from the pericentric heterochromatin, while it is enhanced in rearrangements that move it closer (21). Association of *bw*^D with the pericentric heterochromatin appears to 'drag' its paired homologue into heterochromatic regions in the nucleus, resulting in silencing (22–24).

2.2 A mass-action assembly model

How does heterochromatin exert a repressing effect on genes located many kilobases away? Genetic and cytological evidence (see below) indicates that heterochromatin is composed of a protein–DNA complex involving both histones and non-histone chromosomal proteins. The initial site of complex deposition may be specified by a DNA sequence or structure, from which the complex is proposed to spread in a cooperative and sequence-independent fashion until some boundary or limit is encountered (25). In variegating chromosome rearrangements, gene inactivation is thought to occur when heterochromatin spreads across the heterochromatin–euchromatin junction and extends the heterochromatin domain (Fig. 2). This model implies the presence of a discrete boundary between heterochromatic and euchromatic domains in the normal state. While such a boundary has not yet been identified in *Drosophila*, the presence of interspersed heterochromatic and euchromatic domains along the small fourth chromosome, observed by monitoring expression of a transgene inserted at different sites, implies that such a boundary may be identified (26). (See Chapter 13 for a discussion of chromatin boundaries.)

If heterochromatin-mediated gene silencing reflects the formation and stability of a protein–DNA complex, the levels of the proteins involved should influence the probability that a variegating gene will be silenced (27). Chance differences in protein concentration between cells may account for the mosaic nature of PEV. Mutations reducing the amount of any structural heterochromatin protein would reduce the extent of heterochromatin assembly, and thus the probability of silencing.

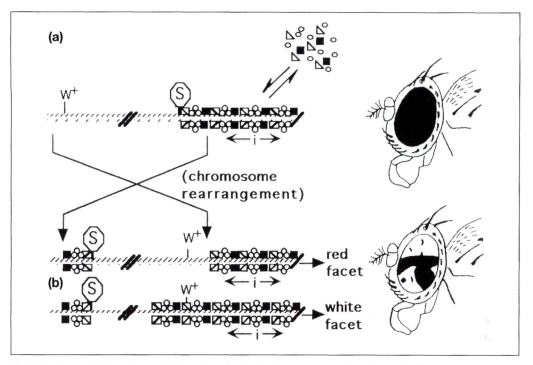

Fig. 2 A schematic illustration of *white* variegation in the X chromosome inversion *In(1)w^{m4}*. The *white* locus (*w*) is located in the distal euchromatin (thin line) of the wild-type X chromosome. It provides a function essential to the normal red pigmentation of the *Drosophila* eye. The inverted X chromosome, *In(1)w^{m4}*, was the result of chromosomal breaks which occurred adjacent to the *white* locus and within the pericentric heterochromatin of the X chromosome (decorated region). Thus, the *white* locus has come to lie *c.* 25 kb away from the heterochromatic breakpoint in this inversion. This abnormal juxtaposition gives rise to flies with mottled or variegated, compound eyes composed of red (*white* gene is active) and white (*white* gene is inactive) eye facts. (a) In the normal chromosome, heterochromatin-specific proteins (represented by geometric symbols) assemble at initiation sites 'i' and cooperatively propagate a condensed heterochromatic structure in *cis* until a terminator or stop (S) is encountered. (b) A chromosome rearrangement that has a breakpoint between the 'i' and 'S' signals permits the spread of heterochromatin into a normally euchromatic region of the chromosome. When the extent of spreading in a given cell includes the gene *w^+*, the gene is inactivated; when the spreading terminates before reaching gene *w^+*, the gene remains active in that cell.

Conversely, mutations reducing the amount of euchromatic proteins required to maintain gene activity would enhance the probability of heterochromatin assembly (Fig. 2). Genetically, the former class of mutations would behave as 'suppressors of variegation', while the latter would be 'enhancers of variegation'. Mutations that alter function or stability of the structure as a whole could also appear as modifiers of PEV. Genes that normally reside in heterochromatin, dependent on that structure for normal function, might be expected to show converse effects in response to modifiers of PEV. The mass action assembly model implies that genes encoding heterochromatin structural proteins will behave, when duplicated, as enhancers of PEV and, when deleted, as suppressors of PEV. Similarly, a structural component of euchromatin that antagonizes heterochromatin spreading would behave, when duplicated, as a suppressor of PEV and, when deleted, as an enhancer of PEV. Loci displaying these properties (termed 'antipodal effects' by Locke (27)) are candidates for genes which encode structural components of heterochromatin and euchromatin; several such loci have been identified.

2.3 Modifiers of heterochromatic silencing

Identification and characterization of modifiers of PEV in *D. melanogaster* has become an important and fruitful strategy in dissecting the composition and properties of heterochromatin. One of the earliest modifiers recognized was the Y chromosome, which is itself heterochromatic in somatic tissue; more Y chromosome material (e.g. a male with the sex chromosome constitution XYY instead of the normal XY) has the effect of suppressing PEV, whereas less Y chromosome material (for example, an XO male) enhances PEV (5). Later investigations showed that X heterochromatin levels also can titrate PEV levels (5). The effect of X and Y dosage on PEV may be understood as a competition for assembly of heterochromatin at the relevant sites, reflecting limiting amounts of structural proteins. A similar competition under conditions of limiting amounts of SIR proteins is observed in *S. cerevisiae*, with a loss of silencing at the weaker heterochromatic sites (Chapter 10).

The histone genes in *Drosophila* are present in a tandem array of *c.*100 copies of a repeating unit, which includes one copy of each of the genes for histones H1, H3, H4, H2A, and H2B (28). Partial deletions of this cluster result in suppression of PEV (29), as does growth in the presence of butyrate or trichostatin A, inhibitors of histone deacetylation (30; J. Ma and J. C. Eissenberg, unpublished). These observations implicate histone proteins themselves in the propagation or stabilization of heterochromatin and suggest that the deacetylated form is required for that purpose (Chapter 8). Consistent with this inference, immunolocalization studies using antibodies specific for histone H4 isoforms acetylated at each of four N-terminal lysine residues showed that the pericentric heterochromatin is relatively poor overall in acetylated H4, although it is reported to be relatively enriched in the H4 isoform(s) acetylated at lysine 12 (31). Furthermore, missense mutations in the gene encoding histone deacetylase, HDAC1, are dominant suppressors of PEV (32). Thus, differential histone acetylation could well be involved in partitioning the genome into heterochromatin and euchromatin. However, a direct role for any histone isoform in heterochromatin-mediated gene repression is yet to be demonstrated. The genetic manipulation of histones is complicated in *Drosophila* (and most other eukaryotes except *S. cerevisiae*) by the multiplicity of histone gene copies.

The genetic dissection of PEV has been pursued intensively through the identification, cloning, and characterization of dominant modifiers of PEV. Genetic screens have identified about 50 modifiers on the two major autosomes of *D. melanogaster* (33). For three of these loci (*Su(var)2–5*, *Su(var)3–7*, and *Su(var)3–9*) comparisons of duplications (or transgenes) and deficiencies overlapping the same interval show the antipodal effects anticipated for chromatin structural proteins. Recently, several of these loci (listed in Table 1) have yielded to molecular cloning, and characterization

Table 1 Summary of cloned loci from Drosophila encoding non-histone modifiers of PEV

Locus	Gene product	Reference
Su(var)2–5	Heterochromatin protein 1 (HP1)	38
Su(var)3–7	Heterochromatin-associated zinc-finger-containing protein	Cited in ref. 95
Su(var)3–9	Chromo and SET domain containing chromosomal protein	Cited in ref. 95
Su(var)3–6	Protein phosphatase type I	Cited in ref. 95
Modulo	DNA-binding protein	Cited in ref. 95
Su(z)5	S-adenosylmethionine synthetase	Cited in ref. 95
E(var)3–93D	Protein with N-terminal homology to transcriptional activators	Cited in ref. 95
E(var)3–64BC	*Drosophila* homologue of the yeast histoned deacetylase RPD3	32
Zeste	DNA-binding protein	Cited in ref. 95
Mus209	Processivity factor for DNA polymerase δ and ε	Cited in ref. 95
D-Ubp-64E^{Evar1}	Ubiquitin-deconjugating enzyme	Cited in ref. 95
Hel	Protein homologous to RNA helicases	Cited in ref. 95
E(var)-trl	GAGA factor	Cited in ref. 95
Regena	Protein homologous to the yeast negative transcriptional regulator CDC36/NOT2	96
E(z)	Chromosonal protein with SET domain, member of the Polycomb group	97
Asx	Chromatin protein	98
E(Pc)	Evolutionary conserved chromatin protein	36
L(3)01544	Ribosomal protein PO/apurinic/apyrimidinic endonuclease	99
E(var)3–93E	Homologue of human E2F transcription factor	Cited in ref. 95

of the gene products has uncovered properties suggestive of structural proteins of chromatin or their modifiers. Some modifiers of PEV (*E(var)3–93D, Asx, E(z),* and *E(var)3-trl*) also exhibit dosage-dependent effects on homeotic gene expression, suggesting that the mechanisms of epigenetic silencing by heterochromatin formation may overlap, at least in some respects, with mechanisms of developmentally regulated gene silencing (see below).

Probably the best-characterized heterochromatin-associated chromosomal protein in *Drosophila* is heterochromatin protein 1 (HP1). Antibodies specific for HP1 localize to the pericentric β-heterochromatin, in a banded pattern on the fourth chromosome, at telomeres, and at a small set of euchromatic sites in indirect immunofluorescent staining of fixed polytene chromosomes (34, 35). The cytological map position of the HP1 gene at 29A coincided with the genetic map position of *Su(var)205*, a dominant suppressor of PEV showing antipodal effects (36) Subsequent molecular studies have identified five mutant alleles of the HP1 gene (now called *Su(var)2–5*), all both dominant suppressors of PEV and recessive lethals (37–40). The *light* and *rolled* genes, which map within heterochromatin, are dependent on normal levels of HP1, as their activity is depressed in *Su(var)2–5* lines (40).

HP1-like proteins are found in fission yeast, protozoa, invertebrates, and vertebrates (41). In the few cases examined, these homologues have been found associated with heterochromatin and have the genetic property of gene silencing. The conserved structural features of all HP1-like proteins are an N-terminal 'chromo domain' (for *ch*romatin *o*rganization *mo*difier) and a C-terminal 'chromo-shadow domain', joined by a variable-length linker peptide (42). In *Drosophila*, these domains function to target heterochromatin, probably through protein–protein interactions with other heterochromatin-associated proteins (39, 43). Both domains have been found in otherwise unrelated proteins; the chromo domain of *Polycomb* also appears to mediate chromatin targeting through protein–protein interactions, playing a critical role in gene silencing in that system (see below, Fig. 4) (44, 45). The physical association of HP1 protein with cytological heterochromatin and the genetic data showing that the gene encoding HP1 is a dosage-dependent modifier of PEV argue strongly for a structural role for HP1 in epigenetic silencing by heterochromatin. This inference is further supported by experiments using HP1 antiserum to stain the polytene chromosomes of a variegating stock: where the variegating locus appears to be part of the condensed β-heterochromatin, it is stained; while in those nuclei in which it is packaged in the banded, polytenized, euchromatic homologue, it is unstained (46).

Interaction with HP1 has been used to implicate other nuclear factors in the mechanism of heterochromatin assembly and gene silencing (41). Reported interactions include binding to the *o*rigin *r*ecognition *c*omplex (ORC) (47) and *c*hromatin *a*ssembly *f*actor 1 (CAF-1) (48). ORC has also been implicated in assembly of heterochromatin at the HMR/HML loci in yeast, acting as a platform to recruit the silencing protein Sir1p (Chapter 10). However, other factors must also be involved in directing heterochromatin assembly, as ORC and CAF-1 play general roles in chromatin replication. Interestingly, mammalian HP1 is reported to interact with the lamin B receptor (49). Such an interaction might anchor heterochromatin to the nuclear lamina; anchoring to the nuclear periphery has been shown to contribute to telomeric silencing in yeast (50). While genetic evidence shows that the chromo domain plays a critical role in silencing, and it has been demonstrated that the chromo domain can direct association with heterochromatin, most of the studies screening for associating proteins have reported interactions involving the chromo-shadow domain. The chromo-shadow domain appears to self-associate in a dimer, creating a binding site recognized by some interacting proteins (51). These patterns of association suggest that HP1 may play an important role in assembling heterochromatin, serving as an *in vivo* bifunctional cross-linking reagent .

2.4 Organization of heterochromatin

What defines *Drosophila* heterochromatin at the DNA sequence level? What signals—sequences, or arrangements of repetitious sequences, perhaps—dictate the characteristic packaging, and/or drive gene silencing? Detailed mapping of the sequence organization of heterochromatin in *Drosophila* has been complicated both by the presence of repetitious DNA and by the paucity of unique markers. Mapping by *in situ* hybridization has indicated that blocks of satellite DNA are interrupted by blocks of middle-repetitive elements (52). In the case of the X chromosome, fine-structure analysis of the basal heterochromatin reveals that it is composed primarily of satellite DNA and transposable elements that are neither exclusively X-linked nor exclusively heterochromatic (53). While the lack of specific markers has made

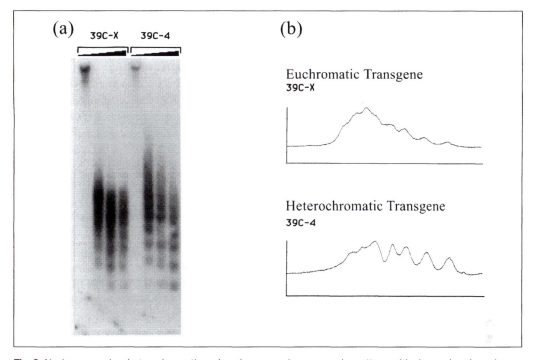

Fig. 3 Nucleosomes in a heterochromatic region show a much more regular pattern, with sharper bands and more uniform spacing, as shown by digestion of the chromatin with micrococcal nuclease. Line 39C-X carries a tagged *hsp26* transgene at a euchromatic site, while line 39C-4 carries the same trangene at a heterochromatic site. (a) Nuclei from third-instar larvae were isolated and incubated with increasing amounts of micrococcal nuclease. The DNA was purified and separated by size on a 1.2% agarose gel, transferred to membrane, and probed with a labelled unique DNA fragment that recognizes the tagged *hsp26* transgene. (b) Densitometric scans of the most digested lanes for each sample shown in (a) (*top* to *bottom* of each lane is *left* to *right* along the *x* axis). Figure adapted from ref. 9, reproduced by permission from Cold Spring Harbor Laboratory Press.

chromatin structure analysis in this region difficult, the $1.688\,\text{g}/\text{cm}^3$ satellite DNA has been shown to be packaged in a highly regular nucleosome array, with two nucleosomes occupying each 359 bp sequence repeat (54). Recent investigations of silenced transgenes localized within heterochromatic domains have also shown packaging in a very uniform nucleosome array (9, 55). Digestion with micrococcal nuclease results in a nucleosome pattern with sharper bands (Fig. 3), and an array that allows distinction of the 10-mer, rather than the 5- to 6-mer typically seen on observing the genome as a whole (55). Such regularity in the nucleosome array could facilitate packaging in a higher-order structure such as a solenoid.

Structural analysis of three variegating breakpoints (25) found middle repetitive elements, structurally related to transposons, at the heterochromatic end of each breakpoint. Reinversions of one of the variegating rearrangements restored wild-type gene expression, yet left the heterochromatic sequences in the immediate vicinity of the breakpoint intact (25), suggesting that the efficient propagation of heterochromatin-mediated silencing activity requires more than the sequences at the breakpoint itself, either a larger mass of heterochromatin or an 'initiator' sequence for heterochromatin formation. No simple rules have been discerned that might predict the effectiveness of a given repetitious sequence in participating in heterochromatin formation. A variety of unique and repetitious sequences (*Doc, hoppel*) have been found adjacent to variegating transgenes in pericentric heterochromatin (56); the data suggest that the immediately adjacent sequence is not a critical parameter. In fact, arrays of P-elements containing exclusively euchromatic DNA have been shown to form ectopic heterochromatin in some cases (57). It has been suggested that some feature of the repeat structure *per se* may be able to nucleate heterochromatin formation (4).

The ability to cause efficient transposition of visible markers to new chromosomal sites in the *Drosophila* genome using a P-element vector has uncovered a wide variety of position effects. In instances where variegated expression of the transposon marker has been encountered, the insertion site has been found to be a telomere, the pericentric heterochromatin, or a subset of sites on the fourth chromosome (9, 26, 58). Since the marker being used and the amount and sequence of DNA in the P-element are a constant, transposons will be especially effective probes for detecting structural and functional heterogeneity at these sites. For example, in contrast to what has been observed in yeast, variegating inserts in the telomeres do not respond to classical modifiers of PEV that do affect silencing of inserts at the chromocentre (9, 15). It is likely that the protein composition and details of chromatin structure vary in different heterochromatin domains (9, 26, 59).

Several lines of evidence suggest that template accessibility is reduced at a locus undergoing heterochromatin-induced silencing. The efficiency of restriction endonuclease cleavage of a target site within a variegating locus is significantly reduced compared to the same sequence in a euchromatic site (9). Importantly, this reduced accessibility is partially relieved by the inclusion of a mutant *Su(var)2–5* allele (56), implicating HP1 in the control of template accessibility. The functional significance of template occlusion is suggested by the observation that footprints for three promoter-binding factors—GAGA factor, TFIID, and RNA polymerase—are significantly reduced at a silenced promoter (60). While these data rely on measurements of template accessibility using isolated nuclei, they are supported by the corresponding reduction in accessibility of a silenced locus to *E. coli dam* DNA methyltransferase *in vivo* (61). This reduced accessibility may reflect the more ordered nature of nucleosome arrays observed at silenced loci (Fig. 3), or might reflect the association of the locus with a nuclear compartment that is enriched for HP1 and relatively depleted in chromatin remodelling factors and transcriptional machinery (15, 55).

3. Maintaining gene expression patterns by the mechanism of cellular memory

The process of pattern formation determines the function and structure that a cell and its descendants attain in the developed body. Differential patterns of gene expression form the basis for the creation of cellular diversity. In *Drosophila*, many of the factors that govern this process have been identified. Master regulatory genes, such as the families of homeotic genes in the bithorax complex (BX-C) and the *Antennapedia* complex (ANT-C), play a fundamental role in defining positional values and establishing (in a combinatorial manner) the determined state of a cell (62).

The regulation of homeotic gene expression can be subdivided into two distinct phases. An early initiation phase establishes the basic body plan during embryogenesis and sets the spatial boundaries of homeotic gene expression. In the subsequent phase, this expression pattern needs to be remembered over many cell generations such that during the differentiation process, the information provided by the homeotic gene products is translated into the appropriate pattern of gene expression to achieve the appropriate morphological structures. This latter mechanism, which maintains the determined state of a cell, has been termed 'cellular memory'. In *Drosophila* the first phase, resulting in the differential expression pattern of homeotic genes, is controlled by a cascade of diffusible transcription factors encoded by the maternal and segmentation genes. In the following phase, two groups of genes are needed to maintain the 'on' and 'off' states, respectively, in a mitotically stable and heritable manner.

3.1 The *Polycomb* group and antipodal *trithorax* group of chromosomal proteins

Genetic analysis has identified a class of mutations, unlinked to the clustered homeotic genes, which none the less displays complex homeotic phenotypes (63, 64). The observed homeotic transformations are the result of an ectopic expression of the homeotic genes in incorrect segments. Thus, in its normal function, this class of regulators, collectively termed the *Polycomb* group (PcG), was considered to be necessary for repressing homeotic gene expression, maintaining the 'off' state. Once molecular markers for homeotic gene products became available, the genetic predictions could be substantiated by a visual assessment of the ectopic expression patterns seen in *PcG* mutants (65, 66). Surprisingly, however, the early pattern of homeotic products was indistinguishable in the mutant embryos from wild type. Only at the stage of germ band retraction did ectopic expression start to emerge. This result demonstrated that the *PcG* was not involved in the establishment of the homeotic pattern, but rather in its maintenance. The products of the *PcG* apparently recognize the inactive expression state of a particular homeotic gene, as defined by the early initiation system, and transmit this repressed state in a stable and heritable form through the many cell divisions occurring during the growth phases of development. Clonal analysis has shown that the repressive role of *PcG* genes is required continuously not only during differentiation of larval tissue, but also for development of adult structures (67, 68).

Subsequent analysis of the regulation exerted by the *PcG* genes, either by closer inspection of the mutant phenotypes or by molecular characterization, revealed that the homeotic genes were only a small subset of the target genes. Many of the other identified targets turned out to encode factors important for regulating developmental processes. Thus, the primary task of the *PcG* is to ensure a faithful and permanent inactivation of developmental regulators in defined domains of the body. Indeed, the phenotypic consequences of *PcG* mutations indicate that the continued maintenance of silencing supported by the *PcG* factors is an important and fundamental regulatory mechanism of the developmental process.

Several members of the *PcG* have now been analysed at the molecular level. *Polycomb* (*Pc*) encodes a 390 amino-acid nuclear protein (69). Both the transcript and the protein are essentially homogeneously distributed in most tissues during development, although there is a preferential accumulation in tissues with high proliferative activity. In polytene chromosomes, the protein is found to be associated with the homeotic gene clusters of the *ANT-C* and *BX-C* loci, as anticipated (70). However, approximately 100 additional sites have been identified, supporting the notion that the *PcG* is responsible for repressing a more extended set of target genes. Interestingly, in this tissue other *PcG* loci were identified as targets of the PC protein, suggesting a cross-regulatory activity within the class. In transgenic lines carrying an isolated copy of a *cis*-regulatory element of a homeotic gene, the PC protein was

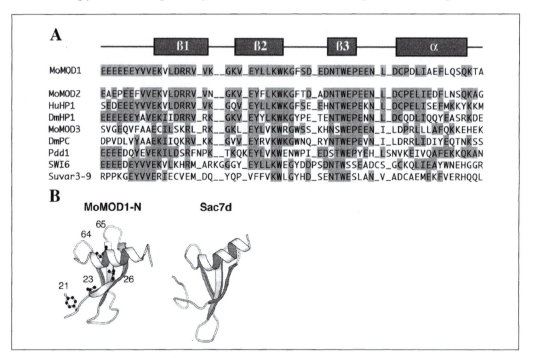

Fig. 4 Conservation and structure of the chromo domain. (A) Alignment of several chromo domains. The location of the secondary structural features as determined from the MoMOD1 chromo domain is shown on the top. β1, β2, and β3 refer to the three strands of the anti-parallel β-sheet, while α indicates the carboxy-terminal α-helix. (B) Comparison of the structure of the MoMOD1 chromo domain with that of the DNA-binding domain from the archaebacterial protein Sac7d. Side-chain residues where mutations affect the function of the chromo domain are indicated with the appropriate residue number. (From ref. 71 and 86; figure reproduced by permission from Elsevier Publishing and Oxford University Press.)

found associated with the transgene at the site of integration. This reflects an obvious specificity in the association of the protein with defined chromosomal sites. A 48 amino-acid chromo domain, with homology to HP1, has been identified near the amino terminus (Fig. 4) (69, 71). In addition to producing the homeotic transformation typical of the *PcG* genes, mutations in the *polyhomeotic* (*ph*) gene cause extensive cell death in the central epidermis, an abnormal pattern of axonal pathways in the central nervous system, and abnormalities in the distribution patterns of the segmentation gene products, indicating that *ph* has a widespread function. The 169 kDa PH protein contains a single zinc-finger motif, a serine–threonine-rich region, and glutamine repeats, suggestive of a potential DNA-binding protein (72).

The embryonic phenotype of *Posterior sex combs* (*Psc*) mutants shows only weak homeotic transformations, due to a strong maternal contribution of protein. Extensive maternal deposition of gene products into the unfertilized egg characterizes the entire class and probably reflects the need for large quantities of these proteins during embryonic development. The *Psc* gene flanks the *Su(z)2* gene, sharing extensive sequence homology in the protein-coding region. Interestingly, some alleles of *Psc* and *Su(z)2* act as suppressors of telomere position effect (15). The developmental profile of PSC protein expression coincides with the previously described patterns of PC and PH proteins, being essentially homogeneous in all tissues tested during most of the embryonic and larval stages. The sequence of the protein reveals the most prominent motif to be a scattered arrangement of zinc fingers. Interestingly, a 200 amino-acid stretch encompassing these zinc fingers has been conserved during evolution and was also found in the mouse Bmi-1 protein and a related Mel-18 protein (73). The murine *bmi-1* gene has been implicated in processes of oncogenesis and cell-cycle control (74). Knock-out mutations in mice also result in homeotic transformations, indicating a functional conservation of the *Bmi-1/Psc* roles. Indeed, functional homologues of many *Drosophila PcG* members have been identified in a variety of organisms (75, 76). This suggests that the function of the *PcG*, maintaining the silenced expression state of master regulatory genes, is fundamental and highly conserved in developing organisms.

In parallel to the genetic identification of the *PcG*, the *trithorax* group (*trxG*) has been recognized as a class of genes necessary for counteracting *PcG* silencing (77). Mutations in members of the *trxG* show a loss-of-function phenotype for homeotic genes, indicating that the products of the *trxG* are needed to maintain the activity of the homeotic genes and other regulators during development. Molecular analyses have shown several members of the *trxG* to be part of chromatin-remodelling complexes. A well-characterized example is the *brahma* product, related to the yeast SWI2/SNF2 factor involved in opening repressive chromatin structures in an energy-dependent manner (78; see Chapters 5, 6).

3.2 *PcG* multimeric protein complexes and the chromo domain

Double *PcG* mutant combinations result in stronger homeotic transformations than the sum of the single phenotypic effects (64). For example, the *engrailed* gene becomes extensively derepressed during embryonic development in double *PcG* mutants, but only weakly derepressed in lines with a single *PcG* mutation (79). In addition, the phenotypic consequences vary with different doses of the target genes. These synergistic effects suggest that the *PcG* proteins cooperate in a common mechanism, and that this role might even extend to an interaction at the molecular level. Taken together, the findings are reminiscent of the cooperative effects that have been proposed for the components of heterochromatin and suggest that *PcG* silencing imposes heterochromatin-like packaging of the nucleosomal fibre to prevent gene activation.

Molecular analyses have also provided evidence that many *PcG* proteins are co-localized in large multiprotein complexes, interacting with different nuclear components (80–82). *In vitro*, the PC protein has been found to bind to nucleosomes through a conserved C-terminal domain (83). This interaction could potentially explain the binding of PC and associated *PcG* components with extended chromosomal domains. Biochemical purification of *PcG* protein complexes has led to the demonstration that *in vitro* the *PcG* activity is able to block the remodelling activity of SWI/SNF complexes on a nucleosomal template (84). These *PcG* functions seem to be accompanied by enzymatic activities that modify histones and correspondingly affect the overall structure and function of the nucleosomal fibre (85).

The conserved chromo domain in the PC protein plays an important role in holding *PcG* complexes together. Mutations in the PC chromo domain result in a disintegration of the *PcG* complex; this has been demonstrated in mutant cells, by yeast two-hybrid analyses and by immunoprecipitation experiments. Resolution of the three-dimensional structure by nuclear magnetic resonance (NMR) shows a folded amino-terminal three-stranded anti-parallel β-sheet laying against a carboxy-terminal α-helix (86) (Fig. 4). Despite knowledge of its structure, the exact molecular function of the chromo domain remains an enigma. However, there is increasing evidence that the chromo domain acts as a protein–protein interaction motif involved in targeting. Mutations in the PC chromo domain completely abolish the binding of the protein to its over 100 binding sites on polytene chromosomes. A chimeric HP1 protein, having a PC chromo domain while retaining its own chromo-shadow domain, locates to the endogenous PC binding sites in addition to its normal position in heterochromatin (40). Interestingly, the HP1 chimera recruits other *PcG* proteins to heterochromatin, indicating that specific chromo domains are sufficient to generate an affinity for the corresponding silencing complex. Additionally, flies expressing this chimeric protein can rescue the loss of silencing induced by mutations in the HP1-encoding gene *Su(var)2–5*, supporting the notion that heterochromatin formation and homeotic gene silencing are mechanistically related (40).

3.3 *PcG* response elements and DNA-binding specificity

A puzzling aspect, contrasting with the observation that *PcG* silencing is gene-specific, is the finding that all members of *PcG* identified so far seem to be homogeneously expressed in all cells. A ubiquitous requirement for function is further supported by the widespread and pleiotropic mutant phenotypes. A sequence-specific DNA-binding factor, the product of *pleiohomeotic*, has been recently identified as a *PcG* member (87). However, repressive effects on specific genes can vary substantially between cells, depending on their position in the embryo. What tags a gene to be inactivated by the *PcG* remains unclear. That there is a definite DNA specificity imposed on the formation of *PcG* complexes is demonstrated by the protein patterns found in polytene chromosomes (Fig. 1). Different methods have been applied to identify the *cis*-acting elements used by the *PcG* to achieve regulation. Reporter gene constructs containing various *cis*-regulatory elements from potential *PcG* binding sites were found to be differentially repressed during embryonic and larval development. In many cases, it was found that larger DNA fragments maintained the repressive effect better than smaller constructs; strong position effects were often observed. This suggested that the larger the size of the silencing complex established by the *PcG* proteins, the better the repression is sustained over developmental time. The DNA elements utilized by the *PcG* to mediate regulation have been termed PREs (*PcG* response *e*lements) (88). That a direct interaction of *PcG* proteins with the defined regulatory elements does exist is demonstrated by immunostaining polytene chromosomes of transgenic lines; at the integration site of the transgene containing a particular *cis*-regulatory element, a new *PcG* protein-binding site can be mapped. In order to keep appropriate segment-specific silencing, the interaction between *PcG* proteins and the respective PREs needs to be maintained throughout the entire development of the fly (68).

A high-resolution map of PREs has been obtained in the bithorax complex (BX-C), containing several homeotic genes, by applying a chromatin cross-linking strategy (89, 90). Tissue-culture cells or embryos are fixed *in vivo* with formaldehyde, and chromatin isolated and reduced in size by shearing. Because of the cross-linking, proteins as well as DNA fragments bound to *PcG* proteins can be purified by immunoprecipitation with specific antibodies. The recovered DNA fragments, enriched in PREs, were PCR-amplified and used as a probe to a genomic walk across the BX-C (Fig. 5). The distributions of PC, PH, and PSC in the BX-C show an extended interaction of these *PcG* proteins centred around single PREs, suggesting a spreading of the *PcG* protein complexes along the DNA over several kilobases. This profile probably reflects the need for an extended interaction of the *PcG* complexes with the nucleosomal backbone, perhaps generating higher-order chromatin structures. The distribution of the *PcG* proteins contrasts with the narrow binding profile of the sequence-specific DNA-binding GAGA factor, a member of the *trxG* (Fig. 5).

Interestingly, components of the repressing *PcG* and of the activating *trxG* have been found to bind and act through the same chromosomal elements (90, 91). These elements may represent the basic *cis*-acting control units of cellular memory, initiating and maintaining the neighbouring chromatin structures in either a silenced or activated mode. In this sense such *PcG*/*trxG*-regulated elements can be considered 'switch elements' that are set during embryogenesis by the patterning system. Indeed, in a transgenic system such an element of the BX-C, *Fab-7*, can be switched from a silenced state into an activated state by an early embryonic pulse of transcription

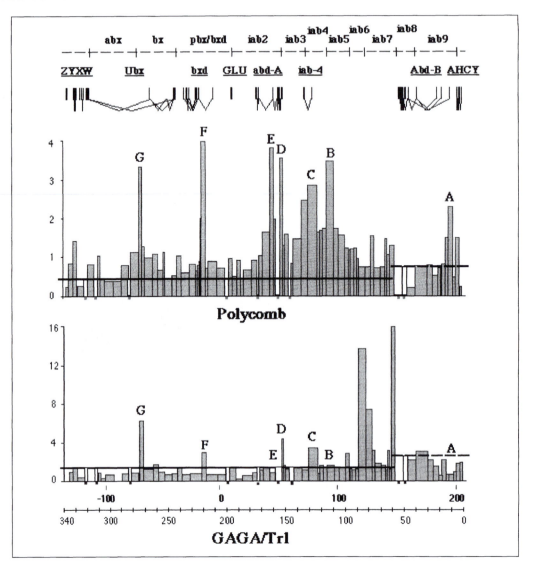

Fig. 5 Distribution of POLYCOMB (*PcG*) and GAGA factor (*trxG*) on the *bithorax* complex (*BX-C*) of *Drosophila*. The various regulatory regions and the organization of the transcription units are indicated at the top of the figure. Binding profiles of PC and GAGA were deduced from chromatin-immunoprecipitation experiments. Restriction fragments showing strong PC binding (PREs) are labelled A–G on both profiles. The thick black line across each profile represents the approximate background level of hybridization. (From ref. 90; figure reproduced by permission of Oxford University Press.)

(92). The activated *Fab-7* is able to maintain the expression of a nearby reporter gene in a mitotically stable manner. This maintenance function becomes independent of the primary transcription factor activating the reporter gene, probably by utilizing the general activating factors of the cell during the maintenance phase. Surprisingly, this activated *Fab-7* state is also meiotically stable and transmissible to the next generation in a substantial proportion of progeny. The activated *Fab-7* is associated with a local hyperacetylation of histone H4, suggesting that this histone modification acts as a epigenetic mark (93). It will be interesting to define elements like *Fab-7* (termed cellular memory modules) in other organisms, as sites where chromatin-modelling *trxG* and *PcG* protein complexes act, maintaining open chromatin structures (allowing continuous expression) or silenced chromatin structures (preventing ectopic expression) for the regulated gene.

4. Final comments: important questions and discussion

While we can now begin to identify pieces of the puzzle, much remains to be learned before one can assemble a model of the chromatin structures capable of maintaining silencing in the *Drosophila* genome. As yet, we do not know what DNA sequences trigger formation of a heterochromatic domain. No particular *cis*-acting elements have been identified; indeed, the evidence at hand suggests that patterns of repetitious elements may be critical. Recognition and silencing of repetitious elements has been suggested to be a defence mechanism against invading transposable elements (4). The effectiveness of silencing appears to be dependent both on the immediate sequence pattern (in some undefined way) and on proximity to the mass of pericentric heterochromatin. This implies that both the observed assembly into a more ordered nucleosome array, and association with a 'heterochromatin compartment', may play a role in silencing. While silencing can be inferred to result from a removal of the gene from the pool that can be accessed by the chromatin-remodelling machinery, how this is accomplished remains a mystery. The shift in nucleosome pattern may block access to the promoter, but other changes in packaging appear to be at work as well (55).

The observation that certain genes map within pericentric heterochromatin seems paradoxical. A simple model might suggest that such genes survive on islands of euchromatin within an ocean of heterochromatin; indeed, there is evidence for the interspersion of blocks of euchromatin and heterochromatin along the small fourth chromosome (26). However, the heterochromatic genes are distinguished by their response to HP1 (and potentially to other modifiers of PEV) in that they are dependent on HP1 for full expression at their endogenous site (40), and function poorly when moved to a euchromatic domain. One way to rationalize how HP1 can be both a silencer of euchromatic genes and an activator of heterochromatic genes is to posit that HP1 functions as an 'organizer' of a condensed chromatin structure. This state of chromatin folding would interfere with the activation of most euchromatic genes, perhaps by blocking access to regulatory regions. Heterochromatic genes may have evolved to depend on HP1-mediated condensation for their activity, perhaps exploiting this chromatin condensation to facilitate enhancer–promoter communication. Chromatin immunoprecipitation data from fission yeast show that domains of binding for the yeast HP1-like protein Swi6 are not co-extensive with pericentric heterochromatin (94); similarly, mosaicism for HP1-associated and HP1-free domains may also exist in *Drosophila* heterochromatin. If heterochromatic gene enhancers and their target promoters occupy separate HP1-free domains, but are linked by an HP1-associated domain, expression could become dependent on HP1-mediated condensation.

In contrast to our current understanding of heterochromatin, we have made significant advances in identifying the *cis*-acting DNA elements (the PREs) that place a gene under the regulation of the *PcG/trxG* system. One possible anchor (the *plieohomeotic* gene product) has been identified. However, it is also clear that others must exist, to explain the complex binding patterns of the PcG proteins observed at PREs. PREs appear to nucleate assembly of the appropriate multiprotein complex; direct *in vitro* assessment has shown that the *PcG* complex can block remodelling of the underlying nucleosomes, a status which should maintain the inactive state. However, the PRE can interact with both *PcG* and *trxG*, and we do not know how the balance is set to determine whether the regulated gene will be maintained in the 'off' or 'on' state. Once established, the activity state is remembered through multiple rounds of mitosis during development; in an artificial construct, the memory has even been maintained in the following generation. Current findings suggest that the histone acetylation state may serve as an epigenetic mark for the 'cellular memory modules,' but again the details of such a mechanism remain to be established. We have much to learn to understand how this multicomponent, delicately balanced mechanism can generate the robust signals required to maintain the differentiated state.

Acknowledgements

We thank our colleagues and fellow authors for comments on the manuscript. Work herein from the Eissenberg lab has been supported by NIH grant GM57005; from the Paro lab by the DFG, HFSP, and the DKFZ-MOS Cooperation; and from the Elgin lab by NIH grants HD23844 and GM31532.

References

1. Urieli-Shoval, S., Gruenbaum, Y., Sedat, J., and Razin, A. (1982) The absence of detectable methylated bases in *Drosophila melanogaster* DNA. *FEBS Lett.*, **146**, 148.
2. Heitz, E. (1928) Das Heterochromatin der Moose. *Jb. Wiss. Bot.*, **69**, 728.
3. Weiler, K. S. and Wakimoto, B. T. (1995) Heterochromatin and gene expression in *Drosophila*. *Annu. Rev. Genet.*, **29**, 557.
4. Henikoff, S. (2000) Heterochromatin function in complex genomes. *Biochim. Biophys. Acta*, **1470**, 1.
5. Spofford, J. B. (1976) Position–effect variegation in *Drosophila*. In *The genetics and biology of Drosophila*, Vol 1c. (ed. M. Ashburner and E. Novitski), p. 955. Academic Press, New York.
6. Henikoff, S. (1981) Position–effect variegation and chromosome structure of a heat shock puff in *Drosophila*. *Chromosoma*, **83**, 381.
7. Reuter, H., Werner, W., and Hoffmann, H. J. (1982) mutations affecting position–effect heterochromatinization in *Drosophila melanogaster*. *Chromosoma*, **85**, 539.

8. Belyaeva, E. S. and Zhimulev, I. F. (1991) Cytogenetic and molecular aspects of position–effect variegation in *Drosophila melanogaster*. III. Continuous and discontinuous compaction of chromosomal material is a result of position effect variegation. *Chromosoma*, **100**, 453.

9. Wallrath, L. L. and Elgin, S. C. R. (1995) Position effect variegation in *Drosophila* is associated with an altered chromatin structure. *Genes Develop.*, **9**, 1263.

10. Lu, B. Y., Ma, J., and Eissenberg, J. C. (1998) Developmental regulation of heterochromatin-mediated gene silencing in *Drosophila*. *Development*, **125**, 2223.

11. Laird, C. D., Chooi, W. Y., Cohen, E. H., Dickson, E., Hutchinson, N., and Turner, S. H. (1973) Organization and transcription of DNA in chromosomes and mitochondria of *Drosophila*. *Cold Spring Harbor Symp. Quant. Biol.*, **38**, 311.

12. Karpen, G. H. and Spradling, A. C. (1990) Reduced DNA polytenization of a mini-chromosome region undergoing position–effect variegation in *Drosophila*. *Cell*, **63**, 97.

13. Rushlow, C. A., Bender, W., and Chovnick, A. (1984) Studies on the mechanism of heterochromatin position effect at the *rosy* locus of *Drosophila melanogaster*. *Genetics*, **108**, 603.

14. Wallrath, L. L., Guntur, V. P., Rosman, L. E., and Elgin, S. C. R. (1996) DNA representation of variegating heterochromatic P-element inserts in diploid and polytene tissues of *Drosophila melanogaster*. *Chromosoma*, **104**, 519.

15. Cryderman, D. E., Morris, E. J., Biessmann, H., Elgin, S. C. R., and Wallrath, L. L. (1999) Silencing at *Drosophila* telomeres: nuclear organization and chromatin structure play critical roles. *EMBO J.*, **18**, 3724.

16. Hilliker, A. J. (1976) Genetic analysis of the centromeric heterochromatin of chromosome 2 of *Drosophila melanogaster*: deficiency mapping of EMS-induced lethal complementation groups. *Genetics*, **83**, 765.

17. Wakimoto, B. T. and Hearn, M. G. (1990) The effects of chromosome rearrangements on the expression of heterochromatic genes in chromosme 2L of *Drosophila melanogaster*. *Genetics*, **125**, 141.

18. Eberl, D. F., Duyf, B. J., and Hilliker, A. J. (1993) The role of heterochromatin in the expression of a heterochromatic gene, the *rolled* locus of *Drosophila melanogaster*. *Genetics*, **134**, 227.

19. Henikoff, S. and Dreesen, T. D. (1989) Trans-inactivation of the *Drosophila brown* gene: evidence for transcriptional repression and somatic pairing dependence. *Proc. Natl Acad. Sci., USA*, **86**, 6704.

20. Dreesen, T. D., Henikoff, S., and Loughney, K. (1991) A pairing-sensitive element that mediates *trans*-inactivation is associated with the *Drosophila brown* gene. *Genes Develop.*, **5**, 331.

21. Platero, J. S., Csink, A. K., Quintanilla, A., and Henikoff, S. (1998) Changes in chromosomal localization of heterochromatin-binding proteins during the cell cycle in *Drosophila*. *J. Cell Biol.*, **140**, 1297.

22. Talbert, P. B., LeCiel, C. D. S., and Henikoff, S. (1994) Modification of the *Drosophila* heterochromatic mutation *brown*[Dominant] by linkage alterations. *Genetics*, **136**, 559.

23. Csink, A. K. and Henikoff, S. (1996) Genetic modification of heterochromatic associations and nuclear organization in *Drosophila*. *Nature*, **381**, 529.

24. Dernburg, A. F., Borman, K. W., Fung, J. C., Marshall, W. F., Philips, J., Agard, D., and Sedat, J. W. (1996) Perturbation of nuclear architecture by long-distance chromosome interactions. *Cell*, **85**, 745.

25. Tartof, K. D., Hobbs, C., and Jones, J. (1984) A structural basis for variegating position effects. *Cell*, **37**, 869.

26. Sun, F.-L., Cuaycong, M. H., Craig, C. A., Wallrath, L. L., Locke, J., and Elgin, S. C. R. (2000) The fourth chromosome of *Drosophila melanogaster*: interspersed euchromatic and heterochromatic domains. *Proc. Natl Acad. Sci., USA*, **97**, 5340.

27. Locke, J., Kotarski, M. A., and Tartof, K. D. (1988) Dosage dependent modifiers of position effect variegation in *Drosophila* and a mass action model that explains their effect. *Genetics*, **120**, 181.

28. Lifton, R. P., Goldberg, M. L., Karp, R. W., and Hogness, D. S. (1978) The organization of the histone genes in *Drosophila melanogaster*: functional and evolutionary implications. *Cold Spring Harbor Symp. Quant. Biol.*, **42**, 1047.

29. Moore, G. D., Sinclair, D. A. R., and Grigliatti, T. A. (1983) Histone gene multiplication and position effect variegation in *Drosophila melanogaster*. *Genetics*, **105**, 327.

30. Mottus, R., Reeves, R., and Grigliatti, T. A. (1980) Butyrate suppression of position-effect variegation in *Drosophila melanogaster*. *Mol. Gen. Genet.*, **178**, 465.

31. Turner, B. M., Birley, A. J., and Lavender, J. (1992) Histone H4 isoforms acetylated at specific lysine residues define individual chromosomes and chromatin domains in *Drosophila* polytene nuclei. *Cell*, **69**, 375.

32. Mottus, R., Sobel, R. E., and Grigliatti, T. A. (2000) Mutational analysis of a histone de-acetylase in *Drosophila melanogaster*: Missense mutations suppress gene silencing associated with position effect variegation. *Genetics*, **154**, 657.

33. Reuter, G. and Spierer, P. (1992) Position effect variegation and chromatin proteins. *BioEssays*, **14**, 605.

34. James, T. C. and Elgin, S. C. R. (1986) Identification of a nonhistone chromosomal protein associated with heterochromatin in *Drosophila* and its gene. *Mol. Cell. Biol.*, **6**, 3862.

35. James, T. C., Eissenberg, J. C., Craig, C., Dietrich, V., Hobson, A., and Elgin, S. C. R. (1989) Distribution patterns of HP1, a heterochromatin-associated nonhistone chromosomal protein of *Drosophila*. *Eur. J. Cell Biol.*, **50**, 170.

36. Sinclair, D. A., Clegg, N. J., Antonchuk, J., Milne, T. A., Stankunas, K., Ruse, C., Grigliatti, T. A., Kassis, J. A., and Brock, H. W. (1998) *Enhancer of Polycomb* is a suppressor of position-effect variegation in *Drosophila melanogaster*. *Genetics*, **148**, 211.

37. Eissenberg, J. C., James, T. C., Foster-Hartnett, D. M., Hartnett, T., Ngan, V., and Elgin, S. C. R. (1990) Mutation in a heterochromatin-specific chromosomal protein is associated with suppression of position–effect variegation in *Drosophila melanogaster*. *Proc. Natl Acad. Sci., USA*, **87**, 9923.

38. Eissenberg, J. C., Morris, G. D., Reuter, G., and Hartnett, T. (1992) The heterochromatin-associated protein HP1 is an essential protein in *Drosophila* with dosage-dependent effects on position effect variegation. *Genetics*, **131**, 345.

39. Platero, J. S., Hartnett, T., and Eissenberg, J. C. (1995) Functional analysis of the chromo domain of HP1. *EMBO J.*, **14**, 3977.

40. Lu, B. Y., Emtage, P. C. R., Duyf, B. J., Hilliker, A. J., and Eissenberg, J. C. (2000) Hetero-chromatin protein 1 is required for the normal expression of two heterochromatin genes in *Drosophila*. *Genetics*, **155**, 699.

41. Eissenberg, J. C. and Elgin, S. C. R. (2000) The HP1 protein family: getting a grip on chromatin. *Curr. Opin. Genet. Devel.*, **10**, 204.

42. Aasland R. and Stewart, A. F. (1995) The chromo shadow domain, a second chromo domain in heterochromatin-binding protein 1, HP1. *Nucleic Acids Res.*, **23**, 3163.

43. Powers, J. A. and Eissenberg, J. C. (1993) Overlapping domains of the heterochromatin-associated protein HP1 mediate nuclear localization and heterochromatin binding. *J. Cell Biol.*, **120**, 291.

44. Messmer, S., Franke, A., and Paro R. (1992) Analysis of the funcitonal role of the *Polycomb* chromo domain in *Drosophila melanogaster*. *Genes Develop.*, **6**, 1241.

45. Platero, J. S., Sharp, E. J., Adler, P. N., and Eissenberg, J. C. (1996) *In vivo* assay for protein–protein interactions using *Drosophila* chromosomes. *Chromosoma*, **104**, 393.

46. Belyaeva, E. S., Demakova, O. V., Umbetova, G. H., and Zhimulev, I. F. (1993) Cytogenetic and molecular aspects of position–effect variegation in *Drosophila melanogaster*. V. Heterochromatin-associated protein HP1 appears in euchromatic chromosomal regions that are inactivated as a result of position effect variegation. *Chromosoma*, **102**, 583.

47. Pak, D. T. S., Pflumm, M., Chesnodov, I., Huang, D. W., Kellum, R., Marr, J., Romanowski, P., and Botchan, M. R. (1997) Association of the origin recognition complex with heterochromatin and HP1 in higher eukaryotes. *Cell*, **91**, 311.

48. Murzina, N., Verreault, A., Laue, E., and Stillman, B. (1999) Heterochromatin dynamics in mouse cells: interaction between chromatin assembly factor 1 and HP1 proteins. *Mol. Cell*, **4**, 529.

49. Ye, Q. and Worman, H. J. (1996) Interaction between an integral protein of the muclear envelope inner membrane and human chromodomain proteins homologous to *Drosophila* HP1. *J. Biol. Chem.*, **271**, 14653.

50. Andrulis, E. D., Neiman, A. M., Zappulla, D. C., and Sternglanz, R. (1998) Perinuclear localization of chromatin facilitates transcriptional silencing. *Nature*, **394**, 592.

51. Brasher, S. V., Smith, B. O., Fogh, R. H., Nietlispach, D., Thiru, A., Nielsen, P. R., Broadhurst, R. W., Ball, L. J., Murzina, N. V., and Laue, E. D. (2000) The structure of mouse HP1 suggests a unique mode of single peptide recognition by the shadow chromo domain dimer. *EMBO J.*, **19**, 1587.

52. Pimpinelli, S., Berloco, M., Fanti, L., Dimitri, P., Bonaccorsi, S., Marchetti, E., Caizzi, R., Caggese, C., and Gatti, M. (1995) Transposable elements are stable structural components of *Drosophila melanogaster* heterochromatin. *Proc. Natl Acad. Sci., USA*, **92**, 3804.

53. Sun, X., Wahlstrom, J., and Karpen, G. (1997) Molecular structure of a functional *Drosophila* centromere. *Cell*, **91**, 1007.

54. Cartwright, I. L., Hertzberg, R. P., Dervan, P. B., and Elgin S. C. R. (1983) Recognition of the nucleosomal structure of chromatin by a cleavage reagent with low sequence preference: (methidium propyl-EDTA) iron (II). *Proc. Natl Acad. Sci., USA*, **80**, 3212.

55. Sun, F.-L., Cuaycong, M. H., and Elgin, S. C. R. (2000) Long-range nucleosome ordering is associated with gene silencing in pericentric heterochromatin, in preparation.

56. Cryderman, D. E., Cuaycong, M. H., Elgin, S. C. R., and Wallrath, L. L. (1998) Characterization of sequences associated with position–effect variegation at pericentric sites in *Drosophila* heterochromatin. *Chromosoma*, **107**, 277.

57. Dorer, D. R. and Henikoff, S. (1994) Expansions of transgenic repeats cause heterochromatin formation and gene silencing in *Drosophila. Cell*, **77**, 993.

58. Hazelrigg, T., Levis, R., and Rubin, G. M. (1984) Transformation of *white* locus DNA in *Drosophila*: dosage compensation, *zeste* interaction, and position effects. *Cell*, **36**, 469.

59. Bishop, C. P. (1992) Evidence for intrinsic differences in the formation of chromatin domains in *Drosophila melanogaster. Genetics*, **132**, 1063.

60. Cryderman, D. E., Tang, H. B., Bell, C., Gilmour, D. S., and Wallrath, L. L. (1999) Heterochromatic silencing of *Drosophila* heat shock genes acts at the level of promoter potentiation. *Nucleic Acids Res.*, **27**, 3364.

61. Boivin, A., and Dura, J.-M. (1998) *In vivo* chromatin accessibility correlates with gene silencing in *Drosophila. Genetics*, **150**, 1539.

62. Lewis, E. B. (1978) A gene complex controlling segmentation in *Drosophila. Nature*, **276**, 565.

63. Duncan, I. M. and Lewis, E. B. (1982) Genetic control of body segment differentiation in *Drosophila*. In *Developmental order: Its origin and regulation*, (ed. S. Subtelny and P. B. Green), p. 533. Liss, New York.

64. Jürgens, G. (1985) A group of genes controlling the spatial expression of the bithorax complex in *Drosophila*. *Nature*, **316**, 153.

65. Struhl, G. and Akam, M. (1985) Altered distributions of *Ultrabithorax* transcripts in *extra sex combs* mutant embryos of *Drosophila*. *EMBO J.*, **4**, 3259.

66. Simon, J., Chiang, A., and Bender, W. (1992) Ten different *Polycomb* group genes are required for spatial control of *abdA* and *AbdB* homeotic products. *Development*, **114**, 493.

67. Struhl, G. (1981) A gene product required for correct initiation of segmental determination in *Drosophila*. *Nature*, **293**, 36.

68. Busturia, A., Wightman, C. D., and Sakonju, S. (1997) A silencer is required for maintenance of transcriptional repression throughout *Drosophila* development. *Development*, **124**, 4343.

69. Paro, R. and Hogness, D. (1991) The Polycomb protein shares a homologous domain with a heterochromatin-associated protein of *Drosophila*. *Proc. Natl Acad. Sci., USA*, **88**, 263.

70. Zink, B. and Paro, R. (1989) *In vivo* binding pattern of a trans-regulator of homoeotic genes in *Drosophila melanogaster*. *Nature*, **337**, 468.

71. Cavalli, G. and Paro, R. (1998) Chromo-domain proteins: linking chromatin structure to epigenetic regulation. *Curr. Opin. Cell Biol.*, **10**, 354.

72. Kyba, M. and Brock, H. W. (1998) The *Drosophila* Polycomb group protein PSC contacts ph and Pc through specific conserved domains. *Mol. Cell. Biol.* **18**, 2712.

73. van Lohuizen, M. (1999) The *trithorax*-group and *Polycomb*-group chromatin modifiers: implications for disease. *Curr. Opin. Genet. Dev.*, **9**, 355.

74. Jacobs, J. J., Kieboom, K., Marino, S., DePinho, R. A., and van Lohuizen, M. (1999) The oncogene and Polycomb-group gene *bmi-1* regulates cell proliferation and senescence through the *ink4a* locus. *Nature*, **397**, 164.

75. Schumacher, A. and Magnuson, T. (1997) Murine *Polycomb*- and *trithorax*-group genes regulate homeotic pathways and beyond. *Trends Genet.*, **13**, 167.

76. Ma, H. (1997) Polycomb in plants. *Trends Genet.*, **13**, 167.

77. Kennison, J. A. (1995) The Polycomb and trithorax group proteins of *Drosophila*: *trans*-regulators of homeotic gene function. *Annu. Rev. Genet.*, **29**, 289.

78. Peterson, C. L. and Tamkun, J. W. (1995) The SWI/SNF complex: a chromatin remodeling machine? *Trends Biochem. Sci.*, **20**, 143.

79. Moazed, D. and O'Farrell, P. H. (1992) Maintenance of the *engrailed* expression pattern by *Polycomb* group genes in *Drosophila*. *Development*, **116**, 805.

80. Franke, A., DeCamillis, M., Zink, D., Cheng, N., Brock, H. W., and Paro, R. (1992) *Polycomb* and *polyhomeotic* are constituents of a multimeric protein complex in chromatin of *Drosophila melanogaster*. *EMBO J.*, **11**, 2941.

81. Peterson, A. J., Kyba, M., Bornemann, D., Morgan, K., Brock, H. W., and Simon, J. (1997) A domain shared by the *Polycomb* group proteins Scm and ph mediates heterotypic and homotypic interactions. *Mol. Cell. Biol.*, **17**, 6683.

82. Sewalt, R. G., Gunster, M. J., van der Vlag, J., Satijn, D. P., and Otte, A. P. (1999) C-Terminal binding protein is a transcriptional repressor that interacts with a specific class of vertebrate Polycomb proteins. *Mol. Cell. Biol.*, **19**, 777.

83. Breiling, A., Bonte, E., Ferrari, S., Becker, P. B., and Paro, R. (1999) The *Drosophila Polycomb* protein interacts with nucleosomal core particles *in vitro* via its repression domain. *Mol. Cell. Biol.*, **19**, 8451.

84. Shao, Z., Raible, F., Mollaaghababa, R., Guyon, J. R., Wu, C. T., Bender, W., and Kingston, R. E. (1999) Stabilization of chromatin structure by PRC1, a Polycomb complex. *Cell*, **98**, 37.

85. Kehle, J., Beuchle, D., Treuheit, S., Christen, B., Kennison, J. A., Bienz, M., and Muller, J. (1998) dMi-2, a hunchback-interacting protein that functions in *Polycomb* repression. *Science*, **282**, 1897.

86. Ball, L. J., Murzina, N. V., Broadhurst, R. W., Raine, A. R. C., Archer, S. J., Stott, F. J., Murzin, A. G., Singh, P. B., Domaille, P. J., and Laue, E. D. (1997) Structure of the chromatin binding (chromo) domain from mouse modifier protein 1. *EMBO J.*, **16**, 2473.

87. Brown, J. L., Mucci, D., Whiteley, M., Dirksen, M. L., and Kassis, J. A. (1998) The *Drosophila Polycomb* group gene *pleiohomeotic* encodes a DNA binding protein with homology to the transcription factor YY1. *Mol. Cell*, **1**, 1057.

88. Simon, J., Chiang, A., Bender, W., Shimell, M.J., and O'Connor, M. (1993) Elements of the *Drosophila bithorax* complex that mediate repression by *Polycomb* group products. *Dev. Biol.*, **158**, 131.

89. Orlando, V. and Paro, R. (1993) Mapping Polycomb-repressed domains in the bithorax complex using *in vivo* formaldehyde cross-linked chromatin. *Cell*, **75**, 1187.

90. Strutt, H., Cavalli, G., and Paro, R. (1997) Co-localization of Polycomb protein and GAGA factor on regulatory elements responsible for the maintenance of homeotic gene expression. *EMBO J.*, **16**, 3621.

91. Chang, Y. L., King, B. O., O'Connor, M., Mazo, A., and Huang, D. H. (1995) Functional reconstruction of *trans* regulation of the *Ultrabithorax* promoter by the products of two antagonistic genes, *trithorax* and *Polycomb*. *Mol. Cell. Biol.*, **15**, 6601.

92. Cavalli, G. and Paro, R. (1998) The *Drosophila* Fab-7 chromosomal element conveys epigenetic inheritance during mitosis and meiosis. *Cell*, **93**, 505.

93. Cavalli, G. and Paro, R. (1999) Epigenetic inheritance of active chromatin after removal of the main transactivator. *Science*, **286**, 955.

94. Partridge, J. F., Borgstrom, B., and Allshire, R. C. (2000) Distinct protein interaction domains and protein spreading in a complex centromere. *Genes Develop.*, **14**, 783.

95. Wallrath, L. L. (1998) Unfolding the mysteries of heterochromatin. *Curr. Opin. Genet. Devel.*, **8**, 147.

96. Frolov, M. V., Benevolenskaya, E. V., and Birchler J. A. (1998) *Regena* (*Rga*), a *Drosophila* homolog of the global negative transcriptional regulator *CDC36* (*NOT2*) from yeast, modifies gene expression and suppresses position effect variegation. *Genetics*, **148**, 317.

97. Laible, G., Wolf, A., Dorn, R., Reuter, G., Nislow, C., Lebersorger, A., Popkin, D., Pillus, L., and Jenuwein, T. (1997) Mammalian homologues of the *Polycomb*-group gene *Enhancer of zeste* mediate gene silencing in *Drosophila* heterochromatin and at *S. cerevisiae* telomeres. *EMBO J.*, **16**, 3219.

98. Sinclair, D., Milne, T., Hodgson, J., Shellard, J., Salinas, C., Kyba, M., Randazzo, F., and Brock, H. (1998) The *additional sex combs* gene of *Drosophila* encodes a chromatin protein that binds to shared and unique Polycomb group sites on polytene chromosomes. *Development*, **125**, 1207.

99. Frolov, M. V. and Birchler, J. A. (1998) Mutation in P0, a dual function ribosomal protein/apurinic/apyrimidinic endonuclease, modifies gene expression and position effect variegation in *Drosophila*. *Genetics*, **150**, 1487.

12 | Epigenetics in mammals

CHRISTOPHER J. SCHOENHERR and SHIRLEY M. TILGHMAN

1. Introduction

It is usually taken for granted in a diploid organism that the two parental alleles of genes that reside on homologous chromosomes function identically. Aside from polymorphic differences between alleles, this assumption is generally true. There are, however, several important instances in mammalian biology where this assumption does not hold. These exceptions are examples of epigenetic regulation that include both random events and highly regulated processes. In this chapter, the discussion of epigenetic regulation will largely be restricted to processes that result in different stable states of transcription between alleles. This allelic discrimination is not determined by DNA sequence alone, but is governed by both covalent modifications, such as DNA methylation, and by different chromatin structures that assemble on the genome.

We describe two examples of determinate epigenetic regulation in mammals—genomic imprinting and X-chromosome inactivation. Both require a complex system to establish and maintain allelic discrimination. Although the basis for discrimination is not completely understood, there is evidence for both novel and established regulatory mechanisms that involve alteration of long- and short-range chromatin organization.

2. Genomic imprinting

2.1 The discovery of genomic imprinting

Genomic imprinting in mammals refers to the unequal expression of autosomal genes based on their parent of origin. These parent-of-origin effects are thought to be exerted at the level of transcription initiation, although other regulatory pathways have not been ruled out. This non-Mendelian pattern of gene expression renders the organism functionally hemizygous for all imprinted genes, putting it at increased genetic risk.

The functional non-equivalence of the maternal and paternal genomes was first suspected from the inviability of uniparental diploid embryos generated by nuclear transplantation (1, 2). In these experiments, one of the two pronuclei was removed

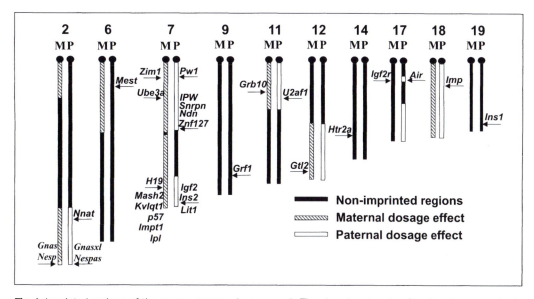

Fig. 1 Imprinted regions of the mouse genome (autosomes). The closed segments of each autosome refer to regions that display no deleterious phenotypes upon maternal duplication/paternal deficiency; open regions show a maternal duplication/paternal deficiency phenotype, and hatched regions show paternal duplication/maternal deficiency effects. The approximate map positions of imprinted genes are indicated, with maternally expressed genes on the left of each autosome and the paternally expressed genes on the right. Adapted from Beechey (http://www.mgu.har.mrc.ac.uk). Following standard convention, mouse genes have initial capitals and are in italics, and human genes are in all capital letters and italics.

from a fertilized egg, and replaced with a pronucleus from a second egg. When the reconstituted embryo contained both a male and a female pronucleus, normal development was observed. However, when both pronuclei were either male or female derived, the embryos developed abnormally. The developmental failure of these uniparental embryos was interpreted to suggest that there must be a gene or genes that are only expressed when inherited from one parent or the other, a process that was termed 'genomic imprinting.'

Further proof that such genes existed came from the study of mice that had developmental defects when either entire chromosomes or subchromosomal regions were inherited from one parent but not the other (see ref. 3 for a review). A nearly complete scan of the mouse genome using both deletions and uniparental duplications narrowed down the genomic regions that might harbour imprinted genes to 11 (Fig. 1 and http://www.mgu.har.mrc.ac.uk/imprinting/all_impmaps.html). It should be noted, however, that the list of regions is not exhaustive, as the screen only scored visible phenotypes. Indeed, imprinted genes have recently been identified in regions that went undetected in this screen.

The search for imprinted genes, which began in 1991 with the discovery of the paternal-specific expression of the insulin-like growth factor II (*Igf2*) gene (4), has identified more than 30 genes to date (Fig. 1). The list includes genes with many different biochemical functions, including growth factors, transcription factors, cell-cycle regulators, a potassium channel, and a surprising number of non-coding RNAs. A substantial number of these genes are expressed during embryogenesis in both the embryo and placenta, and function to regulate fetal growth. Finally it is often the case that they are clustered in the genome.

2.2 The epigenetic mark in genomic imprinting

The fundamental issues in genomic imprinting are how the cell discriminates between two identical genes and how that discrimination results in somatically heritable differences in gene expression. The solution must be a 'mark' or 'imprint' that is placed on one of the two parental chromosomes during gametogenesis or during the early cell divisions of the embryo, where it has recently been shown that the two parental genomes are maintained in separate compartments (5). This epigenetic mark would serve as the basis for a mechanism that leads to differences in gene expression.

The epigenetic mark must fulfil two important criteria. First, it must be capable of being propagated through many cell divisions and must remain restricted to one chromosome. Secondly, the mark must be labile at some point in germ-cell development, so that it can be erased and reset to the appropriate parental state (Fig. 2).

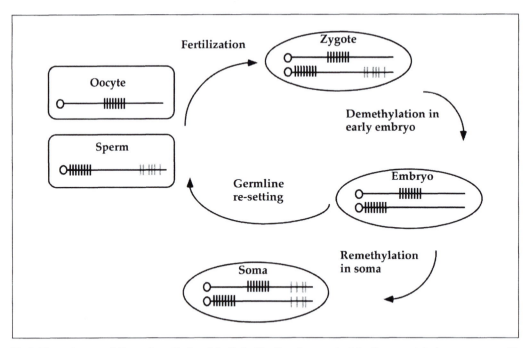

Fig 2 The ontogeny of the epigenetic imprint. At fertilization the zygote inherits parental chromosomes with gametic methylation imprints (vertical black bars) as well as other DNA methylation that is removed in the early embryo (grey bars). The embryonic cells are distributed between the soma, which is re-methylated, and the germline, where methylation is erased and reset in an allele-specific manner.

The ability to reset the mark to match the sex of the parent is essential for proper gene expression in the next generation (Fig. 2).

The favoured candidate for an epigenetic mark is the reversible covalent modification of DNA by methylation, where the cytosine residue in a CpG dinucleotide is methylated at the N_7 position of the pyrimidine ring. CpG methylation is a particularly attractive means to achieve allelic differences because it can readily fulfil the theoretical criteria listed above. It is known to be stably propagated through many cell divisions by virtue of a maintenance DNA methyltransferase, the preferred substrate of which is hemi-methylated DNA at replication forks (6). Conversely, DNA methylation can be erased either by a DNA demethylase, although none has been isolated to date, or by simple dilution through DNA replication in the absence of remethylation.

DNA methylation has another attractive feature as an epigenetic mark, as it has been implicated in gene silencing, based on the correlation between reduced gene transcription and DNA methylation (7). Many genes are more heavily methylated when they are in a silent configuration than when they are actively transcribed. In addition, the DNA in mammalian heterochromatin is heavily methylated. Several mechanisms for DNA-mediated gene silencing have been proposed. The most straightforward is the direct inhibition of the binding of specific transcription factors (8). More recently it has been proposed that DNA methylation acts as a primary signal to assemble chromatin that inhibits gene expression (9, 10). This is thought to occur via the binding of the methyl-binding protein MeCP2, which is part of a histone deacetylase complex (Chapter 8). Thus methylation can serve as both the epigenetic mark that distinguishes the alleles as well as part of the silencing mechanism itself.

Indeed, for a number of imprinted genes, DNA methylation has been shown to be required for the maintenance of imprinting in the embryo. Li *et al.* (11) used a mutation in *Dnmt1* (*DNA methyltransferase 1*), the gene encoding the maintenance DNA methylase, to show that homozygous mutant embryos could not sustain mono-allelic expression of *H19*, *Igf2*, and *Igf2r* in the absence of methylation. Similar findings have now been reported for *Snrpn*, *Kvlqt1*, and *p57Kip2* (12, 13). Interestingly, the effects of loss of methylation were not uniform. For some genes, including *H19* and *Snrpn*, loss of methylation led to reactivation of the silent allele, suggesting that methylation is required to maintain repression of some imprinted genes. In contrast, loss of methylation repressed *Igf2* and *Igf2r*, suggesting that the expression of those genes requires methylation. The significance of this difference will be discussed below.

The response of imprinted genes to loss of DNA methylation is consistent with the common finding that imprinted genes have in their vicinity regions of DNA where one allele is methylated and the other is not. The presence of differential methylation, however, does not guarantee that it represents an epigenetic mark. For a *differentially methylated region* (DMR) to be classified as a true epigenetic mark, the methylation difference must be present in the gametes and survive two critical events: the genome-wide demethylation that occurs shortly after fertilization and a genome-wide remethylation that coincides with implantation (Fig. 2) (14, 15). It has been observed at several imprinted loci that differential methylation at some sites is acquired later in development as a consequence, rather than the cause, of allelic expression differences (e.g. see ref. 16).

DNA methylation may not be the only epigenetic mark for imprinting. For example, loss of DNA methylation has no effect on the imprinting of *Mash2* (12, 17). In such cases, differential protein binding to the parental chromosomes could act as the epigenetic mark. To maintain the difference, however, the protein needs to bind cooperatively to DNA, possibly as an oligomer that can be segregated symmetrically at replication forks. Of course, protein and methylation-based marks are not mutually exclusive. Furthermore, they may be used in succession, with one to initiate imprinting early in development and the other to maintain the imprint for the remainder of the organism's life.

The most obvious consequence of an imprinting mark is the alteration of gene expression. However, other epigenetic differences are consistently observed at imprinted loci. These include differences in the timing of DNA replication, with the paternal alleles of imprinted loci tending to replicate earlier than the maternal alleles (18, 19). This difference reflects the imprinting status of the chromosomes, as shown by changes in the asymmetry of replication timing on chromosomes that have mutations that disrupt imprinting (20, 21). Furthermore, imprinted genes show an unusual association between the two parental chromosomes during late S phase (22). Finally, the rate of meiotic recombination at imprinted loci has been shown to differ between parental chromosomes (23). None of these epigenetic effects are well understood, however.

2.3   The *Ipl-H19* imprinted gene cluster: regulating enhancers

Many imprinted genes are found in clusters that can span over a megabase of DNA. The best studied cluster resides on distal chromosome 7 in the mouse and its syntenic region, 11p15.5, in humans. This region is over one megabase long and contains at least seven maternally expressed and two paternally expressed genes interspersed with several biallelic genes (Fig. 3A). Of the genes in this cluster, the imprinting mechanism of *H19* and *Igf2* has received the most attention. They are adjacent genes, separated by about 90 kb of DNA, and are oppositely imprinted, with *Igf2* paternally and *H19* maternally expressed.

An epigenetic mark has been identified between −2 and −4 kb in the 5′ flanking DNA of the *H19* gene. This element is highly methylated in sperm, but not in oocytes, and this difference is preserved during preimplantation demethylation, as well as in all adult tissues (24). Most importantly deletion of this element leads to the loss of imprinting of both *H19* and *Igf2*, as well as *Ins2*, a gene that is imprinted in the extraembryonic tissues of the mouse, but not the embryo (25, 26). For this reason we refer to this element as an *i*mprinting *c*ontrol *r*egion, or ICR. In contrast, the heavily methylated *H19* promoter and structural gene, as well as two DMRs within the *Igf2* locus, are methylated preferentially on the paternal chromosome, but are not epigenetic marks, based on the observation that their differential methylation patterns are established after fertilization (14, 27).

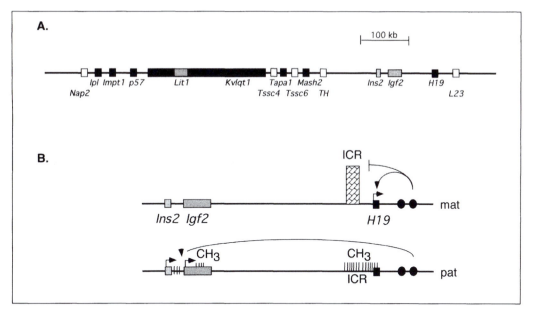

Fig. 3 The structure of the *Ipl–H19* region. (A) The organization of the maternally expressed (black boxes), paternally expressed (grey boxes), and non-imprinted (open boxes) genes in the *Ipl–H19* region in the distal end of mouse chromosome 7. (B) The imprinting of the *Ins2*, *Igf2*, and *H19* genes depends upon the ICR, which is methylated on the paternal chromosome (CH_3) and silences *H19*, and unmethylated on the maternal chromosome where it assembles a chromatin enhancer blocker (rectangular wall). The enhancers are depicted by closed ovals.

The role of the *H19* ICR in establishing and maintaining the imprinting of the three genes in the distal end of the cluster is becoming clear. On the paternal chromosome, where the element is methylated, it is likely that the epigenetic mark serves to nucleate the acquisition of DNA methylation of the region surrounding *H19*, including its promoter. As noted above, loss of DNA methylation results in the reactivation of the paternal *H19* allele, consistent with this model (11). Furthermore the paternal *H19* promoter is resistant to nuclease digestion (28, 29), suggesting that its methylation leads to a closed chromatin structure.

In contrast to *H19*, both the active and the inactive promoters of *Igf2* are unmethylated and hypersensitive to nucleases (30). Furthermore the gene is silenced upon demethylation of the genome, not activated, suggesting that methylation is required to maintain its expression on the paternal chromosome. One model to explain these observations suggested that the methylated DMRs within the locus prevented repressors from binding to the gene (31). A direct test of this idea, however, argued against this model, at least in the early mouse embryo (32). This did not rule out the possibility that paternal *Igf2* methylation could help to maintain differential expression later in development.

A second model proposed a competition between the promoters of *Igf2* and *H19* on the maternal chromosome for enhancers that lie 3′ of *H19* silenced *Igf2* (33, 34). The unmethylated *H19* gene on the maternal chromosome was presumed to win due to a stronger promoter or to its closer proximity to the enhancers (Fig. 3B). This model was eliminated when it was demonstrated that two different deletions of the *H19* promoter showed very little or no effect on *Igf2* imprinting (35, 36).

The most recent proposal suggests that the unmethylated ICR acts as a chromatin boundary element (37, 38). Boundary elements, or insulators, are believed to divide chromosomes into separate functional domains (Chapter 13). They are also capable of blocking the activation of genes if placed between a promoter and an enhancer. The ICR lies between *Igf2* and its enhancers 3' of *H19*, and thus it is in a position to prevent the enhancers from activating *Igf2* on the maternal chromosome (Fig. 3B). A corollary of the boundary model is that methylation of the ICR on the paternal chromosome would inactivate the element and allow the enhancers access to *Igf2*.

Initial biochemical evidence for a boundary at the maternal ICR came from chromatin studies that showed this region was hypersensitive to nucleases (38, 39), a property of boundary elements. Genetic evidence has come from two targeting experiments in mice. In the first, the endodermal enhancers were moved from their normal 3' position and placed midway between *Igf2* and *H19* (37). Upon maternal transmission of this mutation, *Igf2* was now expressed in endodermal tissues, but *H19* was not, even though it remained unmethylated. While other interpretations are possible, this is exactly the result that a boundary model would predict. More direct evidence was obtained by deleting the ICR; this led to *Igf2* expression on the maternal chromosome along with *H19*, albeit both at lower than normal levels (25). Importantly, *H19* was activated on the paternal chromosome with the ICR deletion, indicating that the methylated ICR is required for repression of *H19*. Thus the ICR is a bifunctional element that prevents activation of *Igf2* on the maternal chromosome and, when methylated, of *H19* on the paternal chromosome.

Direct evidence that the *H19* ICR has position-dependent enhancer-blocking activity has come from cell culture and mouse transgenic systems, in which both the mouse and the human ICR were shown to block activation when placed between an enhancer and a promoter (40, 41). In these assays the ICR had no silencing activity, in contrast to experiments in *Drosophila* where the unmethylated ICR showed silencing activity in transgenic flies (42). While the significance of this is unclear, it suggests that *Drosophila* and mammals have different proteins capable of binding to this element.

An excellent candidate for mediating the activity of the ICR is CTCF, a zinc-finger protein that has been implicated in several other vertebrate boundary elements (43) (Chapter 13). CTCF binds *in vitro* to the mouse and human ICRs at sites that coincide with the nuclease hypersensitive sites in chromatin (40, 41). Mutations in these sites eliminate enhancer blocking of the ICR, implying that CTCF is the main mediator of blocking activity *in vivo*. Most importantly, DNA methylation dramatically inhibits the binding activity of CTCF. This strongly suggests that at least one consequence of ICR paternal methylation is to prevent CTCF binding and the formation of the block. Consequently, *Igf2* can be expressed. Thus CTCF is the first protein to be implicated in the activity of an epigenetic mark.

The bifunctional ICR helps explain why the *H19* and *Igf2* genes have remained linked. Yet the ICR has no impact on the imprinting of the more telomeric genes in the cluster, such as *Mash2* and *p57^{Kip2}* (12), leaving open the question of why those genes are clustered with *H19* and *Igf2*. The most convincing evidence for some level of long-range coordinate regulation comes from a subset of humans with Beckwith–Wiedemann syndrome (BWS). BWS is a clinically variable disorder characterized by somatic overgrowth, macroglossia, abdominal wall defects, visceromegaly, and an increased susceptibility to childhood tumours (44). In some patients, the overgrowth phenotype is thought to result from the biallelic expression of *IGF2* that is common in BWS (45). In some cases, the loss of *IGF2* imprinting can be attributed to methylation of both alleles of the *H19* ICR, consistent with the boundary model. Other patients, however, show biallelic expression of *IGF2* but normal ICR methylation, suggesting the existence of an additional imprinting element that also regulates the repression of maternal *IGF2*. Circumstantial evidence for this second element has implicated a DMR in the *KVLQT1* gene (Fig. 3A). This DMR is associated with a paternal-specific transcript, *LIT1*, in both humans and mouse (46, 47) that has not been well characterized. Should *LIT1* have a role in imprinting *IGF2*, it cannot be an essential role, based on the ability of transgenes containing *Igf2* and *H19* to be imprinted when at ectopic sites (48).

2.4 The *Igf2r* gene cluster: regulating antisense transcripts

A second well-studied imprinted region contains the insulin-like growth factor receptor (*Igf2r*) locus (Fig. 4). A DMR is contained within the second intron of the gene, and is methylated on the expressed maternal allele (16). The region, referred to as DMR2, satisfies the criteria for the epigenetic mark as its methylation is present in the oocyte but not in sperm, and remains throughout development. Furthermore DMR2 is required for the imprinting of an *Igf2r* transgene (49). Thus, as is the case for *H19*, an epigenetic mark is coincident with an imprinting element.

The function of DMR2 in imprinting *Igf2r* has not been resolved. By analogy to the *H19* ICR, DMR2 could function as a chromatin boundary for hypothetical down-

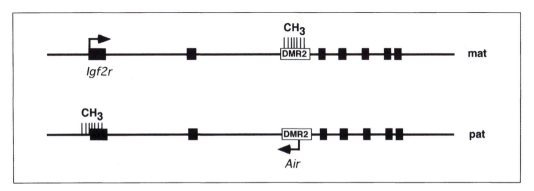

Fig. 4 The structure of the *Igf2r* locus. Arrows indicate the active promoters of the Igf2r and Air transcripts. Exons of the *Igf2r* gene are shown as black boxes. CH₃ indicates DNA methylation.

stream enhancers. Methylation would then inactivate the boundary on the maternal chromosome, allowing *Igf2r* expression. A second class of models involves an unspliced nuclear antisense transcript (*antisense imprinted RNA; Air*) that initiates in the unmethylated DMR2 on the paternal chromosome (49). The first model posits that the *Igf2r* and *Air* promoters compete for enhancers in a manner similar to that originally proposed for *H19* and *Igf2* (see above). In the second model, *Air* transcripts, which overlap *Igf2r* RNA in an antisense direction, are proposed to hybridize to and destabilize paternal *Igf2r* RNA in *cis* (Fig. 4). In both scenarios, methylation of DMR2 inhibits the *Air* promoter, thereby allowing expression of maternal *Igf2r*. This is consistent with the silencing of *Igf2r* in the absence of methylation (11). In addition, *Air* expression is disrupted by the DMR2 deletion that activates paternal *Igf2r*, suggesting that paternal *Air* transcription represses *Igf2r* in *cis*.

Antisense transcripts are frequently observed at imprinted loci. *Gnas*, *Kvlqt1*, and *UBE3A* all have oppositely imprinted, antisense transcripts associated with them (47, 50, 51). This could represent a rather unique mode of regulation for epigenetic loci, as it appears that X inactivation may be regulated by an overlapping, antisense transcript as well (see below). On the other hand, antisense transcripts found at *Igf2* and *Zfp127* are expressed from the same chromosome as the sense RNAs, suggesting that antisense transcripts may commonly be found in transcribed regions (52, 53). Consistent with this possibility, non-coding sense and antisense transcripts have also been found associated with non-imprinted regions such as the β-globin locus (54). At present, an unambiguous function for antisense transcripts in imprinting or X inactivation has not been established.

Recent work on DMR2 has provided insight into the question of how differential methylation may be established and maintained on imprinted genes. Birger *et al.* (55) identified a small sequence in DMR2 that is necessary for the *de novo* methylation of transgenes injected into the female pronucleus. This sequence and its binding protein may represent signals for targeted methylation of imprinted regions. In contrast, a second sequence was found that maintained the unmethylated state of DMR2 transgenes injected into the male pronucleus. Interestingly, this sequence bound a factor that may be paternally expressed, as it was absent from an embryonic cell line made from parthenogenetic embryos. These results imply that the methylation status of both alleles is regulated and neither the methylated nor the unmethylated state can be considered default.

Igf2r is expressed bi-allelically early in development, even though DMR2 is differentially methylated. Mono-allelic expression in the mouse does not begin until after the embryo implants into the uterus, at about 5 days of development (56). This early bi-allelic transcription argues that the presence of a mark is not sufficient for imprinted expression, but that developmentally regulated protein factors are required to interpret the imprinting signal. The need for tissue-specific factors also can be inferred from the observation that several imprinted genes are known to be imprinted in a tissue-specific manner. For example, *Igf2* is expressed bi-allelically in the adult brain (4), whereas *UBE3A* is *only* imprinted in the brain (57, 58).

2.5 The Prader–Willi / Angelman imprinted gene cluster: long-range regulation

A third cluster of imprinted genes on mouse chromosome 7 and human chromosome 15q11–q13 bears both striking similarities as well as differences to the *Ipl/H19* and *Igf2r* regions. This cluster extends over two megabases of DNA, and contains at least four imprinted protein-coding genes and several imprinted non-coding transcripts (59) (Fig. 5). Three of the protein-coding genes, *SNRPN*, *NDN*, and *ZNF127*, are paternally expressed and the remaining gene, *UBE3A*, is maternally expressed. In addition, several paternally expressed non-coding transcripts (*IPW*, *PAR1*, and *PAR5*) have been identified. The function, if any, of these RNAs is unknown.

This region is associated with two distinct human genetic disorders: *Prader–Willi* syndrome (PWS) and *Angelman* syndrome (AS). PWS is characterized by neonatal hypotonia, hypogonadism, obesity, mild dysmorphism, and mild mental retardation. AS children display hyperactivity, ataxia, seizures, and severe mental retardation. While both diseases map to the same region, PWS is associated with large paternal deletions and maternal disomies, whereas AS is associated with maternal deletions and paternal disomies. The parent-of-origin effects indicate that PWS is caused by the loss of expression of paternally expressed gene(s) and AS, the loss of expression of maternally expressed gene(s). Mutations made in mice suggest that the loss of expression of *NDN* and at least one other paternally expressed gene can give rise to PWS (60–62). For AS, mutations in *UBE3A* were found in several AS patients, strongly suggesting that mutations in this gene are solely responsible for AS.

The two parental chromosomes are differentially methylated in the large PWS-AS region, with genes on the maternal chromosome generally more methylated than those on the paternal chromosome (63). This difference is most evident in a region surrounding the *SNRPN* promoter, which is CpG-rich and heavily methylated on the maternal chromosome in both humans and mice. Furthermore, Shemer *et al.* (13)

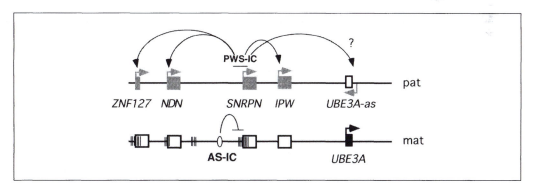

Fig. 5 The structure of the Prader–Willi/Angelman region. The paternally expressed (grey boxes) and maternally expressed (black box) genes in the PWS/AS region are indicated. The PWS-IC (horizontal line) positively regulates the paternally expressed genes while the AS-IC (oval) appears to negatively regulate the PWS-IC on the maternal chromosome. Vertical lines indicate sites of DNA methylation specific to the maternal chromosome. *PAR1* and *PAR5* are not included, as they have not been shown to be independent genes.

have shown that this region of differential methylation has the properties of an epigenetic mark, in that it is established in the gametes and resists alteration after fertilization. Mice that harbour a large deletion that encompasses the differentially methylated region lack paternal expression of *Zfp127*, *Ndn*, and *Ipw*, and manifest several phenotypes common to PWS infants (60).

The mouse studies are also consistent with the phenotypes of PWS and AS patients with small deletions near *SNRPN*. By comparing the extent of microdeletions from different patients, a minimal region of overlap for each disease has been defined (64, 65). For PWS, the region roughly coincides with the differentially methylated region surrounding the *SNRPN* promoter. In AS microdeletion patients, in contrast, the AS minimal deletion was refined to a 1 kb region that lies 40 kb upstream of *SNRPN* and does not show differential methylation (Fig. 5) (66). Paternal transmission of PWS microdeletions is associated with abnormal methylation and repression of several paternal genes (e.g. *ZNF127*, *NDN*, and *IPW*). In contrast, maternal transmission of AS microdeletions results in the loss of methylation and activation of the same paternal genes. Together these regions have been designated *i*mprinting *c*entres (PWS-IC and AS-IC) because they coordinately regulate the imprinting of multiple genes.

The PWS-IC appears to be mechanistically different from the ICR at *H19* and the DMR2 at *Igf2r*, which act locally to negatively regulate the expression of only two or three genes. In contrast, the PWS-IC is necessary for activating or maintaining paternal expression of multiple genes that are spread over a megabase of DNA (67). In addition, there is indirect evidence that the PWS-IC is essential for the resetting of maternal methylation to the unmethylated paternal state in the male germline (59).

The AS-IC is perhaps even more enigmatic. When it is deleted on a maternally inherited chromosome, the PWS-IC becomes unmethylated. This data suggests that the AS-IC is necessary for maintaining maternal methylation of the PWS-IC or that it is necessary for methylation of paternal chromosomes in the female germline. Although preferential nuclease hypersensitivity has been found, no differential methylation has been detected on the human element, making it unclear how the AS-IC might drive allele-specific methylation at such a distance (~40 kb) (66, 68). Although this has not been shown directly, the AS-IC also is likely to positively regulate the expression of maternal *UBE3A*, the AS gene. This activation, however, may require the presence of the PWS-IC: there are microdeletions that remove both ICs that result only in PWS, but no microdeletions have been found that remove both ICs and result in AS. These results suggest that the PWS-IC represses *UBE3A* in *cis*. To explain how the PWS-IC could do this, Brannan and Bartolomei (69) suggested that a recently discovered antisense transcript to *UBE3A* is transcriptionally activated by the PWS-IC. This transcript is paternally expressed and could repress *UBE3A* in a manner analogous to that proposed for *Igf2r/Air* (Fig. 5). Thus, activation of the PWS-IC by deletion of the AS-IC would drive expression of the antisense transcript and down-regulate *UBE3A*.

3. X-chromosome inactivation

3.1 Introduction

Dosage compensation in mammals, a process that evolved to equalize the expression of X-linked genes in XX females and XY males, is achieved by the inactivation of one of the two X chromosomes in females. In eutherian mammals, the choice of which parental X chromosome to inactivate is a random one, and thus female mammals are a mosaic of cells that have inactivated either the male-inherited X chromosome or the female-inherited X chromosome. This form of dosage compensation is therefore fundamentally different from dosage compensation mechanisms in invertebrates, for example the equal suppression of transcription from both X chromosomes in XX females of *Caenorhabditis elegans*, or the hyper-transcription of the single X chromosome in male *Drosophila* embryos (70). These organisms, therefore, are not mosaics, and all somatic cells are equivalent in their expression of X-linked genes. The readers are referred to an excellent and comprehensive review that covers aspects of X-chromosome inactivation that are not discussed in this chapter (71).

The process of X inactivation is a remarkable one, as it results in the silencing of an entire chromosome. The inactivated X chromosome is cytologically distinguishable in mitotic chromosome spreads as a darkly staining, highly condensed structure termed the Barr body, after its discoverer (72). The first insights into the mechanism of X inactivation came from the study of humans and mice with translocations involving the X chromosome (73, 74). It was noted that in reciprocal translocations with autosomes, only one of the two reciprocal pairs displayed any spreading of the condensed structure into the adjacent autosome. By comparing large numbers of such individuals, it was possible to identify a region of the X chromosome that was always present on the partially inactivated chromosomes. This was termed the *XIC/Xic*, for X chromosome *i*nactivation *c*entre. It was hypothesized that the *XIC/Xic* was a site on the X chromosome that was required to initiate X-chromosome inactivation.

Significant progress in understanding the function of the *XIC* came with the discovery of *XIST*, the first gene to map to the *XIC* in both humans and mice (75, 76). *XIST* is expressed exclusively from the inactive X chromosome, unlike all other X-linked genes, and is readily detected in all female cells, but not in male cells. Furthermore the large (17 000 bases long) transcript of the human gene is spliced and polyadenylated but does not contain any open reading frame in common with the mouse *Xist* gene. The RNA product is retained in the nucleus, and in fact, it appears to coat the entire inactive X chromosome in both mouse and humans (77, 78). These properties are sufficiently unusual to make the *Xist* gene a prime candidate for playing a functional role in X-chromosome inactivation.

More recently a new candidate gene for regulating X-chromosome inactivation has been described by Lee and her colleagues (79). The *Tsix* gene encodes an RNA transcript that is initiated within a CpG island 30 kb downstream of *Xist* and proceeds in an antisense direction through the *Xist* transcription unit (Fig. 6). This nucleus-restricted transcript is large, and is transcribed in embryonic cells that have not

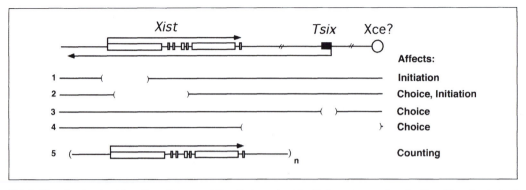

Fig. 6 The structure and function of genes at the *Xic* important for X-chromosome inactivation. The structure of the *Xist* gene, and the relative positions of the *Tsix* promoter and *Xce* are indicated. The deletions that have been made at the locus, and the events in X-chromosome inactivation that are affected are indicated below (1, ref. 81; 2, ref. 90; 3, ref. 92; 4, ref. 91. 5, the structure of transgenes that are counted in embryonic stem cells, ref. 84).

initiated X-chromosome inactivation. Upon X inactivation, transcription of *Tsix* first becomes restricted to the future active X chromosome, and eventually it is repressed on both X chromosomes in differentiated cells. These observations have led to the proposal that *Tsix* is a negative regulator of *Xist* that blocks the accumulation of *Xist* RNA on the future active X chromosome (79).

Xist and *Tsix* have become the focus of attention for those studying the process of X-chromosome inactivation. The process can be divided into a series of discrete steps that have been separated genetically. These include counting the number of X chromosomes to determine whether to inactive one or more of them, choosing which X chromosome to inactivate, initiating the process at the *XIC/Xic* locus, propagating the inactive state throughout the chromosome, and maintaining the silent state through somatic cell divisions.

3.2 Counting

All organisms that perform dosage compensation have a mechanism for sensing or counting the number of sex chromosomes relative to the number of autosomes. In mammals, the counting mechanism must be able to detect the difference between a ratio of 2X:2A in females and 1X:2A in males, where A is the haploid number of autosomes. Female XO animals with a 1X:2A ratio do not initiate X-chromosome inactivation, whereas XXY males with a 2X:2A ratio do, illustrating that the process is not restricted to females. Furthermore, XXX females inactivate two X chromosomes, while tetraploid individuals (4X:4A) inactivate two X chromosomes. These observations have led to the 'n – 1' rule, which posits that in the presence of the normal diploid complement of autosomes, the organism will inactivate all but one X chromosome.

The elements on the X chromosome and on the autosomes that are counted have not been precisely defined. However, an experimental approach to the problem is provided by embryonic stem (ES) cells, pluripotent cells that are derived from mouse embryos before X inactivation has occurred. Female ES cells have two active X chromosomes, but upon *in vitro* differentiation, they randomly inactivate one (80, 81). Male ES cells, as expected, show no X chromosome inactivation upon differentiation. When a yeast artificial chromosome-derived transgene that spanned the *Xist* locus was introduced into male ES cells, the endogenous X chromosome initiated inactivation, suggesting that the transgene was being counted as an X chromosome according to the '*n* – 1' rule (82). As one would predict, inactivation of the endogenous X chromosome resulted in cell death. Using this approach with progressively smaller transgenes, several groups have narrowed down the region required for initiating X inactivation to the *Xist* gene along with 9 kb of upstream and 6 kb of downstream sequence (Fig. 6) (83, 84). Thus the *Xist* gene itself may be a component of the counting mechanism. The *Tsix* gene, however, was excluded, as it is absent from some constructs that are counted.

3.3 Choice

In most mammals, the soma consists of *c*. 50:50 mixtures of cells that have inactivated one of the two parental X chromosomes, implying that each chromosome has an equal likelihood of being chosen for inactivation. However, complete randomness is not always observed. In crosses between distantly related species of laboratory mice, the choice can be skewed as much as 65:35 (85), and interspecific crosses of a second mouse genus, *Peromyscus*, results in a skew of 85:15 (86). Genetic analysis has shown that skewing is controlled by an X-linked locus called *Xce*, for X-controlling *e*lement (87). The *Xce* affects the primary choice decision, rather than the survival of cells that have made each choice, as demonstrated by Rastan (88) in early mouse embryos.

The *Xce* lies very close to the *Xist* gene itself, but can be separated from it genetically (85). The current status of mapping suggests that *Xce* lies distal to both *Xist* and *Tsix*, and may, in fact, be a regulatory element controlling the expression level of the *Xist* gene (Fig. 6). However, the level of *Xist* RNA expression in various mouse species does not correlate with the strength of their *Xce* alleles. Others have noted a difference in the degree of DNA methylation of the *Tsix* promoter in different *Xce* strains, suggesting that *Xce* could act through epigenetic control of *Xist*/*Tsix* expression (89).

Targeted mutations in the *Xist* gene have greatly strengthened the conclusion that choice is affected by the expression of *Xist* in *cis*. Penny *et al.* (81) first reported that a deletion of the minimal promoter of the *Xist* gene along with 7 kb of the first exon resulted in preferential inactivation of the non-mutated X chromosome in female ES cells (Fig. 6, line 1). These authors suggested that *Xist* functioned downstream of the choice mechanism, based on their finding that differentiated heterozygous female ES cells consisted of two cell populations: those that had inactivated the wild-type X chromosome and those that had two active Xs. The latter population was interpreted to suggest that the *Xist* deletion had not affected choice, but had affected the ability of the mutant chromosome to inactivate. However Marahrens *et al.* (90), analysing a larger deletion of the structural *Xist* gene itself (Fig. 6, line 2), observed non-random inactivation of the wild-type X chromosome in heterozygous females, leading them to conclude that *Xist* is part of the choice mechanism, but not the counting mechanism. These findings can be reconciled if choice requires sequences that are deleted in the Marahrens *et al.* allele (line 2) but not in the Penny *et al.* allele (line 1); that is, sequences within the *Xist* gene itself (Fig. 6). What is clear from both reports, however, is that in the absence of *Xist* expression, X-chromosome inactivation is not initiated in *cis*.

The reciprocal finding was obtained with mutations in the antisense *Tsix* gene. A targeted 65 kb deletion downstream of the *Xist* gene that encompasses the *Tsix* gene and possibly the *Xce* (Fig. 6, line 4) resulted in constitutive inactivation of the deleted X chromosome (91). Thus the downstream region must act to repress inactivation. This activity maps to the *Tsix* gene itself, as a much more restricted deletion of its promoter yielded the same constitutive inactivation of the mutant chromosome in female but not male mutant ES cells (Fig. 6, line 3) (92). This finding has reinforced the notion that *Tsix* negatively regulates *Xist* expression, possibly by blocking the stabilization of *Xist* RNA on the future active X chromosome. In its absence, *Xist* cannot be down-regulated, and its continued expression marks the future inactive X chromosome. This result also demonstrates that *Tsix*, unlike *Xist*, is not required for X chromosome inactivation, but it is a component of the choice mechanism.

3.4 Initiation

The events associated with the initiation of X-chromosome inactivation have been studied primarily in ES cells undergoing differentiation (Fig. 7). Although the temporal order of events described below has not been rigorously determined, the analysis of populations of early differentiating cells implies that the following events occur. Undifferentiated male and female cells co-express *Xist* RNA and *Tsix* RNA at low levels on all X chromosomes (79). Upon differentiation, down-regulation of *Tsix* on the future inactive X chromosome is the first change that can be detected, followed

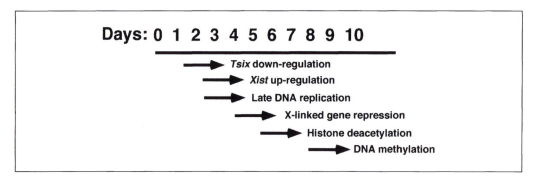

Fig. 7 The temporal order of events in X-chromosome inactivation. X inactivation is initiated in ES cells at day 0. The temporal order of the events that follow are indicated by the positions of the arrow tails.

by *Xist* RNA accumulation along the chromosome in *cis* (78). Recently it has been shown that the accumulation of *Xist* RNA is brought about by stabilization of the transcript, and not by changes in the rate of transcription (93, 94). At a slightly later time, the future active X chromosome turns off transcription of the unstable *Xist* RNA, and finally the low level of *Tsix* transcription on that chromosome is lost.

Consideration of the '*n* – 1' rule has led to a number of general models for the initiation of X-chromosome inactivation. In one model, a hypothetical limiting 'initiation factor' is produced that binds to the future inactive X chromosome. This model suffers from its inability to explain the '*n* – 1' rule, as a limiting initiation factor cannot explain the inactivation of multiple X chromosomes in XXX females. A more attractive model proposes a limited hypothetical 'blocking factor' that binds to the future active X chromosome, and prevents its inactivation (Fig. 8) (95). If the blocking factor is encoded by an autosome, it would explain the activity of two X chromosomes in tetraploid females. The ability of multi-copy *Xist* transgenes to initiate X-chromosome inactivation in male ES cells (84) suggests that they can titrate out this factor and that the binding site must lie within 15 kb of the *Xist* gene itself (see line 5, Fig. 6). More complex models invoking both positive and negative factors have also been proposed (92).

These models suffer from the requirement for a precisely limiting factor, possibly present at just one molecule per cell. This is inherently problematic, unless one allows for considerable error leading to cell death at the outset. Cells that fail to inactivate will die, as will those that inactivate both X chromosomes. A less problematic model suggests that the limiting factor is not a diffusible molecule but a site in the nucleus

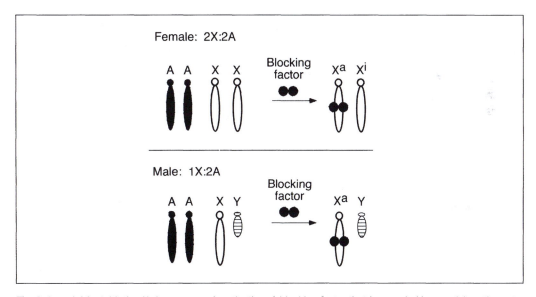

Fig. 8 A model for initiating X-chromosome inactivation. A blocking factor that is encoded by gene(s) on the auto-somes is produced in limiting concentration in both males and females, so as to block a single X chromosome from inactivation. In females, the second X chromosome is then inactivated upon differentiation.

that binds the future active X chromosome. A second possible solution is the generation of a feedback loop that rapidly signals from the chosen active X to all other X chromosomes to initiate their inactivation.

The *Xic* transgenes and targeted mutations described above shed some light on the requirements for initiation. It is clear that the *Xist* gene itself is required in *cis* for initiation, but that the *Tsix* gene is not. Rather, the down-regulation of *Tsix* appears to be required for inactivation, based on the observation that an X chromosome carrying the *Tsix* deletion is always inactivated. How might *Tsix* regulate *Xist*? It has been proposed that *Tsix* may act through its RNA product to base pair with and mask functional domains in *Xist* RNA, or to destabilize the *Xist* RNA. Alternatively *Tsix* transcription may directly affect *Xist* transcription through competition for regulatory elements, or it may change the chromatin structure at the *Xist* promoter (92).

3.5 Propagation and maintenance of X-chromosome inactivation

It is generally thought, based on the single dominant activity of the *Xic*, that X inactivation spreads or propagates from the *Xic* bi-directionally throughout the chromosome. The *Xist* gene is thought to be both necessary and sufficient for propagation of the inactive state. This is argued from transgene studies, in which large *Xist* transgenes integrated into ectopic sites on autosomes are able to initiate the spread of an inactive state on the autosomes (82). This activity is reminiscent of the spread of heterochromatin into autosomal DNA in X : autosome translocations. However, the repressive effects of the transgene require the co-integration of multiple copies of the transgene. Single-copy inserts do not display this property (96). Thus it may be that other sequences on the X chromosome are able to amplify a single *Xist* signal, but autosomes require a high level of *Xist* DNA or RNA.

The molecular mechanism of the activity of *Xist* in propagation is unknown. Both the remarkable 'painting' of the inactive X chromosome by *Xist* RNA and the deletion of *Xist* that prevents X inactivation in *cis* imply a direct role for the RNA. It has been widely hypothesized that *Xist* RNA interacts with proteins that condense chromatin and nucleate the spread of a heterochromatic structure bi-directionally on the chromosome. Consistent with this model, *Xist* RNA spreads along the chromosome prior to gene inactivation, and this binding is resistant to DNAase digestion, implying that it interacts with the protein scaffold of the chromosome (77). Because the X acquires its late-replicating status prior to gene silencing, an alternative function for *Xist* RNA may be to delay DNA replication on the future inactive X, thereby ensuring that the chromosome acquires the condensed structure characteristic of late-replicating DNA (95).

After the propagation of inactivation is complete, several changes occur in the X chromosome. The inactive X chromosome is rich in hypoacetylated histones H2A, H3, H4, and macroH2A1.2, an H2A variant histone (97–99). Furthermore the DNA is heavily methylated even within CpG islands that normally resist DNA methylation (100, 101). All of these chromatin modifications are thought to contribute to the maintenance of the inactivated state, but whether they also participate in its initial spreading remains unclear. The fact that gene silencing occurs prior to the underacetylation of the inactive chromosome argues against a role in propagation (Fig. 7) (102). Likewise DNA methylation occurs too late in development to be a component of the propagation process (103).

The maintenance of the inactive state, however, is not dependent upon the continual expression of the *Xist* gene. This was first demonstrated in cell lines in which the *Xist* gene was deleted and in leukaemia cells with X-chromosomal translocations (104). More recently, Csankovszki *et al.* (105) generated a conditional deletion of *Xist*, and showed that, in embryonic fibroblasts, the inactive state is maintained in the absence of *Xist* expression. They did note, however, that the inactive X was no longer associated with macroH2A1.2, and argued that *Xist* RNA may have a role in re-modelling chromatin.

3.6   Stability of the inactive X chromosome

It has been known for some time that X inactivation is not complete, and that some genes are able to escape inactivation. Carrel *et al.* (106) evaluated 224 X-linked genes and found that 34 appeared to escape inactivation. The genes were distributed non-randomly along the X; 31 of 34 map to Xp, including three that map to the pseudo-autosomal region. Xp is the region of the X chromosome that has most recently (in evolution) been part of an autosome, and thus the escape frequency may reflect the history of the X chromosome. In mouse, by contrast, many fewer genes are thought to escape inactivation, based on the relative viability of mouse XO conceptuses (107). However, fewer genes have been examined to date. The ability of some genes to escape inactivation is reminiscent of active genes present in heterochromatic regions in *Drosophila* (Chapter 11).

Once established, the inactive status of a gene on the X chromosome is stable in eutherians. Although cells that have reactivated the inactive X have been detected in ageing animals (108), generally the silent state is thought to prevail. This is not the case in marsupials, where the inactivated state is relatively unstable (109). Marsupials also display less CpG methylation on the inactive X chromosome, but whether this explains the instability remains to be determined.

3.7   Imprinted X-chromosome inactivation

Marsupials display imprinted X-chromosome inactivation, in that the paternal X-chromosome is preferentially inactivated in all somatic cells (110). Imprinted X inactivation also occurs in the extra-embryonic lineages of the mouse embryo (111), where only paternal *Xist* RNA is detected (112). Norris *et al.* (113) showed that the promoter of the *Xist* gene is methylated in oocytes but not sperm, and proposed that gamete-specific pre-emptive methylation controlled *Xist* expression in precisely the same manner as is proposed for imprinted genes such as *H19*. This bias must be erased early in development in the cells that form the embryo proper, in order to achieve random choice.

The requirement for paternal X-chromosome inactivation in extra-embryonic lineages was demonstrated by Marahrens *et al.* (90) who showed that an *Xist* gene deletion could be passed from mothers to daughters with no deleterious effects, as the paternal X chromosome in daughters was then inactivated in all embryonic and extra-embryonic cells. Fathers, however, could not transmit the mutation to daughters because the extra-embryonic cells could not undergo dosage compensation, and with overexpression of X-linked genes in those lineages, the embryos died. Note that imprinting apparently overrides the counting mechanism in these lineages.

4. Final comments

4.1 Future directions in genomic imprinting

While the mechanisms that govern somatic silencing of imprinted genes are beginning to be revealed, virtually nothing is known about the mechanisms that establish and erase the epigenetic marks in the germline. Fundamental questions remain regarding the nature of the DNA sequences that constitute epigenetic marks and how they are recognized differently in the two germlines. It is possible, for example, that there are germline-specific methylases with different recognition sequences, as has been proposed for *Dnmt1* (114). The role of the methyltransferases *Dnmt3a* and *Dnmt3b* in establishing methylated epigenetic marks (115) has not been tested, although they are excellent candidates based on their ability to methylate DNA *de novo*. Nor is it clear whether methylation is actively inhibited in the germline. For example, proteins like CTCF, which binds to the unmethylated *H19* ICR in somatic cells, could prevent methylation of the ICR in oocytes. The special properties of epigenetic marks that protect them from demethylation in the early embryo, as well as re-methylation during differentiation, are also unclear. How are these CpG islands differentiated from CpG islands that are normally unmethylated in all cells?

The clustering of imprinted genes in the genome suggests that long-range co-ordinate regulation, such as found in the PWS-AS region, may be common at imprinted loci. It is possible, however, that the clustering is a historical artefact that reflects the evolutionary acquisition of imprinting, and not a requirement at the present time. Given the recent evolution of imprinting in mammals, it is likely that the organism has adopted a variety of gene silencing mechanisms that pre-existed in the progenitor, and a unique imprinting mechanism may not have evolved.

4.2 Future directions in X-chromosome inactivation

Despite significant advances in this field over the past 10 years, a full understanding of the mechanism of X-chromosome inactivation remains a relatively elusive goal. The identification of the *Xist* and *Tsix* genes, and the demonstration that they are involved in some way in the inactivation process, are major steps forward. However, many questions remain to be resolved. The mechanism by which the cell counts the number of X chromosomes is still mysterious, despite the fact that the region on the X chromosome that counts has been narrowed to the *Xist* gene itself. Is there a titration of a limiting factor that designates the future active chromosome? If so, what is the factor and where does it bind? There are few clues as to the role that the autosomes play in the process, despite clear evidence that they, too, are counted. Choice is also affected by sequences within the *Xist* gene itself, as well as within the antisense *Tsix* gene and the *Xce* (further 3′ of both genes). Whether the requirement for *Tsix* in choice is direct, or whether it acts by negatively regulating *Xist*, is still unresolved. Does *Tsix* have a role as an RNA, possibly working as an antisense transcript, or is it transcription of *Tsix* that is important? The evidence that *Xist* RNA itself is an active player in the choice and initiation of X inactivation is based on internal deletions of the gene, as well as the striking way it 'paints' the inactive X chromosome. Finally *Xist*, but not *Tsix*, has clearly been implicated in the initiation, but not the maintenance, of X-chromosome inactivation, but how it acts to silence an entire chromosome is unknown. It is unknown, however, how *Xist* RNA directs the silencing machinery, presumably a combination of DNA methylation and heterochromatin, to the future inactive X chromosome.

4.3   Concluding remarks

One of the most intriguing aspects of epigenetics in mammals is the possible involvement of non-coding RNAs in both imprinting and X-chromosome inactivation. RNAs have also been implicated in dosage compensation in *Drosophila*. Two RNAs, roX1 and roX2, have been identified as male-specific, non-coding RNAs that interact with the male X chromosome in association with the proteins that constitute the dosage compensation machinery, shown to have histone acetylase activity (116). It has recently been suggested that the roX RNAs define, in *cis*, specific chromatin entry sites that nucleate the spread of the dosage compensation complexes (117), and thus, changes in chromatin structure. This could be analogous to the role of *Xist* RNA in mammals. RNAs have also been involved in gene silencing in plants and by RNA interference in an increasing number of organisms (118). In these cases, however, the silencing is not allele-specific, and thus may not be analogous to gene silencing in imprinting or X-chromosome inactivation.

References

1. Surani, M. A. H. and Barton, S. C. (1983) Development of gynogenetic eggs in the mouse: implications for parthenogenetic embryos. *Science*, **222**, 1034.
2. McGrath, J. and Solter, D. (1983). Nuclear transplantation in mouse embryos. *J. Exp. Zool.*, **228**, 355.
3. Cattanach, B. M. and Jones, J. (1994) Genetic imprinting in the mouse: implications for gene regulation. *J. Inherited Metab. Dis.*, **17**, 403.
4. DeChiara, T. M., Robertson, E. J., and Efstratiadis, A. (1991) Parental imprinting of the mouse insulin-like growth factor II gene. *Cell*, **64**, 849.

5. Mayer, W., Niveleau, A., Walter, J., Fundele, R., and Haaf, T. (2000) Demethylation of the zygotic paternal genome. *Nature*, **403**, 501.

6. Leonhardt, H., Page, A. W., Weier, H. U., and Bestor, T. H. (1992) A targeting sequence directs DNA methyltransferase to sites of DNA replication in mammalian nuclei. *Cell*, **71**, 865.

7. Bird, A. P. and Wolffe, A. P. (1999) Methylation-induced repression—belts, braces, and chromatin. *Cell*, **99**, 451.

8. Tate, P. H. and Bird, A. P. (1993) Effects of DNA methylation on DNA-binding proteins and gene expression. *Curr. Opin. Genet. Devel.*, **3**, 226.

9. Nan, X., Ng, H. H., Johnson, C. A., Laherty, C. D., Turner, B. M., Eisenman, R. N., and Bird, A. (1998). Transcriptional repression by the methyl-CpG-binding protein MeCP2 involves a histone deacetylase complex. *Nature*, **393**, 386.

10. Jones, P. L., Veenstra, G. J., Wade, P. A., Vermaak, D., Kass, S. U., Landsberger, N., Strouboulis, J., and Wolffe, A. P. (1998) Methylated DNA and MeCP2 recruit histone deacetylase to repress transcription. *Nature Genet.*, **19**, 187.

11. Li, E., Beard, C., and Jaenisch, R. (1993) The role of DNA methylation in genomic imprinting. *Nature*, **366**, 362.

12. Caspary, T., Cleary, M. A., Baker, C. C., Guan, X.-J., and Tilghman, S. M. (1998) Multiple mechanisms regulate imprinting of the mouse distal chromosome 7 gene cluster. *Mol. Cell. Biol.*, **18**, 3466.

13. Shemer, R., Birger, Y., Riggs, A. D., and Razin, A. (1997) Structure of the imprinted mouse *Snrpn* gene and establishment of its parental-specific methylation pattern. *Proc. Natl Acad. Sci., USA*, **94**, 10267.

14. Brandeis, M., Kafri, T., Ariel, M., Chaillet, J. R., McCarrey, J., Razin, A., and Cedar, H. (1993) The ontogeny of allele-specific methylation associated with imprinted genes in the mouse. *EMBO J.*, **12**, 3669.

15. Monk, M., Boubelik, M., and Lehnert, S. (1987) Temporal and regional changes in DNA methylation in the embryonic, extraembryonic and germ cell lineages during mouse embryo development. *Development*, **99**, 371.

16. Stoger, R., Kubicka, P., Liu, C.-G., Kafri, T., Razin, A., Cedar, H., and Barlow, D. P. (1993) Maternal-specific methylation of the imprinted mouse *Igf2r* locus identifies the expressed locus as carrying the imprinting signal. *Cell*, **73**, 61.

17. Tanaka, M., Puchyr, M., Gertsenstein, M., Harpal, K., Jaenisch, R., Rossant, J., and Nagy, A. (1999) Parental origin-specific expression of Mash2 is established at the time of implantation with its imprinting mechanism highly resistant to genome-wide demethylation. *Mech. Dev.*, **87**, 129.

18. Kitsberg, D., Selig, S., Brandeis, M., Simon, I., Keshet, I., Driscoll, D. J., Nicholls, R. D., and Cedar, H. (1993) Allele-specific replication timing of imprinted gene regions. *Nature*, **364**, 459.

19. Knoll, J. H., Cheng, S. D., and Lalande, M. (1994) Allele specificity of DNA replication timing in the Angelman/Prader–Willi syndrome imprinted chromosomal region. *Nature Genet.*, **6**, 41.

20. White, L. M., Rogan, P. K., Nicholls, R. D., Wu, B. L., Korf, B., and Knoll, J. H. (1996) Allele-specific replication of 15q11–q13 loci: a diagnostic test for detection of uniparental disomy. *Am. J. Hum. Genet.*, **59**, 423.

21. Greally, J. M., Starr, D. J., Hwang, S., Song, L., Jaarola, M., and Zemel, S. (1998) The mouse *H19* locus mediates a transition between imprinted and non-imprinted DNA replication patterns. *Hum. Mol. Genet.*, **7**, 91.

22. LaSalle, J. M. and Lalande, M. (1996) Homologous association of oppositely imprinted chromosomal domains. *Science*, **272**, 725.
23. Paldi, A., Gyapay, G., and Jami, J. (1995) Imprinted chromosomal regions of the human genome display sex-specific meiotic recombination frequencies. *Curr. Biol.*, **5**, 1030.
24. Tremblay, K. D., Saam, J. R., Ingram, R. S., Tilghman, S. M., and Bartolomei, M. S. (1995) A paternal-specific methylation imprint marks the alleles of the mouse *H19* gene. *Nature Genet.*, **9**, 407.
25. Thorvaldson, J. L., Duran, K. L., and Bartolomei, M. S. (1998) Deletion of the *H19* differentially methylated domain results in loss of imprinted expression of *H19* and *Igf2*. *Genes Develop.*, **12**, 3693.
26. Leighton, P. A., Ingram, R. S., Eggenschwiler, J., Efstratiadis, A., and Tilghman, S. M. (1995) Disruption of imprinting caused by deletion of the *H19* gene region in mice. *Nature*, **375**, 34.
27. Feil, R., Walter, J., Allen, N. D., and Reik, W. (1994) Developmental control of allelic methylation in the imprinted mouse *Igf2* and *H19* genes. *Development*, **120**, 2933.
28. Bartolomei, M. S., Webber, A. L., Brunkow, M. E., and Tilghman, S. M. (1993). Epigenetic mechanisms underlying the imprinting of the mouse *H19* gene. *Genes Develop.*, **7**, 1663.
29. Ferguson-Smith, A. C., Sasaki, H., Cattanach, B. M., and Surani, M. A. (1993) Parental-origin-specific epigenetic modifications of the mouse *H19* gene. *Nature*, **362**, 751.
30. Sasaki, H., Jones, P. A., Chaillet, J. R., Ferguson-Smith, A. C., Barton, S., Reik, W., and Surani, M. A. (1992) Parental imprinting: potentially active chromatin of the repressed maternal allele of the mouse insulin-like growth factor (*Igf2*) gene. *Genes Develop.*, **6**, 1843.
31. Walter, J., Allen, N., Kruger, T., Engemann, S., Kelsey, G., Feil, R., Forne, T., and Reik, W. (1996) Genomic imprinting and modifier genes in the mouse. In *Epigenetic mechanisms of gene regulation*, (ed. V.E.A. Russo, R.A. Martienssen, and A.D. Riggs), pp. 195. Cold Spring Harbor Laboratory Press, Cold Spring Harbor.
32. Jones, B. K., Levorse, J., and Tilghman, S. M. (1998). *Igf2* imprinting does not require its own DNA methylation or H19 RNA. *Genes Develop.*, **12**, 2200.
33. Bartolomei, M. S. and Tilghman, S. M. (1992) Parental imprinting of mouse chromosome 7. *Semin. Dev. Biol.*, **3**, 107.
34. Leighton, P. A., Saam, J. R., Ingram, R. S., Stewart, C. L., and Tilghman, S. M. (1995) An enhancer deletion affects both *H19* and *Igf2* expression. *Genes Develop.*, **9**, 2079.
35. Ripoche, M.-A., Kress, C., Poirier, F., and Dandolo, L. (1997) Deletion of the *H19* transcription unit reveals the existence of a putative imprinting control element. *Genes Develop.*, **11**, 1596.
36. Schmidt, J. V., Levorse, J. M., and Tilghman, S. M. (1999) Enhancer competition between *H19* and *Igf2* does not mediate their imprinting. *Proc. Natl Acad. Sci., USA*, **96**, 9733.
37. Webber, A., Ingram, R. I., Levorse, J., and Tilghman, S. M. (1998) Location of enhancers is essential for imprinting of *H19* and *Igf2*. *Nature*, **391**, 711.
38. Hark, A. T. and Tilghman, S. M. (1998) Chromatin conformation of the *H19* epigenetic mark. *Hum. Mol. Genet.*, **7**, 1979.
39. Szabo, P. E., Pfeifer, G. P., and Mann, J. R. (1998) Characterization of novel parent-specific epigenetic modifications upstream of the imprinted mouse *H19* gene. *Mol. Cell. Biol.*, **18**, 6767.
40. Bell, A. C. and Felsenfeld, G. (2000) Modulation of a CTCF-dependent enhancer boundary by DNA methylation control imprinting of the *Igf2* gene. *Nature*, **405**, 482.
41. Hark, A. T., Schoenherr, C. J., Katz, D. J., Ingram, R. S., Levorse, J. M., and Tilghman, S. M. (2000) CTCF mediates methylation-sensitive enhancer blocking activity at the *H19/Igf2* locus. *Nature*, **405**, 486.

42. Lyko, F., Brenton, J. D., Surani, M. A., and Paro, R. (1997) An imprinting element from the mouse *H19* locus functions as a silencer in *Drosophila*. *Nature Genet.*, **16**, 171.

43. Bell, A. C., West, A. G., and Felsenfeld, G. (1999) The protein CTCF is required for the enhancer blocking activity of vertebrate insulators. *Cell*, **98**, 387.

44. Elias, E. R., DeBaun, M. R., and Feinberg, A. P. (1998) Beckwith–Wiedemann Syndrome. In *Principles of molecular medicine*, (ed. J. L. Jamison), p. 1047. Humana Press.

45. Weksberg, R., Shen, D. R., Fei, Y. L., Song, Q. L., and Squire, J. (1993) Disruption of insulin-like growth factor 2 imprinting in Beckwith–Wiedemann syndrome. *Nature Genet.*, **5**, 143.

46. Lee, M. P., DeBaun, M. R., Mitsuya, K., Galonek, H. L., Brandenburg, S., Oshimura, M., and Feinberg, A. P. (1999) Loss of imprinting of a paternally expressed transcript, with antisense orientation to *KVLQT1*, occurs frequently in Beckwith–Wiedemann syndrome and is independent of *insulin-like growth factor II* imprinting. *Proc. Natl Acad. Sci., USA*, **96**, 5203.

47. Smilinich, N. J., Day, C. D., Fitzpatrick, G. V., Caldwell, G. M., Lossie, A. C., Cooper, P. R., Smallwood, A. C., Joyce, J. A., Schofield, P. N., Reik, W., *et al.* (1999) A maternally methylated CpG island in *KVLQT1* is associated with an antisense paternal transcript and loss of imprinting in Beckwith–Wiedemann syndrome. *Proc. Natl Acad. Sci., USA*, **96**, 8064.

48. Ainscough, J. F.-X., Koide, T., Tada, M., Barton, S., and Surani, A. (1997) Imprinting of *Igf2* and *H19* from a 130 kb YAC transgene. *Development*, **124**, 3621.

49. Wutz, A., Smrzka, O. W., Schweifer, N., Schellander, K., Wagner, E. F., and Barlow, D. P. (1997) Imprinted expression of the *Igf2r* gene depends on an intronic CpG island. *Nature*, **389**, 745.

50. Rougeulle, C., Cardoso, C., Fontes, M., Colleaux, L., and Lalande, M. (1998) An imprinted antisense RNA overlaps UBE3A and a second maternally expressed transcript. *Nature Genet.*, **19**, 15.

51. Wroe, S. F., Kelsey, G., Skinner, J. A., Bodle, D., Ball, S. T., Beechey, C. V., Peters, J., and Williamson, C. M. (2000) An imprinted transcript, antisense to *Nesp*, adds complexity to the cluster of imprinted genes at the mouse *Gnas* locus. *Proc. Natl Acad. Sci., USA*, **97**, 3342.

52. Moore, T., Constancia, M., Zubair, M., Bailleul, B., Feil, R., Sasaki, H., and Reik, W. (1997) Multiple imprinted sense and antisense transcripts, differential methylation and tandem repeats in a putative imprinting control region upstream of mouse *Igf2*. *Proc. Natl Acad. Sci., USA*, **94**, 12509.

53. Jong, M. T., Gray, T. A., Ji, Y., Glenn, C. C., Saitoh, S., Driscoll, D. J.. and Nicholls, R. D. (1999) A novel imprinted gene, encoding a RING zinc-finger protein, and overlapping antisense transcript in the Prader–Willi syndrome critical region. *Hum. Mol. Genet.*, **8**, 783.

54. Ashe, H. L., Minks, J., Wijgerde, M., Fraser, P., and Proudfoot, N. J. (1997) Intergenic transcription and transduction of the human β-globin locus. *Genes Develop.*, **11**, 2494.

55. Birger, Y., Shemer, R., Perk, J., and Razin, A. (1999) The imprinting box of the mouse *Igf2r* gene. *Nature*, **397**, 84.

56. Lerchner, W. and Barlow, D. P. (1997) Paternal repression of the imprinted mouse *Igf2r* locus occurs during implantation and is stable in all tissues of the post-implantation mouse embryo. *Mech. Develop.*, **61**, 141.

57. Vu, T. H. and Hoffmann, A. R. (1997) Imprinting of Angelman syndrome gene, *UBE3A*, is restricted to brain. *Nature Genet.*, **17**, 12.

58. Rougeulle, C., Glatt, H., and Lalande, M. (1997) The Angelman syndrome candidate gene, *UBE3A-AP*, is imprinted in brain. *Nature Genet.*, **17**, 14.

59. Nicholls, R. D., Saitoh, S., and Horsthemke, B. (1998) Imprinting in Prader–Willi and Angelman syndromes. *Trends Genet.*, **14**, 194.

60. Yang, T., Adamson, T. E., Resnick, J. L., Leff, S., Wevrick, R., Francke, U., Jenkins, N. A.,

Copeland, N. G., and Brannan, C. I. (1998) A mouse model for Prader–Willi syndrome imprinting-centre mutations. *Nature Genet.*, **19**, 25.

61. Tsai, T. F., Armstrong, D., and Beaudet, A. L. (1999) *Necdin*-deficient mice do not show lethality or the obesity and infertility of Prader–Willi syndrome. *Nature Genet.*, **22**, 15.

62. Gerard, M., Hernandez, L., Wevrick, R., and Stewart, C. L. (1999) Disruption of the mouse *necdin* gene results in early post-natal lethality. *Nature Genet.*, **23**, 199.

63. Glenn, C. C., Porter, K. A., Jong, M. T., Nicholls, R. D., and Driscoll, D. J. (1993) Functional imprinting and epigenetic modification of the human *SNRPN* gene. *Hum. Mol. Genet.*, **2**, 2001.

64. Buiting, K., Saitoh, S., Gross, S., Dittrich, B., Schwartz, S., Nicholls, R. D., and Horsthemke, B. (1995) Inherited microdeletions in the Angelman and Prader–Willi syndromes define an imprinting centre on human chromosome 15. *Nature Genet.*, **9**, 395.

65. Sutcliffe, J. S., Nakao, M., Christian, S., Orstavik, K. H., Tommerup, N., Ledbetter, D. H., and Beaudet, A. L. (1994) Deletions of a differentially methylated CpG island at the *SNRPN* gene define a putative imprinting control region. *Nature Genet.*, **8**, 52.

66. Schumacher, A., Buiting, K., Zeschnigk, M., Doerfler, W., and Horsthemke, B. (1998) Methylation analysis of the PWS/AS region does not support an enhancer-competition model. *Nature Genet.*, **19**, 324.

67. Bielinska, B., Blaydes, S. M., Buiting, K., Yang, T., Krajewska-Walasek, M., Horsthemke, B., and Brannan, C. I. (2000) *De novo* deletions of *SNRPN* exon 1 in early human and mouse embryos result in a paternal to maternal imprint switch. *Nature Genet.*, **25**, 74.

68. Schweizer, J., Zynger, D., and Francke, U. (1999) *In vivo* nuclease hypersensitivity studies reveal multiple sites of parental origin-dependent differential chromatin conformation in the 150 kb *SNRPN* transcription unit. *Hum. Mol. Genet.*, **8**, 555.

69. Brannan, C. I. and Bartolomei, M. S. (1999) Mechanisms of genomic imprinting. *Curr. Opin. Genet. Dev.*, **9**, 164.

70. Cline, T. W. and Meyer, B. J. (1996) Vive la difference: males vs females in flies vs worms. *Annu. Rev. Genet.*, **30**, 637.

71. Heard, E., Clerc, P., and Avner, P. (1997) X-chromosome inactivation in mammals. *Annu. Rev. Genet.*, **31**, 571.

72. Barr, M. L. and Carr, D. H. (1961) Correlations between sex chromatin and sex chromosomes. *Acta Cytol.*, **6**, 34.

73. Russell, L. B. and Cacheiro, N. L. (1978) The use of mouse X-autosome translocations in the study of X-inactivation pathways and nonrandomness. *Basic Life Sci.*, **12**, 393.

74. Therman, E. and Sarto, G. E. (1983) Inactivation center on the human X chromosome. In *Cytogenetics of the mammalian X chromosome. Part A. Basic mechanisms of X chromosome behavior*, (ed. A. A. Sandber), p. 315, Liss, New York.

75. Brown, C. J., Ballabio, A., Rupert, J. L., Lafreniere, R. G., Grompe, M., Tonlorenzi, R., and Willard, H. F. (1991) A gene from the region of the human X chromosome inactivation centre is expressed exclusively from the inactive X chromosome. *Nature*, **349**, 38.

76. Brockdorff, N., Ashworth, A., Kay, G. F., Cooper, P., Smith, S., McCabe, V. M., Norris, D. P., Penny, G. D., Patel, D., and Rastan, S. (1991) Conservation of position and exclusive expression of mouse Xist from the inactive X chromosome. *Nature*, **351**, 329.

77. Clemson, C. M., McNeil, J. A., Willard, H. F., and Lawrence, J. B. (1996) XIST RNA paints the inactive X chromosome at interphase: evidence for a novel RNA involved in nuclear/chromosome structure. *J. Cell Biol.*, **132**, 259.

78. Panning, B. and Jaenisch, R. (1996) DNA hypomethylation can activate *Xist* expression in mouse embryonic cells. *Genes Develop.*, **10**, 1191.

79. Lee, J. T., Davidow, L. S., and Warshawsky, D. (1999) *Tsix*, a gene antisense to *Xist* at the X-inactivation centre. *Nature Genet.*, **21**, 400.

80. Rastan, S. and Robertson, E. J. (1985) X-chromosome deletions in embryo-derived (EK) cell lines associated with lack of X-chromosome inactivation. *J. Embryol. Exp. Morphol.*, **90**, 379.

81. Penny, G. D., Kay, G. F., Sheardown, S. A., Rastan, S., and Brockdorff, N. (1996) Requirement for *Xist* in X chromosome inactivation. *Nature*, **379**, 131.

82. Lee, J. T., Strauss, W. M., Dausman, J. A., and Jaenisch, R. (1996) A 450 kb transgene displays properties of the mammalian X-inactivation center. *Cell*, **86**, 83.

83. Lee, J. T., Lu, N., and Han, Y. (1999) Genetic analysis of the mouse X inactivation center defines an 80-kb multifunction domain. *Proc. Natl Acad. Sci., USA*, **96**, 3836.

84. Herzing, L. B. K., Romer, J. T., Horn, J. M., and Ashworth, A. (1997) *Xist* has properties of the X-chromosome inactivation centre. *Nature*, **386**, 272.

85. Simmler, M. C., Cattanach, B. M., Rasberry, C., Rougeulle, C., and Avner, P. (1993). Mapping the murine Xce locus with (CA)n repeats. *Mamm. Genome*, **4**, 523.

86. Vrana, P. B., Fossella, J. A., Matteson, P., del Rio, T., O'Neill, M. J., and Tilghman, S. M. (2000) Genetic and epigenetic incompatibilities underlie hybrid dysgenesis in *Peromyscus*. *Nature Genet.*, **25**, 120.

87. Cattanach, B. M., Perez, J. N., and Pollard, C. E. (1970) Controlling elements in the mouse X-chromosome. II. Location in the linkage map. *Genet. Res.*, **15**, 183.

88. Rastan, S. (1982) Primary non-random X-inactivation caused by controlling elements in the mouse demonstrated at the cellular level. *Genet. Res.*, **40**, 139.

89. Avner, P., Prissette, M., Arnaud, D., Courtier, B., Cecchi, C., and Heard, E. (1998) Molecular correlates of the murine *Xce* locus. *Genet. Res.*, **72**, 217.

90. Marahrens, Y., Panning, B., Dausman, J., Strauss, W., and Jaenisch, R. (1997) *Xist*-deficient mice are defective in dosage compensation but not spermatogenesis. *Genes Dev.*, **11**, 156.

91. Clerc, P. and Avner, P. (1998) Role of the region 3′ to *Xist* exon 6 in the counting process of X-chromosome inactivation. *Nature Genet.*, **19**, 249.

92. Lee, J. T. and Lu, N. (1999) Targeted mutagenesis of *Tsix* leads to nonrandom X inactivation. *Cell*, **99**, 47.

93. Panning, B., Dausman, J., and Jaenisch, R. (1997) X chromosome inactivation is mediated by *Xist* RNA stabilization. *Cell*, **90**, 907.

94. Sheardown, S. A., Duthie, S. M., Johnston, C. M., Newall, A. E., Formstone, E. J., Arkell, R. M., Nesterova, T. B., Alghisi, G. C., Rastan, S., and Brockdorff, N. (1997) Stabilization of *Xist* RNA mediates initiation of X chromosome inactivation. *Cell*, **91**, 99.

95. Panning, B. and Jaenisch, R. (1998) RNA and the epigenetic regulation of X chromosome inactivation. *Cell*, **93**, 305.

96. Heard, E., Mongelard, F., Arnaud, D., and Avner, P. (1999) *Xist* yeast artificial chromosome transgenes function as X-inactivation centers only in multicopy arrays and not as single copies. *Mol. Cell. Biol.*, **19**, 3156.

97. Jeppesen, P. and Turner, B. M. (1993) The inactive X chromosome in female mammals is distinguished by a lack of histone H4 acetylation, a cytogenetic marker for gene expression. *Cell*, **74**, 281.

98. Belyaev, N., Keohane, A. M., and Turner, B. M. (1996) Differential underacetylation of histones H2A, H3 and H4 on the inactive X chromosome in human female cells. *Hum. Genet.*, **97**, 573.

99. Costanzi, C. and Pehrson, J. R. (1998) Histone macroH2A1 is concentrated in the inactive X chromosome of female mammals. *Nature*, **393**, 599.

100. Norris, D. P., Brockdorff, N., and Rastan, S. (1991) Methylation status of CpG-rich islands on active and inactive mouse X chromosomes. *Mamm. Genome*, **1**, 78.

101. Tribioli, C., Tamanini, F., Patrosso, C., Milanesi, L., Villa, A., Pergolizzi, R., Maestrini, E., Rivella, S., Bione, S., Mancini, M., *et al.* (1992) Methylation and sequence analysis around *EagI* sites: identification of 28 new CpG islands in XQ24-XQ28. *Nucleic Acids Res.*, **20**, 727.

102. Keohane, A. M., O'Neill L, P., Belyaev, N. D., Lavender, J. S., and Turner, B. M. (1996) X-inactivation and histone H4 acetylation in embryonic stem cells. *Dev. Biol.*, **180**, 618.

103. Bartlett, M. H., Adra, C. N., Park, J., Chapman, V. M., and McBurney, M. W. (1991) DNA methylation of two X chromosome genes in female somatic and embryonal carcinoma cells. *Somat. Cell Mol. Genet.*, **17**, 35.

104. Brown, C. J. and Willard, H. F. (1994) The human X-inactivation centre is not required for maintenance of X-chromosome inactivation. *Nature*, **368**, 154.

105. Csankovszki, G., Panning, B., Bates, B., Pehrson, J. R., and Jaenisch, R. (1999) Conditional deletion of *Xist* disrupts histone macroH2A localization but not maintenance of X inactivation. *Nature Genet.*, **22**, 323.

106. Carrel, L., Cottle, A. A., Goglin, K. C., and Willard, H. F. (1999) A first-generation X-inactivation profile of the human X chromosome. *Proc. Natl Acad. Sci., USA*, **96**, 14440.

107. Ashworth, A., Rastan, S., Lovell-Badge, R., and Kay, G. (1991) X-chromosome inactivation may explain the difference in viability of XO humans and mice. *Nature*, **351**, 406.

108. Wareham, K. A., Lyon, M. F., Glenister, P. H., and Williams, E. D. (1987) Age related reactivation of an X-linked gene. *Nature*, **327**, 725.

109. Cooper, D. W., Johnston, P. G., Watson, J. M., and Graves, J. A. M. (1993) X-inactivation in marsupials and monotremes. *Semin. Develop. Biol.*, **4**, 117.

110. Cooper, D. W., VandeBerg, J. L., Sharman, G. B., and Poole, W. E. (1971) Phosphoglycerate kinase polymorphism in kangaroos provides further evidence for paternal X inactivation. *Nature New Biol.*, **230**, 155.

111. Takagi, N. and Sasaki, M. (1975) Preferential inactivation of the paternally derived X chromosome in the extraembryonic membranes of the mouse. *Nature*, **256**, 640.

112. Kay, G. F., Penny, G. D., Patel, D., Ashworth, A., Brockdorff, N., and Rastan, S. (1993) Expression of *Xist* during mouse development suggests a role in the initiation of X chromosome inactivation. *Cell*, **72**, 171.

113. Norris, D. P., Patel, D., Kay, G. F., Penny, G. D., Brockdorff, N., Sheardown, S. A., and Rastan, S. (1994). Evidence that random and imprinted *Xist* expression is controlled by preemptive methylation. *Cell*, **77**, 41.

114. Mertineit, C., Yoder, J. A., Taketo, T., Laird, D. W., Trasler, J. M., and Bestor, T. H. (1998) Sex-specific exons control DNA methyltransferase in mammalian germ cells. *Development*, **125**, 889.

115. Okano, M., Bell, D. W., Haber, D. A., and Li, E. (1999) DNA methyltransferases *Dnmt3a* and *Dnmt3b* are essential for *de novo* methylation and mammalian development. *Cell*, **99**, 247.

116. Stuckenholz, C., Kageyama, Y., and Kuroda, M. I. (1999) Guilt by association: non-coding RNAs, chromosome-specific proteins and dosage compensation in *Drosophila*. *Trends Genet.*, **15**, 454.

117. Meller, V. H., Gordadze, P. R., Park, Y., Chu, X., Stuckenholz, C., Kelley, R. L., and Kuroda, M. I. (2000) Ordered assembly of roX RNAs into MSL complexes on the dosage-compensated X chromosome in *Drosophila*. *Curr. Biol.*, **10**, 136.

118. Fire, A. (1999) RNA-triggered gene silencing. *Trends Genet.*, **15**, 358.

13 | Chromatin boundaries

VICTOR G. CORCES and GARY FELSENFELD

1. Introduction

Because in any given cell the eukaryotic genome is organized into regions that are transcriptionally active, interspersed with those that are not, there must be *de facto* boundaries between the regions. The boundary could be established merely as the result of some compromise between opposing reactions: those favouring opening of the chromatin structure and those promoting condensation. The location of such a boundary might vary with the level of transcriptional activity or the supply of protein components necessary to stabilize the condensed state.

A second kind of boundary involves distinct DNA sequence elements that are known in most cases to bind specific proteins essential to their function. Unlike the potentially fluctuating boundaries mentioned above, these are fixed in position. The first sequences proposed as fixed boundary elements, scs and scs', were identified morphologically in the 87A7 cytogenetic locus of *Drosophila melanogaster*, surrounding the two *hsp70* genes (1). The elements were initially defined by two sets of closely spaced nuclease-hypersensitive sites arranged around a central nuclease-resistant segment. The presence of hypersensitive sites is a good indication that the binding of regulatory proteins has displaced or disrupted nucleosome structure. As we will discuss below, proteins have been identified that bind to scs and scs'.

Since the discovery of scs/scs', a number of different candidate sequences have been proposed as boundary elements. In a small number of cases in *Drosophila* it has been possible to investigate *in vivo* the effect of deleting, mutating, or moving the element, and to observe directly changes in gene expression arising from altered interactions with (for example) nearby enhancers. In most situations, however, it is necessary to prepare reporter constructs containing the putative boundary element, and introduce them by transient or stable transformation into cell lines where boundary activity can be tested.

The question of how one assays for 'boundary activity' is therefore central to any discussion of boundary elements. In their seminal paper, Kellum and Schedl (1) devised an assay that tested for protection against position effects. They created a transposable element carrying the *white* gene driven by a weakened promoter, so that expression varied from line to line depending on the site of insertion. They then showed that when the reporter was surrounded by scs elements this variability was

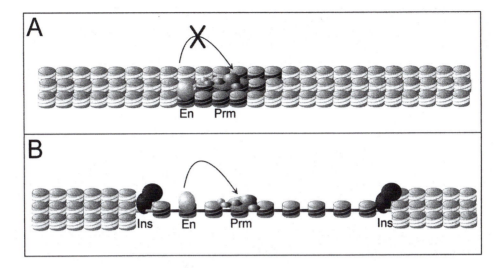

Fig. 1 Insulator elements buffer gene expression from repressive effects of adjacent chromatin. En represents an enhancer with an associated transcription factor bound to nucleosomal DNA. Prm is the promoter of the gene where the different components of the transcription complex are present. Ins is an insulator element with its associated proteins. Solid arrows indicate a positive activation of transcription by the enhancer element; an X on the arrows indicates the inability of the enhancer to activate transcription. (A) A transgene (represented by the dark DNA) integrated in the chromosome in a region of condensed chromatin is not properly expressed; the repressive chromatin structure of the surrounding region presumably spreads into transgene sequences, inhibiting enhancer–promoter interactions. (B) If the transgene is flanked by insulator elements, these sequences inhibit the spreading of the repressive chromatin conformation, allowing an open chromatin conformation and normal transcription of the gene.

suppressed. This is plausible behaviour for a putative boundary element, because one might expect that at least in some circumstances it must protect a locus against the encroachment of silencing or activating signals coming from outside the locus (Fig. 1).

A different boundary assay measures the ability of the element to block activation when it is placed between an enhancer and a promoter (Fig. 2). This blocking activity is distinguishable from normal silencing because no effect is observed when the element is placed elsewhere. Versions of this assay have been used in a wide variety of experiments to assess the ability of an element to insulate promoters from enhancers. The term 'insulator' has been used more or less interchangeably with 'boundary element'. It seems reasonable that a boundary should have the ability to block effects of distal enhancers, but positive results in this assay, as well as in the position effect assay described above, do not guarantee that the element serves as a boundary in its natural position in the genome. In a number of cases, however, the genomic position of the element relative to genes or gene clusters provides strong indirect evidence for such a role.

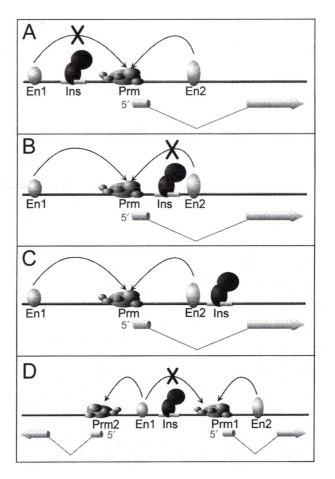

Fig. 2 Polar effect of an insulator on enhancer–promoter interactions. Symbols are as in Fig. 1. (A) An insulator located in the 5′ region of the gene inhibits its transcriptional activation by an upstream enhancer (En1) without affecting the function of a second enhancer (En2) located in the intron of the gene. (B) When the insulator is located in the intron, expression from the downstream enhancer (En2) is blocked, whereas the upstream enhancer (En1) is active. (C) When the insulator is located in the intron but distal to the En2 enhancer, both enhancers are active and transcription of the gene is normal. This property distinguishes an insulator from a typical silencer. (D) If a second gene is located upstream of the En1 enhancer, although this enhancer cannot act on the Prm1 promoter, it is still functional and able to activate transcription from the upstream Prm2 promoter.

2. Boundary elements: specific examples

2.1 *Drosophila* insulator elements in the *hsp70* heat-shock locus

A variety of chromatin boundaries or insulator elements have been described in *Drosophila*, including the *Mcp*, *Fab-6*, *Fab-7*, and *Fab-8* elements present in the bithorax complex, the scs and scs′ elements flanking the 87A7 heat-shock gene locus, the *gypsy* insulator present in the *gypsy* retrotransposon, and an insulator present in the *even-*

skipped (*eve*) promoter which contains a binding site for the GAGA protein. These insulators have some common characteristics that might suggest shared mechanisms of action, while at the same time they display idiosyncratic properties suggestive of particular roles in chromatin organization and regulation of gene expression. The first experimental evidence of a specific DNA sequence having insulator activity was obtained with the identification of the scs (which stands for 'specialized chromatin structure') and scs' elements of *Drosophila*. These sequences were identified and became prominent because of their location at the borders of the 87A7 chromomere (2). The location of the scs and scs' sequences, flanking the puff that arises upon induction of gene expression by heat shock, suggests that they might demarcate the extent of chromatin that decondenses consequent to induction of transcription by temperature elevation. This hypothesis is supported by the chromatin organization of the region containing these sequences: two strong nuclease-hypersensitive sites surrounding a nuclease-resistant core. This central structure is flanked by additional weaker nuclease cleavage sites present at intervals corresponding to the length of a nucleosome (2). A similar pattern of strong hypersensitive sites at the location of the proposed boundary elements is observed at the sites of the chicken β-globin 5′ boundary (3) as well as other insulators, and suggests that this chromatin organization might play a role in boundary function. The properties of the scs and scs′ sequences have been confirmed by studying their ability to confer position-independent expression when flanking two different types of *white* reporter genes (Fig. 1) (1), and to insulate the *Drosophila hsp70* promoter from the effect of the yolk protein *yp1* enhancer in a hybrid *hsp70–lacZ* reporter gene construct; activation of transcription of the *lacZ* gene by the *yp1* enhancer in the female fat body is abolished when scs sequences are interposed between the enhancer and the promoter (Fig. 2) (4).

The roles of specific sequences and particular chromatin structures contained within the originally defined scs element in providing boundary function have been determined by testing the ability of specific subfragments to block the activation of a downstream promoter by the eye and testis enhancers of the *white* gene (5). Results from this analysis have led to the conclusion that the sequences associated with DNAase I hypersensitive sites are essential for complete blocking activity of enhancer function, whereas the central nuclease-resistant A/T-rich region is dispensable for this effect. Deletion of sequences associated with some hypersensitive sites leads to a reduction in enhancer blocking, whereas multimerization of subfragments with partial activity restores full boundary function (5). Further insights into the specific sequences required for boundary function have come from the identification of SBP (scs *b*inding *p*rotein) as the product of the *zeste-white 5* (*zw5*) gene (6). SBP binds to a 24 bp sequence of scs *in vitro*, and multiple copies of this sequence have insulator activity, as determined by their ability to block enhancer–promoter interactions *in vivo*. Mutations in the sequence that disrupt SBP binding also disrupt insulator function. In addition, mutations in the *zw5* gene decrease the enhancer-blocking activity of these sequences. The ZW5 protein contains zinc-finger motifs and is essential for cell viability. Many mutations in the gene are recessive lethal, but hypomorphic alleles display a variety of pleiotropic effects on wing, bristle, and eye development consistent with a role for this protein in chromatin organization (6).

Sequences responsible for the boundary function of the scs' element have also been characterized in more detail with the identification of binding proteins (see below). A series of CGATA repeats that interact with the BEAF-32 proteins are responsible for the insulator activity of the scs' sequences (7, 8). Mutations in this sequence that interfere with binding of the BEAF-32 protein also abolish insulator activity, whereas multimers containing several copies of the sequence display boundary function. The latter results are similar to those obtained with the *Drosophila* scs sequences and the chicken β-globin insulator, and suggest that the effect of boundary elements on transcription might require the binding of a critical number of proteins that somehow cause chromatin alterations as a consequence of their interaction with DNA.

Two related 32 kDa proteins, termed BEAF-32A and BEAF-32B (for *b*oundary *e*lement *a*ssociated *f*actor of 32 kDa), have been purified from nuclei of a *Drosophila* cell line and found to interact with scs' sequences (7, 9). These proteins bind with high affinity to a site containing three copies of the CGATA motif that flanks the two hypersensitive regions in the scs' sequence. The DNA-binding activity resides in the amino-terminal region, which is different in the two proteins; the carboxy terminus is shared and it is involved in heterocomplex formation between the two proteins. The sequence containing BEAF-32 binding sites acts as a typical boundary element in an enhancer-blocking assay involving stable transformation into cultured cells of a reporter containing a chloramphenicol acetyltransferase gene under the control of ecdysone response and heat shock regulatory elements. The BEAF-32 binding site blocks the activity of both heat-shock and ecdysone-responsive enhancers in stably transfected cells (7). Immunolocalization of BEAF-32 using antibodies shows the presence of this protein in specific subnuclear regions and its exclusion from the nucleolus. BEAF-32 is present in the interband region that separates the highly reproducible and characteristic polytene bands of *Drosophila* third instar larval chromosomes (Chapter 11). Interbands contain lower amounts of DNA than bands, and are presumed to be regions of partial unfolding of the 30 nm chromatin fibre. As expected, BEAF-32 is present at the scs'-containing border of the 87A7 chromomere, and is also found at the edges of many developmental puffs typically seen in polytene chromosomes at this stage of larval development (7). This observation suggests that BEAF-32 might have general structural and functional roles in defining many boundary elements throughout the *Drosophila* genome.

Recently, a second protein capable of interacting with endogenous BEAF insulators has been identified (10). This protein is the transcription factor DREF (*D*NA *r*eplication-related *e*lement-binding *f*actor); it binds to a sequence overlapping that recognized by BEAF, suggesting that the two proteins might compete for DNA binding *in vivo*. DREF participates in the regulation of genes encoding proteins required for DNA replication and cell proliferation, suggesting that displacement of BEAF by binding of DREF might occur during rapid proliferation in the affected cells. Competition for binding to insulator sites would open up a possibility for regulation of boundary function.

2.2   *Drosophila* insulator elements: the bithorax complex

The *Ultrabithorax* (*Ubx*), *Abdominal-A* (*Abd-A*), and *Abdominal-B* (*Abd-B*) genes of the bithorax complex are expressed in a parasegmental-specific pattern dictated by a complex set of regulatory sequences arranged over 300 kb of DNA in a linear fashion, corresponding to the order of expression along the anterior–posterior axis. These parasegment-specific regulatory sequences appear to be separated by boundaries initially identified due to the dominant gain-of-function phenotypes observed in 'boundary deletion mutants' that result in the fusion of two adjacent parasegment-specific regulatory elements into one single functional unit (11). The best studied of these boundaries is the *Fab-7* element located between the *iab-6* and *iab-7* regulatory sequences that control expression of the *Abd-B* gene in parasegments PS11 and PS12. The *Fab-7* region contains a *Polycomb* response element (PRE, see Chapter 11) and a sequence that has the properties of a chromatin boundary or insulator (12–14). The latter sequences behave as a typical insulator in an enhancer-blocking assay; when the *Fab-7* element is placed between the eye- and testes-specific enhancers and the promoter of the *Drosophila white* gene, transcription is blocked in these two tissues. In addition, deletion of the insulator in the chromosomal DNA results in cross-talk between the *iab-6* and *iab-7* regulatory regions, causing homeotic phenotypes in the adult fly. These results indicate that the *Fab-7* region contains an insulator element that is involved in the normal regulation of the *Abd-B* gene. The location of the insulator has been narrowed down to a 1.2 kb DNA fragment using the enhancer-blocking assay described above. This fragment contains one weak and two strong DNAase I hypersensitive sites (12, 14).

In the case of the bithorax complex, the role of the insulators that separate different parasegment-specific regulatory sequences is to avoid interactions between these sequences and to maintain proper segmental expression of the genes. This organization nevertheless poses the problem of how these regulatory elements can overcome the effect of the insulators to activate transcription of the *Abd-B* gene when appropriate. A solution to this problem might lie in a recently described sequence named the PTS (promoter-targeting sequence). This sequence, found within the *Fab-8* element, which also contains an insulator, allows distal enhancers to overcome the blocking effects of the *Fab-8* insulator (15, 16).

2.3   *Drosophila* insulator elements: the *gypsy* retrovirus

Another insulator element found in *Drosophila* is associated with the *gypsy* retrotransposon. This insulator is 350 bp in length and is located in the 5' transcribed, untranslated region of *gypsy*, upstream from the start of the *gag* open reading frame (reviewed in 17). The *gypsy* insulator was originally identified by the nature of the mutations induced by insertion of this retroelement. In most instances studied, *gypsy* has inserted into non-coding regions of genes and causes a tissue-specific mutant phenotype due to the inactivation of specific enhancers. The inactivated enhancers are always located distal to the insertion site with respect to the promoter. This polar effect on transcription regulation is one of the characteristic properties of insulator elements, and the mutagenic effect of *gypsy* on adjacent genes can be explained exclusively by the presence of the internal insulator element (18–20). The *gypsy* insulator does not inactivate the adjacent enhancer, as this can still activate transcription of a gene located on the same side as the enhancer (21, 22) (see Fig. 2). The *gypsy* insulator can also buffer the expression of a transgene from position effects due to adjacent sequences in the genome (23), and it protects the replication origin of the *Drosophila* chorion genes from similar position effects (24). This insulator contains 12 copies of a 26 bp sequence, including a core element similar to the octamer motif found in various vertebrate enhancers and promoters; this core element is flanked by an AT-rich region that induces DNA bending and is required for insulator function. Mutations that interfere with bending of the DNA also impair the ability of the insulator to interfere with enhancer–promoter interactions. The strength of the insulator depends on the number of copies of the 26 bp basic motif; one copy causes a very small effect on enhancer activation of transcription, while additional copies result in a stronger effect, with an apparent linear relationship between number of copies and enhancer blocking (25, 26). As in other boundary elements, the *gypsy* insulator also contains a series of five strong DNAase I hypersensitive sites indicative of a special chromatin organization (27).

The *gypsy* insulator is perhaps the best studied with respect to the characterization of protein components that interact with insulator DNA. One of these components, su(Hw), the suppressor of *Hairy-wing* protein, was originally identified on the basis of the observation that mutations in the *su(Hw)* gene reverse the phenotypic effect of *gypsy*-induced mutations. This observation can now be interpreted in light of the fact that *gypsy*-induced mutations result from the presence of the insulator and its effect on enhancer–promoter interactions; in the absence of su(Hw) protein, the insulator is not able to block enhancer function, suggesting that su(Hw) is an essential component of the insulator. The su(Hw) protein contains 12 zinc fingers involved in DNA binding; interaction of the su(Hw) protein with its target sequence is a prerequisite for proper insulator function, as mutations that disrupt the zinc fingers hamper the ability of the insulator to block enhancer–promoter interactions (28). In addition, su(Hw) contains two acidic domains, located in the amino- and carboxy-terminal ends of the protein. These two domains are dispensable for the effect of the insulator on enhancer function and for its ability to buffer against chromosomal position effects. An α-helical region homologous to the second helix–coiled coil region of basic HLH-zip proteins is absolutely required for insulator function (28). Since leucine zipper regions usually mediate protein–protein interactions, and su(Hw) does not interact with itself, this observation suggests the involvement of other proteins, in addition to su(Hw), in the formation of the boundary element.

A second component of the *gypsy* insulator has been identified by searching for mutations that alter *gypsy*-induced phenotypes (29). Mutations in the *mod(mdg4)* *(modifier of mdg4)* gene reverse the effect of the insulator on enhancer function. The *mod(mdg4)* gene encodes several different proteins arising from alternatively spliced RNAs, all of which contain a BTB domain (30–32). The BTB domain is present in various transcription factors, including GAGA factor, a transcriptional activator encoded by the *Trithorax-like* (*Trl*) gene (33). Genetic and molecular analyses suggest that mod(mdg4) proteins interact directly with su(Hw) and therefore constitute a second component of the *gypsy* insulator (31).

Several general conclusions can be drawn from the data accumulated so far on the structure of insulator DNAs and associated proteins. Insulators are composed of relatively large DNA sequences containing binding sites for multiple proteins or tandem arrays of a binding site for a single specific protein. These sequences determine the establishment of a specific chromatin structure that is manifested by the formation of multiple DNAase I hypersensitive sites. Whether this chromatin conformation is required for, or is simply a consequence of, insulator function is not known at this time.

2.4 Vertebrate insulator elements: globin locus boundary elements, BEAD, RO, MARs, and others

The first insulator element discovered in vertebrates is located near the 5′ end of the chicken β-globin locus. It is marked by a strong DNAase I hypersensitive site (5′HS4) present in all cells and tissues that have been examined (34; see Chapter 9). Attention was originally focused on the site because it seemed possible that it marked the 5′ boundary of the open β-globin chromatin domain. Subsequent studies of the general (low-level) DNAase I sensitivity and histone acetylation patterns in this region confirmed that suggestion; a fairly sharp transition occurs just 5′ of 5′HS4 between an open (DNAase I-sensitive, hyperacetylated) structure and a condensed (insensitive, less acetylated) structure (35). This, of course, does not prove that there is a cause-and-effect relationship, merely a correlation. Enhancer-blocking assays in stably transformed cell lines did, however, confirm that there is a positional enhancer-blocking element within a 1.2 kb DNA segment that contains 5′HS4 (3; Fig. 3). Further studies showed that most of this activity was contained in a 250 bp 'core' sequence within the 1.2 kb segment (36). Subsequent experiments identified a single binding site for the protein CTCF (*CCTC-binding factor*) that was sufficient to confer enhancer-blocking activity (37). CTCF contains 11 zinc fingers and, in other contexts, had been reported to act as a repressor or activator of transcription (38).

Relatively few other elements with enhancer-blocking activity have been identified in vertebrates. The human T-cell receptor α/δ locus contains a 1.6 kb sequence designated BEAD-1 (*blocking element alpha/delta 1*), which prevents action of an enhancer on a promoter only when placed between them (39). It has been proposed that BEAD-1 prevents a δ-specific enhancer from acting on the α genes early in T-cell development. A binding site for CTCF has been detected within BEAD-1, and deletion of this site largely abolishes enhancer-blocking effects (37). Similarly, a site within the *Xenopus* ribosomal RNA gene repeat, the 'repeat organizer' (RO), which has been shown to have limited enhancer-blocking activity when assayed in *Xenopus* oocytes (40), has been identified as a CTCF binding site (37).

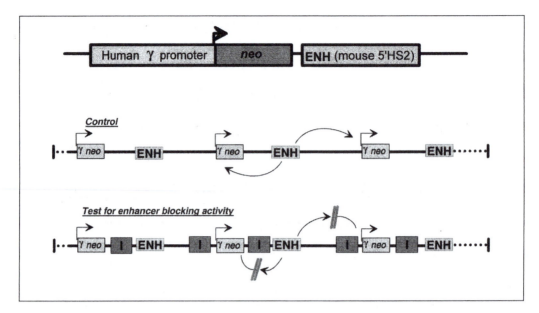

Fig. 3 Assay used to detect enhancer-blocking activity. A reporter is used that contains a strong human erythroid (γ-globin) promoter and mouse erythroid enhancer (ENH, from the β-globin locus control region) driving expression of a gene for neomycin resistance. The reporter is stably transformed into a human erythroleukaemia line (K562) and the number of colonies resistant to G418 is measured. The presence of an insulator (I) surrounding each enhancer is expected to block expression and reduce the number of colonies. See ref. 3 for details.

Quite recently it has been reported that the 3' boundary of the chicken β-globin locus (as determined by DNAase I sensitivity) is also marked by a CTCF-binding site with positional enhancer-blocking activity (41). This is strong evidence in support of the idea that both the 3' and 5' sites are indeed boundary elements, i.e. that they play some role in the establishment and maintenance of the globin locus as a unit of transcriptional regulation. Located not far beyond those boundaries are other genes with quite different patterns of expression. Beyond the 5' boundary (in the 5' direction) a region of condensed chromatin extends for about 16 kb, and beyond that is a gene coding for a folate receptor that is expressed only early in erythroid development, before globin gene expression is switched on (42, 43). The presence of this nearby gene, with its own programme of expression, raises the possibility that the boundary element may serve to prevent inappropriate cross-reaction between regulatory elements for the folate receptor gene and those for the globin genes. Similarly, at the 3' end, beyond the CTCF site, there is a gene coding for an odorant receptor, again with an expression pattern distinct from that of the nearby globin genes (41).

The enhancer-blocking activity of CTCF allows, in principle, for the establishment of a permanent boundary in which enhancers on one side of the CTCF-binding site are permanently blocked from activating promoters on the other side. Recent results (44, 45; Fig. 4) with the imprinted *Igf2/H19* locus in mouse, rat, and human, however,

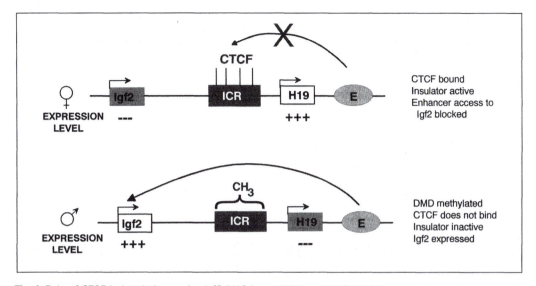

Fig. 4 Role of CTCF in imprinting at the *Igf2/H19* locus. Within the *Igf2/H19* locus in mouse, the maternally transmitted allele expresses *H19* but not *Igf2*. Expression patterns are reversed in the paternally transmitted allele, which is also methylated in the region marked ICR (imprinting control region, see text). The behaviour of *Igf2* can be accounted for by the presence of four binding sites (or a larger number of sites in the human locus) for CTCF within the ICR, which confer enhancer-blocking activity that prevents the enhancer E from acting on the *Igf2* promoter (44, 45). In the paternal allele, methylation of the ICR prevents binding of CTCF and abolishes the insulating activity, allowing activation of the *Igf2* promoter. It should be noted that *H19* expression is not controlled by the same mechanism.

suggest that there are ways in which this blocking activity can be regulated. Earlier work had led to the speculation (46) that an unspecified insulator element situated between the two genes was responsible for blocking activation of *Igf2* by *H19* enhancers in the maternally transmitted allele, but that in the paternally transmitted allele, methylated in part of the region between the two genes, this insulator was inactive. This would, in principle, account for the silencing of expression of *Igf2* from the maternal allele, and its expression from the paternal allele (Chapter 12). The region targeted for methylation has now been shown to contain a series of CTCF-binding sites that possess strong insulating activity (44, 45). Furthermore, methylation of these sites abolishes CTCF binding; mutation of the sites, which also prevents binding, abolishes the enhancer-blocking activity. The imprinting phenomenon at this locus is thus closely connected to insulator activity, and this activity can be controlled by methylation. It seems reasonable to suppose that other sites will be discovered at which insulator activity can be modulated, to provide fine control of enhancer action during the cell cycle or as part of a developmental process.

The chicken 5′ β-globin boundary element also has been tested for its ability to protect against position effects (47). In these experiments an erythroid-specific reporter gene expressing a cell-surface marker is introduced by stable transformation into a pre-erythroid chicken cell line in which the promoter and enhancer used in the construct are only moderately active. Under such circumstances expression of the reporter is quite variable from line to line, and after withdrawal of hygromycin, used for selection of transformed cells, there is in most cases a gradual extinction of expression over a period of 40–80 days. In contrast, when two copies of the complete 1.2 kb β-globin boundary element surround the reporter, expression is quite uniform among lines and no extinction of expression is observed. These are both manifestations of protection against position effects. Recent data (Recillas-Targa *et al.*, unpublished) show that two copies of the 250 bp core work as well as two copies of the full 1.2 kb fragment in this assay. It should be noted, however, that this property is not associated with CTCF binding, but with the presence of other DNA sequence elements within the boundary. Thus the chicken 5' β-globin boundary involves a multicomponent complex (Recillas-Targa *et al.*, unpublished), as does the *gypsy* insulator of *Drosophila*.

As had been shown earlier in a different reporter system, position effects of this kind are associated with loss of core histone acetylation (48). Similarly, in the case of the cell-surface marker reporter gene discussed above, histones H3 and H4 covering the body of the gene become deacetylated in the absence of the insulator sequences. In the presence of insulator elements, deacetylation does not occur (47). This suggests that the boundary elements either promote acetylation of the protected region or prevent the action of histone deacetylases. Inactivation is also accompanied by DNA methylation, but the correlation between activity and methylation levels is more complicated and awaits detailed analysis of the roles of individual methylation sites.

Whatever the mechanism involved in protection against position effects, the 1.2 kb β-globin boundary element, as well as the 250 bp core sequence within it, have proven useful in creating stably transformed animals (49, 50) and cell lines (51, 52) in which expression of reporters is uniform and at high levels from line to line. These boundary elements work well with retroviral vectors and should therefore be useful in gene therapy.

There is no reason to believe that the CTCF-based enhancer-blocking elements are the only ones capable of this activity in vertebrates. It has been suggested that *m*atrix *a*ttachment *r*egions (MARs) may also have this property. MARs or SARs (*s*caffold *a*ttachment *r*egions) are AT-rich DNA sequences, often containing topoisomerase II cleavage sites, that mediate the anchoring of the chromatin fibre to the chromosome scaffold or nuclear matrix and delimit the boundaries of discrete and topologically independent higher-order domains (53). A MAR derived from the chicken lysozyme locus has been reported to contain an enhancer-blocking element (54), but recent results indicate that the element actually resides upstream of the AT-rich region characteristic of MARs, and involves CTCF-binding sites (37). This does not exclude the possibility that MAR elements in other contexts can mark boundaries (see below). A distinctly different, 265 bp DNA sequence (the *sns* sequence) with enhancer-blocking activity has been observed near the 3' end of the sea urchin early H2A histone gene (55). It contains multiple CTC repeats, which may resemble in significant ways the TTCCC repeats earlier reported to have a similar activity (56). In the latter case, activity was attributed to the formation of triple-strand H-DNA. Whether unusual

DNA structures or sequence-specific binding of proteins (or both) are involved in this activity, it is likely to involve a different mechanism of action from that of CTCF.

2.5 Yeast boundary elements

The exploration of boundaries in yeast has focused on the regulation of silencing in the neighbourhood of telomeres and the mating-type loci (Chapter 10). This kind of silencing arises from a condensed chromatin domain; under appropriate circumstances it encroaches on a potentially active gene. Aparicio and Gottschling (57) showed that if a gene is placed adjacent to the TG_{1-3} tract of a telomere from which subtelomeric regions have been removed, it will exhibit position effect variegation. This tendency to silencing can be overcome by addition of a transactivator protein that increases expression of the gene. The choice between states is thus the result of a competition, and is made in G_2/metaphase. Beyond that point in the cell cycle, an inactivated test gene cannot be reactivated even by addition of the transactivator, until the cell has again replicated and divided.

The 'boundary' between active and inactive chromatin in this case is clearly not fixed from one cell to another. The condensed chromatin structure at yeast telomeres involves the RAP1 protein as an 'initiator' which binds to the telomeric repeats, leading to the formation of an extended complex of variable length that includes SIR2, SIR3, and SIR4, and histones H3 and H4 (58). A related (but not identical) complex forms at the yeast mating-type loci *HML* and *HMR*, and in those cases there has been considerable interest in determining what limits the extension of the silencing structure over adjacent sequences. Recent studies show that there are at least two ways in which this can be accomplished. The yeast *HML-I* silencer establishes a boundary between active and inactive chromatin by virtue of its polarity: it organizes a condensed chromatin structure in only one direction (59). The biochemical basis of this polarity is not known, but is at least suggestive of the presence of a barrier at one end of the element that prevents spreading of silencing activity in that direction. More direct evidence for such a boundary comes from work with the *HMR-E* silencer (60). Silenced chromatin structures propagate from *HMR-E* both inward to repress the *HMR* domain, and outward to repress expression of test genes located 475 bp, but not 2840 bp, to the left of *HMR-E*. Further examination has revealed the existence of an insulator that acts as a barrier to silencing. The element contains both a TY1 LTR and a tRNA gene. The insulator activity is impaired by mutations in *smc1* and *smc3*, which code for chromosomal structural proteins, establishing a connection between chromatin structure and boundary function.

Other factors also can prevent the propagation of a silencing structure. Experiments with *HML-E* show that silencing can be interfered with by insertion of multiple binding sites for RAP1, which occur naturally in the regulatory regions of certain genes (61). This presumably involves a kind of 'chain termination' process that blocks the further polymerization of silencing factors on chromatin.

Yeast subtelomeric regions also contain repetitive DNA sequences that are capable of preventing the propagation of silencing structures. These elements (subtelomeric *a*ntisilencing *r*egions, or STARS) can prevent the kind of telomeric silencing described above (57) when inserted between TG_{1-3} repeats and a reporter gene (62). Binding sites for two proteins, TBF1 and REB1, have been implicated in this insulator activity. Within the subtelomeric region they are found in combination with elements that reinforce silencing, located in such a way as to permit alternating silenced and expressed domains.

3. Mechanisms

Given the variety of phenomena that are gathered together under the terms 'boundary' and 'insulator', it would be surprising if a single mechanism were involved in all of them. In particular, the enhancer-blocking assays and the position effect assays may detect quite different kinds of interactions. In yeast the boundary elements so far detected interfere with the progress of a condensation reaction. Although the details of condensed chromatin structure in yeast and higher eukaryotes are certainly different, it seems likely that mechanisms to inhibit the propagation of such structures exist, for example, in *Drosophila* as well, perhaps to provide boundaries for regions condensed under the influence of the Polycomb group (PcG) proteins. Such a role has recently been suggested for su(Hw) protein in its action at the *cut* locus (63, 64).

Many but not all of the insulator elements so far described are capable not only of protecting against position effects but also of positional enhancer blocking. To the extent that both properties have to do with shielding the reporter gene from external activating influences, this is reasonable. It is less clear why an enhancer blocker should shield against the encroachment of a condensed chromatin structure. In fact, it has been shown in constructs designed to test the enhancer-blocking function of scs and of the *gypsy* insulator that their action does not depend (at least in these constructs) on unidirectional establishment of a condensed chromatin structure (21).

3.1 Relationship to mechanisms of enhancer action

Many of the proposed mechanisms of insulator action are in fact attempted explanations of positional enhancer-blocking activity. They are therefore intimately connected with models of enhancer action. Since the earliest recognition that transcriptional activation could be mediated by enhancers located some distance from their target promoters, two kinds of models have been proposed. Data in support of both kinds of models exist in the literature. The first class can be referred to as 'tracking' models, the second as 'looping' or direct contact models. Tracking models suppose that there is a signal sent from enhancer to promoter that must actually traverse the DNA linking the two. In the simplest version of this model, the enhancer binds activating proteins that then transfer to the adjacent DNA and diffuse, maintaining contact with the DNA (or chromatin), until they reach the promoter. Alternatively, the enhancer-bound complex might loop over to make contact with the DNA, and then the complex could track to the promoter as the loop enlarges. Evidence at least consistent with such a mechanism has been reported (65).

Tracking models have the great advantage that they easily explain positional enhancer blocking: the role of the insulator is somehow to derail the tracking mechanism, and it can do this only if it lies between enhancer and promoter, in the path of the advancing signal. There are, however, some results that are difficult to explain in this way. For example, experiments in *Xenopus* oocytes have shown that an enhancer can activate a promoter when the two are on separate but interlinked closed circular plasmids (66). Furthermore, surrounding the enhancer or the promoter in such linked plasmids with insulators is sufficient to block enhancer action. Although it is possible to imagine tracking models that would account for this behaviour, it does make such explanations more difficult to accept.

3.2 Loop or domain models

An alternative view suggests that insulators exert their effects on transcription through changes in higher-order chromatin structure (1). This model is supported by the observation that insulators are usually associated with strong DNAase I-hypersensitive sites and tend to separate chromatin domains with different degrees of condensation (2, 43, 59, 60). A role of insulators in chromatin organization is also supported by the properties of one of the protein components of the *gypsy* insulator. The *mod(mdg4)* (*modifier of mdg4*) gene acts as a classical enhancer of position effect variegation and has the properties characteristic of *trithorax group* (*trxG*) genes (Chapter 11) (30, 31, 67). These observations suggest that insulators might affect gene expression by establishing higher-order domains of chromatin organization (Fig. 5).

Much of the direct evidence supporting this type of model comes from analysis of the subnuclear distribution of protein components of the *gypsy* insulator. Results from immunofluorescence experiments using antibodies against su(Hw) and mod(mdg4) indicate that these proteins are present at hundreds of sites in polytene chromosomes from salivary glands. These sites do not contain copies of the *gypsy* retrotransposon and are presumed to be endogenous insulators, similar to the one found in *gypsy*, that play a role in the normal regulation of gene expression in *Drosophila*. The mod(mdg4) protein is located in approximately 500 sites and overlaps with su(Hw) protein at all sites where this protein is present (67). Polytene chromosomes can be considered to be at interphase, and the observed distribution of insulator proteins suggests that they should bind at more or less regular intervals in the chromosomes of interphase diploid cells. Given the large number of sites and their regular distribution along the chromosome arms, one would expect to observe a diffuse homogeneous scattering of insulator sites in the nuclei of interphase diploid cells. Surprisingly, this is not the case; instead, *gypsy* insulator proteins accumulate at a small number of nuclear locations. This has led to the suggestion that each of the locations where su(Hw) and mod(mdg4) proteins accumulate in the nucleus is made up of several individual sites that come together, perhaps through interactions among protein components of the insulator. Interestingly, the locations where individual insulator sites appear to aggregate in the nucleus are not random. Analyses of the distribution of *gypsy* insulator aggregates in three-dimensional reconstructions of nuclei from diploid cells

Fig. 5 Schematic model explaining the role of su(Hw), mod(mdg4) and trxG and PcG proteins in the function of the gypsy insulator. The diagram represents a section though a cell with a nucleus surrounded by the nuclear membrane and the nuclear lamina. The chromatin fibre is represented as a long cylindrical wire; proteins are represented as spheres located in the nuclear periphery (su(Hw) and mod(mdg4)) or ovals located in the central region of the nucleus (trxG and PcG).

in interphase indicate that, although not all the aggregate sites are present in the nuclear periphery, approximately 75% of them are present immediately adjacent to the nuclear lamina (68). This finding suggests that the formation of *gypsy* insulator aggregates may require a substrate for attachment, and that physical attachment might play a role in the mechanism by which this insulator affects enhancer–promoter interactions (Fig. 5). The nuclear lamina itself might serve as a substrate for attachment, perhaps through interactions between lamin and protein components of the insulator. The nature of the substrate involved in the attachment of the aggregate sites found in the interior of the nucleus is unknown, but it is interesting that a lamin network has also been detected inside of the nucleus (69).

A model suggesting a role for *gypsy* insulators in the subnuclear organization of the chromatin fibre has several testable predictions. For example, this model predicts that a DNA sequence usually located randomly with respect to the periphery versus interior compartments in the nucleus should become preferentially located in the periphery when a copy of the *gypsy* insulator is present in this sequence. In addition, a DNA sequence should be located at the site of insulator aggregates when the sequence contains an insulator, but should be present in a different nuclear location when the insulator is not located within the sequence. Finally, two different sequences located far apart in a chromosome might come together at the same nuclear location when insulators are present within these sequences. Experimental results have confirmed all three predictions, supporting the idea that *gypsy* insulators play a role in the higher-order domain organization of the chromatin fibre (68).

Since the *gypsy* insulator aggregates appear to be present preferentially in the nuclear periphery, insulator proteins might physically attach to a solid support such as the nuclear lamina. This attachment might impose a topological or physical constraint on the DNA that interferes with the transmission of a signal from an enhancer located in one domain to a promoter located in an adjacent one. The preferential aggregation of insulator sites at the nuclear periphery and the possibility that this targeting might take place through interactions with the nuclear lamina led to the idea that the *gypsy* insulator might be equivalent to MARs/SARs (67). This hypothesis is directly supported by the finding of MAR activity within the DNA sequences containing the *gypsy* insulator (70). An important question arising from these results is whether the organization imposed by the *gypsy* insulator is static, and has a mostly structural role, or whether the organization is dynamic and has a direct regulatory significance. If the latter is the case, one would expect a correlation between the pattern of nuclear organization of insulator bodies and the transcriptional state of the cell. This is an important topic for future studies.

4. Future questions

The study of the properties of chromatin boundaries/insulators should lead to a better understanding of the mechanisms by which enhancers activate transcription in eukaryotes and of the role of complex levels of chromatin organization in the control of gene expression. In the following sections we speculate on some of the important questions that will need to be addressed in the immediate future.

4.1 How does activation and deactivation over long distances work in higher eukaryotes?

Transcriptional activation in eukaryotic organisms involves changes in chromatin structure that are probably a prerequisite for the ensuing interactions of enhancer-bound transcription factors with the transcription complex present at the promoter. These changes in chromatin structure probably involve alterations of higher-order levels of organization as well as changes in nucleosome structure or organization in the primary chromatin fibre involving histone acetylases and deacetylases or other chromatin remodelling complexes (reviewed in 71; see also earlier chapters). Many of the available results on these issues come from studies carried out in yeast, where upstream activating sequences are located relatively close to the promoter. But in most eukaryotes, including *Drosophila*, enhancer elements are located tens or even hundreds of kilobases away from the promoters of genes. How do eukaryotic enhancers activate transcription over such long distances? Since insulators regulate this interaction, studies on the mechanisms of insulator function should give insights into how enhancers activate transcription over long distances (72). Studies of the effects of the *gypsy* insulator on the regulation of the *cut* gene by the wing margin enhancer have already led to the identification of Chip, a protein that appears to regulate enhancer–promoter interactions (63). Chip is a homologue of the mouse Nli/Ldb1/ Clim-2 family and, as can other members, it can also interact with nuclear LIM domain proteins. Chip is widely distributed on *Drosophila* polytene chromosomes and it is required for the expression of many genes, although it does not participate directly in transcriptional activation. These results have led to the suggestion that Chip facilitates enhancer–promoter interactions by stabilizing the formation of chromatin structures that bring enhancers located far upstream in close proximity with the promoter (72). The analysis of Chip and Nipped-B (73), found to affect insulator function, will shed light on the mechanisms of long-range interactions between enhancers and promoters.

4.2 Are there any common patterns of action shared by the insulator elements so far identified? Are there many more kinds of insulator elements?

It may be that the problem in providing a single explanation for insulator activity is that we are actually looking at a series of distinct functions. As mentioned above, there is no reason why elements found *in vivo* should necessarily be capable of both enhancer blocking and protection against position effects. Furthermore, enhancer blocking may be accomplished in different ways by different combinations of binding sites and proteins. It is evident that quite different proteins are bound, for example, at scs, scs', *gypsy*, *Fab-7*, and 5'HS4. So far the only protein associated with enhancer blocking to be identified in vertebrates is CTCF. In sea urchin, there is at least one other kind of site (in the histone gene locus), which must bind something else, and in all probability other kinds of sites are as yet undetected. Like all regulatory elements, insulators seem to exert their effect by binding proteins that, in turn, recruit a host of others that are essential to function. It is likely that the great diversity of properties and regulated activities that such mechanisms impart to enhancers and promoters will also be conferred on insulators.

4.3 What cofactors are involved in establishing boundaries, and what is their relationship to chromatin structure?

A connection between insulators and other proteins involved in the establishment of particular chromatin structures was made by the observation that the mod(mdg4) protein of the *gypsy* insulator has properties of E(var) and trxG proteins. This connection is supported by the recent finding of an insulator in the promoter of the *eve* gene containing a binding site for the trG protein GAGA (74). Hypomorphic mutations in *mod(mdg4)* have a specific effect on the activity of the *gypsy* insulator, and this effect can be transmitted maternally (i.e. wild-type progeny of *mod(mdg4)* mutant mothers still show a partially inactive insulator). Interestingly, maternal transmission is affected by mutations in *trxG* and *PcG* genes. This genetic interaction correlates with changes in the ability of *gypsy* insulator sites to form aggregates in the nuclei of interphase diploid cells; in the background of mutations in *trxG* and *PcG* genes, these aggregates fail to form and the insulator sites appear to be distributed throughout the nucleus. These observations have been interpreted in the context of a model in which trxG and PcG proteins participate and help insulator proteins in the establishment and maintenance of higher-order chromatin domains (Fig. 5) (67).

Other factors that are more directly involved in regulating insulator activity must be present in the nucleus. If insulators play a role in establishing higher-order domains of chromatin organization, their activity might be modulated both during cell division and cell differentiation. There must then be proteins that are either constitutive insulator components or are functionally linked to alter the properties of insulators by modifying their protein components. Such proteins have not been identified as yet; their existence would lend support to the idea that insulators play important roles in global aspects of gene regulation.

References

1. Kellum, R. and Schedl, P. (1991) A position-effect assay for boundaries of higher order chromatin domains. *Cell*, **64**, 941.
2. Udvardy, A., Maine, E., and Schedl, P. (1985) The 87A7 chromomere. Identification of novel chromatin structures flanking the heat shock locus that may define the boundaries of higher order domains. *J. Mol. Biol.*, **185**, 341.
3. Chung, J. H., Whiteley, M., and Felsenfeld, G. (1993) A 5′ element of the chicken beta-globin domain serves as an insulator in human erythroid cells and protects against position effects in *Drosophila*. *Cell*, **74**, 505.
4. Kellum, R. and Schedl, P. (1992) A group of scs elements function as domain boundaries in an enhancer-blocking assay. *Mol. Cell. Biol.*,**12**, 2424.
5. Vazquez, J. and Schedl, P. (1994) Sequences required for enhancer blocking activity of scs are located within two nuclease-hypersensitive regions. *EMBO J.*, **13**, 5984.
6. Gaszner, M., Vazquez, J., and Schedl, P. (1999) The Zw5 protein, a component of the scs chromatin domain boundary, is able to block enhancer–promoter interaction. *Genes Develop.*, **13**, 2098.
7. Zhao, K., Hart, C. M., and Laemmli, U. K. (1995) Visualization of chromosomal domains with boundary element-associated factor BEAF-32. *Cell*, **81**, 879.
8. Cuvier, O., Hart, C. M., and Laemmli, U. K. (1998) Identification of a class of chromatin boundary elements. *Mol. Cell. Biol.*, **18**, 7478.
9. Hart, C., Zhao, K., and Laemmli, U. K. (1997) The scs′ boundary element: characterization of boundary element-associated factors. *Mol. Cell. Biol.*, **17**, 999.
10. Hart, C. M., Cuvier, O., and Laemmli, U. K. (1999) Evidence for an antagonistic relationship between the boundary element-associated factor BEAF and the transcription factor DREF. *Chromosoma*, **108**, 375.

11. Mishra, R. and Karch, F. (1999) Boundaries that demarcate structural and functional domains of chromatin. *J. Biosci.*, **24**, 377.

12. Hagstrom, K., Muller, M., and Schedl, P. (1996) *Fab-7* functions as a chromatin domain boundary to ensure proper segment specification by the *Drosophila bithorax* complex. *Genes Develop.*, **10**, 3202.

13. Zhou, J., Barolo, S., Szymanski, P., and Levine, M. (1996) The Fab-7 element of the *bithorax* complex attenuates enhancer–promoter interactions in the *Drosophila* embryo. *Genes Develop.*, **10**, 3195.

14. Mihaly, J., Hogga, I., Gausz, J., Gyurkovies, H., and Karch, F. (1997) *In situ* dissection of the *Fab-7* region of the *bithorax* complex into a chromatin domain boundary and a Polycomb-response element. *Development*, **124**, 1809.

15. Zhou, J. and Levine, M. (1999) A novel *cis*-regulatory element, the PTS, mediates an anti-insulator activity in the *Drosophila* embryo. *Cell*, **99**, 567.

16. Barges, S., Mihaly, J., Galloni, M., Hagstrom, K., Muller, M., Shanower, G., Schedl, P., Gyurkovics, H., and Karch, F. (2000) The *Fab-8* boundary element defines the distal limit of the bithorax complex *iab-7* domain and insulates *iab-7* from initiation elements and a PRE in the adjacent *iab-8* domain. *Development*, **127**, 779.

17. Gdula, D. A., Gerasimova, T. I., and Corces, V. G. (1996) Genetic and molecular analysis of the *gypsy* chromatin insulator of *Drosophila*. *Proc. Natl Acad. Sci., USA*, **93**, 9378.

18. Holdridge, C. and Dorsett, D. (1991) Repression of *hsp70* heat shock gene transcription by the *suppressor of Hairy-wing* protein of *Drosophila melanogaster*. *Mol. Cell. Biol.*, **11**, 1894.

19. Jack, J., Dorsett, D., Delotto, Y., and Liu, S. (1991) Expression of the *cut* locus in the *Drosophila* wing margin is required for cell type specification and is regulated by a distal enhancer. *Development*, **113**, 735.

20. Geyer, P. K. and Corces, V. G. (1992) DNA position-specific repression of transcription by a *Drosophila* zinc finger protein. *Genes Develop.*, **6**, 1865.

21. Cai, H. and Levine, M. (1995) Modulation of enhancer–promoter interactions by insulators in the *Drosophila* embryo. *Nature*, **376**, 533.

22. Scott, K. S. and Geyer, P. M. (1995) Effects of the *Drosophila* su(Hw) insulator protein on the expression of the divergently transcribed yolk protein genes. *EMBO J.*, **14**, 6258.

23. Roseman, R. R., Pirrotta, V., and Geyer, P. K. (1993) The su(Hw) protein insulates expression of the *Drosophila melanogaster white* gene from chromosomal position-effects. *EMBO J.*, **12**, 435.

24. Lu, L. and Tower, J. (1997) A transcriptional insulator element, the su(Hw) binding site, protects a chromosomal DNA replication origin from position effects. *Mol. Cell. Biol.*, **17**, 2202.

25. Spana, C. and Corces, V. G. (1990) DNA bending is a determinant of binding specificity for a *Drosophila* zinc finger protein. *Genes Develop.*, **4**, 1505.

26. Scott, K. C., Taubman, A. D., and Geyer, P. K. (1999) Enhancer blocking by the *Drosophila gypsy* insulator depends upon insulator anatomy and enhancer strength. *Genetics*, **153**, 787.

27. Chen, S. (2000) The effect of the *gypsy* insulator on chromatin structure. Ph.D. Thesis, The Johns Hopkins University, Baltimore, MD.

28. Harrison, D. A., Gdula, D. A., Coyne, R. S., and Corces, V. G. (1993) A leucine zipper domain of the *suppressor of Hairy-wing* protein mediates its repressive effect on enhancer function. *Genes Develop.*, **7**, 1966.

29. Georgiev, P. G. and Gerasimova, T. I. (1989) Novel genes influencing the expression of the *yellow* locus and mdg4 (*gypsy*) in *Drosophila melanogaster*. *Mol. Gen. Genet.*, **220**, 121.

30. Dorn, R., Krauss, V., Reuter, G., and Saumweber, H. (1993) The enhancer of position-effect variegation of *Drosophila E(var)3–93D* codes for a chromatin protein containing a conserved domain common to several transcriptional regulators. *Proc. Natl Acad. Sci., USA,* **90,** 11376.

31. Gerasimova, T. I., Gdula, D. A., Gerasimov, D. V., Simonova, O., and Corces, V. G. (1995) A *Drosophila* protein that imparts directionality on a chromatin insulator is an enhancer of position-effect variegation. *Cell,* **82,** 587.

32. Büchner, K., Roth, P., Schotta, G., Kraus, V., Saumweber, H., Reuter, G., and Dorn, R. (2000) Genetic and molecular complexity of the position effect variegation modifier *mod(mdg4)* in *Drosophila. Genetics,* **155,** 141.

33. Farkas, G., Gausz, J., Galloni, M., Reuter, G., Gyurkovics, H., and Karch, F. (1994) The *Trithorax-like* gene encodes the *Drosophila* GAGA factor. *Nature,* **371,** 806.

34. Reitman, M. and Felsenfeld, G. (1990) Developmental regulation of topoisomerase II sites and DNase I hypersensitive sites in the chicken beta-globin locus. *Mol. Cell. Biol.,* **10,** 2774.

35. Hebbes, T. R., Clayton, A. L., Thorne, A. W., and Crane-Robinson, C. (1994) Core histone hyperacetylation co-maps with generalized DNase I sensitivity in the chicken beta-globin chromosomal domain. *EMBO J.,* **13,** 1823.

36. Chung, J. H., Bell, A. C., and Felsenfeld, G. (1997) Characterization of the chicken beta-globin insulator. *Proc. Natl Acad. Sci., USA,* **94,** 575.

37. Bell, A. C., West, A. G., and Felsenfeld, G. (1999) The protein CTCF is required for the enhancer-blocking activity of vertebrate insulators. *Cell,* **98,** 387.

38. Filippova, G. N., Fagerlie, S., Klenova, E. M, Myers, C., Dehner, Y., Goodwin, G., Neiman, P. E., Collins, S. J., and Lobanenkov, V. V. (1996) An exceptionally conserved transcriptional repressor, CTCF, employs different combinations of zinc fingers to bind diverged promoter sequences of avian and mammalian c-myc oncogenes. *Mol. Cell. Biol.,* **16,** 2802.

39. Zhong, X. P. and Krangel, M. S. (1997) An enhancer-blocking element between alpha and delta gene segments within the human T-cell receptor alpha/delta locus. *Proc. Natl Acad. Sci., USA,* **94,** 5219.

40. Robinett, C. C., O'Connor, A., and Dunaway, M. (1997) The repeat organizer, a specialized insulator element within the intergenic spacer of the *Xenopus* rRNA genes. *Mol. Cell. Biol.,* **17,** 2866.

41. Saitoh, N., Bell, A., Recillas-Targa, F., West, A., Simpson, M., Pikaart, M., and Felsenfeld, G. (2000) Structural and functional conservation at the boundaries of the chicken beta-globin domain. *EMBO J.,* **19,** 2315.

42. Bulger, M., von Doorninck, H., Saitoh, N., Telling, A., Farrell, C., Bender, M.A., Felsenfeld, G., Axel R., and Groudine, M. (1999) Conservation of sequence and structure flanking the mouse and human beta-globin loci: the beta-globin genes are embedded within an array of odorant receptor genes. *Proc. Natl Acad. Sci., USA,* **96,** 5129.

43. Prioleau, M.-N., Nony, P., Simpson, M., and Felsenfeld, G. (1999) An insulator element and condensed chromatin region separate the chicken β-globin locus from an independently regulated erythroid-specific folate receptor gene. *EMBO J.,* **18,** 4035.

44. Bell, A. and Felsenfeld, G. (2000) Methylation of a CTCF-dependent boundary controls imprinted expression of the *Igf2* gene. *Nature,* **405,** 482.

45. Hark, A. T., Schoenherr, C. J., Katz, D. J., Ingram, R. S., Levorse, J. M., and Tilghman, S. M. (2000) CTCF mediates methylation-sensitive enhancer blocking activity at the *H19/Igf2* locus. *Nature,* **405,** 486.

46. Leighton, P. A., Ingram, R. S., Eggenschwiler, J., Efstratiadis, A., and Tilghman, S. M. (1995) Disruption of imprinting caused by deletion of the *H19* gene region in mice. *Nature,* **375,** 34.

47. Pikaart, M. J., Recillas-Targa, F., and Felsenfeld, G. (1998) Loss of transcriptional activity of a transgene is accompanied by DNA methylation and histone deacetylation, and is prevented by insulators. *Genes Develop.*, **12**, 2852.

48. Chen, W. Y., Bailey, E. C., McCune, S. L., Dong, J. Y., and Townes, T. M. (1997) Reactivation of silenced, virally transduced genes by inhibitors of histone deacetylase. *Proc. Natl Acad. Sci., USA*, **94**, 5798.

49. Wang, Y., DeMayo, F. J., Tsai, S. Y., and O'Malley, B. W. (1997) Ligand-inducible and liver-specific target gene expression in transgenic mice. *Nat. Biotechnol.*, **15**, 329.

50. Taboit-Dameron, F., Malassagne, B., Viglietta, C., Puissant, C., Leroux-Coyau, M., Chereau, C., Attal, J., Weill, B., and Houdebine, L. M. (1999) Association of the 5'HS4 sequence of the chicken beta-globin locus control region with human EF1 alpha gene promoter induces ubiquitous and high expression of human CD55 and CD59 cDNAs in transgenic rabbits. *Transgenic Res.*, **8**, 223.

51. Inoue, T., Yamaza, H., Sakai, Y., Mizuno, S., Ohno, M., Hamasaki, N., and Fukumaki, Y. (1999) Position-independent human beta-globin gene expression mediated by a recombinant adeno-associated virus vector carrying the chicken beta-globin insulator. *J Hum. Genet.*, **44**, 152.

52. Rivella, S., Callegari, J. A., May, C., Tan, C. W., and Sadelain, M. (2000) The cHS4 insulator increases the probability of retroviral expression at random chromosomal integration sites. *J. Virol.*, **74**, 4679.

53. Laemmli, U. K., Käs, E., Poljak, L., and Adachi, Y. (1992) Scaffold-associated regions: cis-acting determinants of chromatin structural loops and functional domains. *Curr. Opin. Genet. Dev.*, **2**, 275.

54. Stief, A., Winter, D. M., Stratling, W. H., and Sippel, A. E. (1989) A DNA attachment element mediates elevated and position-independent gene activity. *Nature*, **341**, 343.

55. Palla, F., Melfi, R., Anello, L., Di Bernardo, M., and Spinelli, G. (1997) Enhancer blocking activity located near the 3' end of the sea urchin early H2A histone gene. *Proc. Natl Acad. Sci., USA*, **94**, 2272.

56. Michel, D., Chatelain, G., Herault, Y., Harper, F., and Brun, G. (1993) H-DNA can act as a transcriptional insulator. *Cell. Mol. Biol. Res.*, **39**, 131.

57. Aparicio, O. M. and Gottschling, D. E. (1994) Overcoming telomeric silencing: a trans-activator competes to establish gene expression in a cell cycle-dependent way. *Genes Develop.*, **8**, 1133.

58. Grunstein, M. (1997) Molecular model for telomeric heterochromatin in yeast. *Curr Opin. Cell. Biol.*, **9**, 383.

59. Bi, X., Braunstein, M., Shei, G. J., and Broach, J. R. (1999) The yeast HML silencer defines a heterochromatin domain boundary by directional establishment of silencing. *Proc. Natl Acad. Sci., USA*, **96**, 11934.

60. Donze, D., Adams, C. R., Rine, J., and Kamakaka, R. T. (1999) The boundaries of the silenced HMR domain in *Saccharomyces cerevisiae*. *Genes Develop.*, **13**, 698.

61. Bi, X. and Broach, J. R. (1999) UASrpg can function as a heterochromatin boundary element in yeast. *Genes Develop.*, **13**, 1089.

62. Fourel, G., Revardel, E., Koering, C., and Gilson, E. (1999) Cohabitation of insulators and silencing elements in yeast subtelomeric regions. *EMBO J.*, **18**, 2522.

63. Morcillo, P., Rosen, C., Baylies, M. K. and Dorsett, D. (1997) Chip, a widely expressed chromosomal protein required for segmentation and activity of a remote wing margin enhancer in *Drosophila*. *Genes Develop.*, **11**, 2729.

64. Bulger, M. and Groudine, M. (1999) Looping versus linking: toward a model for long-distance gene activation. *Genes Develop.*, **13**, 2465.

65. Courey, A. J., Plon, S. E., and Wang J. C. (1986) The use of psoralen-modified DNA to probe the mechanism of enhancer action. *Cell*, **45**, 567.

66. Dunaway, M. and Droge, P. (1989) Transactivation of the *Xenopus* rRNA gene promoter by its enhancer. *Nature*, **341**, 657.

67. Gerasimova, T. I. and Corces, V.G. (1998) Polycomb and trithorax group proteins mediate the function of a chromatin insulator. *Cell*, **92**, 511.

68. Gerasimova T. I. and Corces V.G. (2000) A chromatin insulator determines the nuclear localization of DNA, submitted.

69. Neri, L. M., Raymond, Y., Giordano, A., Capitani, S., and Martelli, A. M. (1999) Lamin A is part of the internal nucleoskeleton of human erythroleukemia cells. *J. Cell. Physiol.*, **178**, 284.

70. Nabirochkin, S., Ossokina, M., and Heidmann, T. (1998) A nuclear matrix/scaffold attachment region co-localizes with the *gypsy* retrotransposon insulator sequence. *J. Biol. Chem.*, **273**, 2473.

71. Blackwood, E. M. and Kadonaga, J. T. (1998) Going the distance: a current view of enhancer action. *Science*, **281**, 61.

72. Dorsett, D. (1999) Distant liaisons: long-range enhancer-promoter interactions in *Drosophila. Curr. Opin. Genet. Dev.*, **9**, 505.

73. Rollins, R. A., Morcillo, P., and Dorsett, D. (1999) *Nipped-B*, a *Drosophila* homologue of chromosomal adherins, participates in activation by remote enhancers in the *cut* and *Ultrabithorax* genes. *Genetics*, **152**, 577.

74. Ohtsuki, S. and Levine, M. (1998) GAGA mediates the enhancer-blocking activity of the *eve* promoter in the *Drosophila* embryo. *Genes Develop.*, **12**, 3325.

14 | Linking large-scale chromatin structure with nuclear function

NICOLA L. MAHY, WENDY A. BICKMORE, TUDORITA TUMBAR, and ANDREW S. BELMONT

1. Introduction: why study large-scale chromatin/chromosome structure?

Simple transcription factor binding to specific DNA sequences is insufficient to explain the control of gene expression. In previous chapters, the close link between small-scale chromatin structure and control of DNA function, particularly transcription, has been established. The discovery of an ever-growing number of modifying factors, including histone acetyl transferases (HATs), histone deacetylases (HDACs), and chromatin re-modelling complexes has considerably strengthened this connection. Beyond the nucleosome, the formation of specialized chromatin structures, e.g. heterochromatin, involves large multisubunit protein complexes. Mechanisms of gene regulation such as dosage compensation, imprinting, locus control region (LCR) function, and boundary elements suggest that many other levels of chromatin structure impinge on transcription and its regulation. Finally, attention has turned to the spatial organization and functional compartmentalization of chromosomes, and of the nucleus itself, in the quest to understand how the expression of complex genomes is regulated. This is the focus of this chapter.

2. Comparing large-scale chromatin structure within mitotic and interphase chromosomes

Within the higher eukaryotic chromosome, DNA is folded through DNA–protein interactions into multiple levels of organization. At the highest level, these yield a compaction ratio of more than 20 000 : 1 in terms of the ratio of linear, B-form DNA to the length of the fully compacted metaphase chromosome.

While the extent of compaction within mitotic chromosomes is well known, less appreciated is the fact that compaction remains extremely high within interphase nuclei. The bulk of genomic DNA in interphase is likely to be packaged within large-scale structures well above the level of the 30 nm chromatin fibre. Fluorescence *in situ* hybridization (FISH) studies suggest a linear DNA compaction ratio of *c*. 200–1000 : 1 in the nucleus, and direct *in vivo* visualization has revealed compact interphase centromere structures not significantly different in size from their metaphase conformation (1). By inserting lac operator direct repeats adjacent to a selectable marker, and combining this with gene amplification, it has been possible to visualize amplified chromosome regions or even entire chromosome arms using green fluorescent protein (GFP)-lac repressor staining in living cells (2). Distinct, large-scale chromatin fibres extending for up to 5 μm can be selectively visualized in this way. Direct analysis of a late replicating, heterochromatic amplified chromosome arm has revealed compaction ratios only several fold lower than metaphase values during late stages of DNA replication (3). After targeting a strong transcriptional activator to this region, large-scale chromatin fibres with compaction ratios near 1000 : 1, which show extremely high levels of transcriptional activity, can be seen within interphase nuclei (4) (Plate 4). What impact this level of organization has on transcriptional activity remains an open question. Chromosomes 18 and 19, which are of a similar size and hence have a similar compact structure at mitosis, have been demonstrated to have highly contrasting structural and transcriptional characteristics in interphase (5).

2.1 Lessons from model, non-mitotic chromosomes

With the exception of a very small number of specialized cells, spatially distinct, 30 nm chromatin fibres cannot be visualized within interphase nuclei by transmission electron microscopy (TEM) (6). Instead, chromatin is packaged primarily into higher-order chromatin domains within which the linear fibre organization is not evident. The exceptions to this technical limitation are special interphase chromosome types: these include lampbrush chromosomes in meiotic nuclei, and polytene chromosomes in many dipterans. The distinct banding pattern of polytene chromosomes, combined with the transient appearance of decondensed 'puffs' associated with high transcriptional activity, immediately suggest the existence of defined genomic regions with different higher-order chromatin compaction levels. Electron microscopy (EM) ultrastructural analysis of a Balbiani Ring puff in *Chironimus* polytene chromosomes has reinforced these earlier light microscopy observations (7). Abrupt transitions in banding patterns and chromatin folding are seen within several kilobases of the 5′ and 3′ ends of the transcription units.

Inferences based on studies of polytene chromosomes have been supported by the visualization of ribonucleoprotein (RNP)-covered loops of highly transcriptionally active loci in lampbrush chromosomes (8). Again, sharp demarcations between the decondensed, transcriptionally active loops and the highly condensed chromatin along the chromosome axis are seen. These observations indicate that very distinct and sudden transitions in higher-order chromatin structure can occur in a sequence-dependent manner. The appearance of lampbrush chromosome loops has reinforced models of decondensed chromatin with looped domains associated with gene activity. Polymerase densities on these lampbrush loops and polytene puffs are often so high that nucleosomes may be substantially lost from the transcribed regions. In contrast, the majority of RNA polymerase II (RNA Pol II)-transcribed genes in somatic cells are expressed with less than one polymerase per transcription unit (9). The organization of these transcribed regions in the context of chromatin is still unknown.

2.2 Models of mitotic and interphase chromosome structure

For technical reasons, most research into chromosome structure has focused on the structure of maximally condensed, metaphase chromosomes. An experimental approach based largely on unfolding chromosome structure through extraction of chromosomal proteins has led to a radial loop model of chromosome structure. In this model, structural proteins, which are resistant to high salt and detergent extraction, anchor the bases of 30 nm chromatin fibre loops (~20–200 kb long) to a chromosome 'scaffold', which itself may be helically coiled (10). Specific SAR/MAR DNA sequences (scaffold *attachment* *regions* or *matrix* *attachment* *regions*) are hypothesized to form the bases of these loops, attached to specific proteins which are predicted to make up the chromosome scaffold (11). Specific sequences are found remaining at the axial core in extracted human metaphase chromosomes (12), but it is not clear whether the same SAR/MAR sequences are attached to an underlying scaffold in both mitotic and interphase chromosomes. FISH on cell nuclei has led to a giant-loop, random walk model for interphase chromosomes, based on statistical analysis of the mean separation between two chromosome sites, as a function of genomic distance (13).

Ideally, any model of large-scale chromatin folding would unify mitotic and interphase chromosome structure and predict the structural transitions accompanying cell-cycle-driven chromosome condensation/decondensation. The radial-loop, helical-coil model of mitotic chromosome structure (Fig. 1a) has been extended to interphase chromosomes. However, this has required postulating a particular loop geometry that might, under special circumstances, give rise to a fibre with an elliptical 60–90 nm cross-section (14, 15). An alternative model proposes a successive, helical coiling of 10 nm chromatin fibres into 30–50 nm tubes, and of these into 200 nm diameter tubes, which coil into c. 600 nm metaphase chromatids (16). Finally, a folded chromonema model is based on *in vivo* light microscopy combined with TEM ultrastructural analysis of folding intermediates during the transition into and out of mitosis. In this model, 10 and 30 nm chromatin fibres fold to form a c. 100 nm diameter chromonema fibre, which then folds into a 200–300 nm diameter prophase chromatid, which itself coils to form the metaphase chromosome (17) (Fig. 1b). It is still unclear how these structural models of mitotic and interphase chromosome structure integrate with the underlying biochemistry responsible for chromosome condensation. The two chief

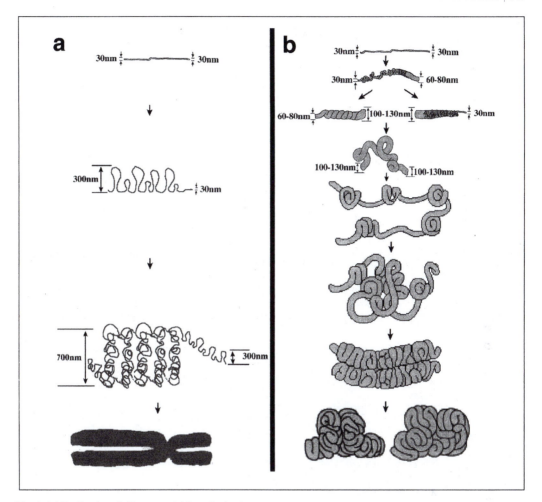

Fig. 1 (a) Textbook radial-loop model for mitotic chromosome structure. A looping of the 30 nm fibre gives rise to a 300 nm structure in which 50–100 kb looped DNA attaches at the base of the loop to a chromosomal scaffold. This structure coils helically to form the metaphase chromosome. (b) Chromonema model of interphase chromatin structure. Progressive levels of coiling of the 30 nm fibre into 60–80 nm and 100–130 nm fibres are depicted. Chromonema fibres kink and coil to form regions of more dispersed or compact chromatin. Extended chromonema fibres predominate in G_1 while more compact structures become abundant during cell-cycle progression. Chromonema folding culminates with the formation of the G_2 chromatid, which coils to form the compact metaphase chromosome.

protein components of the mitotic chromosome scaffold, topoisomerase IIα and SCII, have more clearly identified DNA topological activities than structural roles. SCII is a component of the mitotic condensin complex, which recently has been demonstrated to have the ability to introduce positive supercoils into DNA in the presence of topoisomerase II in a stoichiometric manner (reviewed in 18). SCII also shows a non-ATP-dependent enhancement of re-annealing of complementary DNA strands.

3. Organization of chromosomes within the interphase nucleus

3.1 Chromosome territories and intermingling between chromosomes

Two lines of experimental evidence indicate that interphase chromosomes occupy distinct territories in the nucleus, and that the DNA of different chromosomes does not extensively intermingle. First, DNA damage induced by UV-laser micro-irradiation of Chinese hamster cells has been shown to be restricted to a few chromosomes when the cells are followed to the subsequent mitosis (19). Secondly, defined interphase chromosome domains have been visualized by FISH, first in rodent–human hybrid cells (20, 21) and subsequently in a large number of different animal and plant cells (15, 19, 22).

Probes for individual genes, selected chromosome segments, and entire single chromosomes are now being used to study the three-dimensional organization of chromosome territories in detail (19, 23) (Plate 5). These studies have revealed different levels of organization within chromosome territories: separate domains are formed by individual chromosome arms (24, 25) and by early and late-replicating chromatin (26) representing R- and G-bands, respectively (27). Thus, genome segmentation revealed through metaphase chromosome bands appears to be translated into interphase nuclear space. Spatially distinct, labelled, and compact chromosome territories can also be visualized in living cells by microinjection of fluorescent nucleotides followed by chases over several cell generations (28).

While chromosome territories or chromosomal subdomains appear to be well defined at the resolution of the light microscope, fibre-like structures may be seen embedded in other territories at a small number of sites (29). This is closer to the picture generated using higher-resolution methods such as TEM, in which nuclei contain decondensed chromatin masses, inconsistent with a 'solid' chromosome territory model (30). Further research is required to determine whether there are specific sites of intermingling between chromosomes. One such site has been identified where a loop of several megabases of chromatin extends from the surface of chromosome 6p (31). This loop must be projecting into another chromosome domain, although whether it interacts with specific sequences within that domain remains to be determined. Other sites where DNA probes apparently locate away from their native chromosome have also been identified (Mahy and Bickmore, unpublished).

3.2 Spatial distribution of chromosomes relative to each other

A territorial organization of interphase chromosomes raises the question of whether there are distinct spatial relationships between the territory of one chromosome and that of another. The diffusional constraints on chromatin movement *in vivo* (32) mean that the physical proximity of different chromosomes in interphase, and their interactions with nuclear substructure, may be important in determining the likelihood of any two chromosomes meeting and interacting.

The paradigm for a distinct spatial relationship of one chromosome to another occurs at the nucleolus, a specialized nuclear domain where ribosomal DNA (rDNA) is transcribed and pre-ribosomes are assembled. It is organized by 'nucleolar organizer regions' (NORs) which, in humans, are localized at the short arms ('p' arms) of acrocentric chromosomes (HSA13, 14, 15, 21, and 22) and contain the ribosomal RNA genes (33). In the interphase nucleus rDNA-containing chromosomes are intrinsic to the nucleolus and also remain associated with one another through successive cell cycles (34).

Homologous chromosome alignment is prominent in meiosis (35) but, for most organisms, it is not a normal characteristic of somatic nuclear organization. However, in *Drosophila melanogaster* and other dipteran insects, homologous chromosomes are usually paired at interphase (36). Homologous chromosomes have also been shown to interact genetically in *Drosophila* (37, 38); suppression or enhancement of a phenotype (*trans*-sensing) can be observed when pairing is disrupted by chromosomal rearrangements (39, 40).

Homologous pairing in mammalian somatic nuclei does not normally occur (41, 42), but homologous loci of human and mouse genes subject to imprinting are transiently associated during late S-phase (43). It has been suggested that these *trans*-interactions between oppositely imprinted chromosome regions are important for the maintenance of imprinting (44). Subtelomeric chromosome regions, which contain a high density of genes, also show an elevated incidence of somatic pairing in human nuclei (45).

3.3 Spatial distribution of chromosomes and chromosome domains relative to nuclear landmarks

In 1885, Rabl proposed his theory on the internal structure of the nucleus based on observations in *Salamandra maculata* (46). Today, the 'Rabl configuration' refers to the idea that chromosomes maintain their anaphase–telophase orientation, as well as their individuality, with centromeres localized at one side of the nucleus and telomeres at the other. This organization is apparent in *Drosophila* polytene chromosomes (47) and early embryonic nuclei (48). However, in larval imaginal discs, chromosomes maintain this Rabl orientation for less than 2 hours after mitosis and then reorient, while still maintaining heterochromatic areas at the nuclear periphery (49). A peripheral localization of telomeres has also been observed in *Trypanosoma* (50), fission yeast (51), and some, but not all, plant species (52, 53).

In *Saccharomyces cerevisiae*, telomeres are bound by at least seven proteins, including Rap1p, three Sir proteins, and the yKu complex (54, 55). These proteins, and the telomeres themselves, are concentrated at the nuclear periphery, and even interact with nuclear pore complexes (56). Complex disruption leads to loss of both telomeric gene silencing and peripheral localization (55, 57).

In mammalian somatic cells, centromere and telomere positioning is non-Rabl and is probably cell-type specific (41, 58, 59). Intriguingly, EM reveals human telomeric DNA around the surface of small electron-opaque structures (60). It is not clear whether these structures represent sites in the nucleus where telomeres from different chromosomes cluster together, or whether each structure is derived from the nucleoprotein complex, and the large duplex loop that is present at the end of each individual telomere (61).

Several lines of evidence suggest that the interphase organization of chromosomes with respect to nuclear landmarks is not random, and may be cell-type and cell-cycle specific (15, 19, 62). In addition to the rDNA-containing chromosomes, which are intrinsic to the nucleolus, other human autosomes can also adopt preferred positions in the nucleus: gene-rich human chromosome 19 localizes towards the interior of the nucleus, whereas gene-poor chromosome 18 is positioned towards the periphery (5), and the inactive X (Xi) is consistently positioned against the nuclear membrane in mammalian cells (41). A general trend for the more gene-rich fractions of the genome to localize to the interior volume of the nucleus is supported by the nuclear distribution of the most highly acetylated isoforms of histone H4 (63). In an artificial system, a heterochromatic chromosome arm, engineered by gene amplification, was peripherally located through most of the cell cycle, but targeting a transcription factor to this arm resulted in a shift to an interior location and dramatic uncoiling (4).

How chromosome territories are arranged in a highly defined nuclear architecture remains unclear (64). It is known that rDNA transcription directs the acrocentric chromosomes to the nucleolus but chromosome 19, which also localizes adjacent to the nucleolus (65), is devoid of ribosomal genes and must be positioned by an alternative mechanism. Components of the nuclear lamina have chromatin-binding properties, which could serve to position certain chromosome domains to the nuclear periphery. Specifically, the lamin B receptor (LBR) has been shown to interact with human HP1-type chromodomain proteins (66). It is interesting to note that 15–20 Mb sub-chromosomal regions are sufficient to confer the characteristic subnuclear localization of human chromosomes 18 and 19 (5).

4. Evidence for the role of large-scale chromatin structure in regulation

Evidence is gathering in support of the idea that mammalian chromosomes regularly adopt defined addresses within the nucleus and that chromosomes align to build up higher-order compartments within the interphase nucleus. However, the significance of such an organization for gene and genome function is, so far, only implied. Previous chapters have described a variety of mechanisms by which small-scale chromatin structure, from the level of the nucleosome to the 30 nm fibre, is re-modelled or modified to enable or repress transcription. But what is the evidence that higher-order levels of chromatin organization play a direct role in gene regulation?

4.1   Chromatin compaction

The most striking example of long-range transcriptional repression coincident with changes in large-scale chromatin structure is the formation of the Barr body during mammalian X-chromosome inactivation. Great advances have been made in deciphering the mechanisms that initiate the events of X inactivation and the resultant changes in histone acetylation and DNA methylation (Chapters 8 and 12). However, we still do not understand the role, if any, that large-scale chromatin structure plays in achieving repression of transcription from the inactive X (Xi). Early cytological studies indicated that facultative and constitutive heterochromatin is more compact than euchromatin in interphase nuclei. The inactive X chromosome (Xi) appears as a dense mass by DNA staining methods both in light microscopy and by examination in the TEM. Using FISH, Xa and Xi interphase chromosomes differ in their shape and surface structure but still have similar volumes (67), suggesting that silencing, in this case, is not simply a matter of total chromosome compaction. Whether this apparent discrepancy is related to resolution problems associated with FISH, or reflects differences in cell type and or cell-cycle stage, remains unclear.

Distinct dosage compensation mechanisms have evolved independently in other species but have a common mechanistic basis; specific molecules decorate the X chromosome and play a role in remodelling the chromatin structure of the dosage-compensated chromosome in one sex, but not in the other. In *Drosophila*, the male X chromosome is hyperactivated by an RNA–protein complex which effects site-specific acetylation of histone H4 and results in the up-regulation of gene transcription twofold relative to each X chromosome in female cells (68). A global change in the chromosome structure of the hyperactivated X is suggested by its more diffuse cytological appearance (69, 70).

Conversely, in nematodes, a distinct protein complex is recruited to both X chromosomes in the interphase nuclei of hermaphrodite worms, which reduces gene transcription levels to those of the single X chromosome in males. This protein complex is related to the 13S condensin complex; key components are DPY-27 (dumpy protein-27) and MIX-1 (mitosis- and X-associated protein-1). Both of these are SMC (structural maintenance of chromosomes) proteins (71, 72), putative ATPases that resemble molecular motors and are implicated in a wide variety of higher-order chromosome dynamics, including chromosome condensation (Fig. 2) (73). Although the role of DPY-27 is specific to dosage compensation, MIX-1 has an essential role in chromosome segregation and localizes to all chromosomes at mitosis. Other components of the worm dosage compensation complex (DPY-26 and -28) also have some homology to non-SMC subunits of the 13S condensin complex. Thus, dosage compensation in nematodes appears to provide a direct link between chromatin compaction and regulation of transcription (Fig. 2).

Normal mitotic chromosome condensation is accompanied by a cessation of transcription, although it is unclear what role compaction of chromatin *per se* has in this. A mitotic-like condensation of interphase X chromosomes may directly decrease the transcription of X-linked genes in hermaphrodite worms, or might act as a trigger

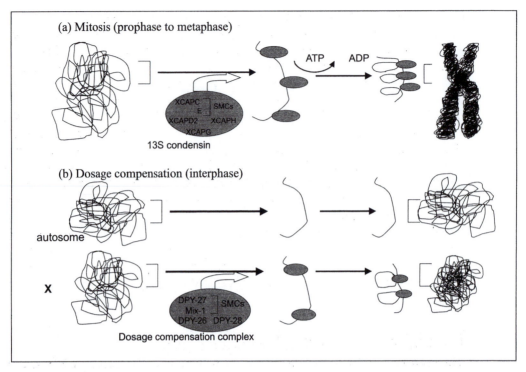

Fig. 2 Dosage compensation in *Caenorhabditis elegans*. (a) The 13S condensin complex is required for mitotic chromosome condensation. Key components of this complex are two SMC proteins (called XCAPC and E in *Xenopus laevis*) that may use the energy generated by ATP hydrolysis to drive coiling of the interphase chromatid fibre, and so condense it for mitosis. (b) In dosage compensation in the nematode worm *C. elegans* a dosage compensation complex is recruited to the X chromosomes, but not the autosomes, of hermaphrodite animals during interphase. Two components of this complex (DPY-27 and Mix-1) are SMC proteins; Mix-1 also plays a role in a condensin complex acting on all chromosomes at mitosis. In the model the SMC proteins in this complex bring about a partial condensation of the X chromosomes that either directly or indirectly dampens transcription of X-linked genes.

to modify the X-linked chromatin in other ways. It will be interesting to see whether gene repression through modulation of higher-order chromatin structure is found at other loci and in other species. For example, there is genetic evidence that SMC proteins play a role in maintaining boundary-insulator element function (Chapter 13) at the silent mating-type loci in yeast (74).

4.2 Does replication timing dictate the formation of distinct chromatin domains?

Mammalian chromosomal domains replicate at defined, developmentally regulated times during S phase. It has been suggested that mammalian genomes are compartmentalized according to the programme of DNA replication timing (75, 76), and that these compartments are established at late telophase/early G_1 (77). Evidence also indicates that the G- and R-bands of mitotic chromosomes persist in interphase as distinct domains, termed subchromosomal foci (27), suggesting that structural, transcriptional, and replicational domains share topographical boundaries as basic units of chromosome organization. Gene-rich sequences cluster within the nuclear interior, giving rise to a compartment containing early replicating DNA that is transcriptionally competent and active. G-band sequences organize a compartment with transcriptionally incompetent and inactive late-replicating DNA at the nuclear periphery (63).

Nuclear envelope assembly is completed prior to the global repositioning of chromosomal domains (77). This is consistent with the 2–3 hours required after mitosis for decondensation and dispersal throughout the nuclear interior of chromonema fibres (17), and for the localization of human chromosome 18 to the nuclear periphery (65). Dimitrova and Gilbert propose that the temporal order of replication is not established until sequences are repositioned within the nucleus (77). However, Bridger *et al.* have evidence that in quiescent human cells, where chromosome 18 has an uncharacteristic positioning in the nuclear interior, the replication timing of this chromosome is unchanged (65), challenging the idea that the correct sequence of replication timing requires a particular configuration of chromosomes within the nucleus.

4.3   Localization of sites of transcription

The observation, by light microscopy, that chromosomes occupy discrete domains in interphase has prompted speculation that transcription, RNA processing, and RNA transport occur in a space between chromosomes which links to the nuclear pores. This space has been called the inter-chromosomal domain (ICD) compartment (19). Accumulations of specific RNAs have been seen at the border of the chromosome territory from which they originated, and components of the splicing machinery have also been reported to be outside of chromosome territories (78, 79).

However, sites of transcription appear to be present throughout chromosome territories, with the obvious exception of the human Xi chromosome (80, 81). Furthermore, TEM localization of uridine or 5-bromouridine 5-triphosphate (BrUTP) incorporation has shown heavy labelling at the edge of condensed, large-scale chromatin domains rather than at the surface of chromosomes *per se* (82, 83). These studies suggest that locally compacted and unfolded regions within an interphase chromosome form distinct subdomains, and that chromatin folding is organized in such a way that transcriptionally active DNA is at the surface of large-scale chromatin fibres. The ICD compartment has thus been extended to include space between surfaces of subchromosomal compartments.

4.4   Are active genes displayed on the surface of chromosome territories?

The localization of a small number of individual genes from different chromosomes has been examined within their chromosome territories. In one such study, three coding regions of the human genome were all found to be located at the surface of their respective chromosome territories, independent of transcriptional status (84); one might infer from this that most genes lie on the surface of chromosome territories. However, three is a small number and these genes are located randomly on different chromosomes. On the X chromosome, an active gene is found in a more peripheral location within the Xa chromosome territory than in its counterpart Xi (85). With the exception of Xi, both early (gene-rich) and late-replicating (gene-poor) DNA are distributed throughout their chromosome territories, implying that gene-rich DNA is also located within the chromosome interior (26), and GC-rich (gene-rich) DNA sequences show no preferential localization within chromosome territories (86). As sites of transcription also localize throughout chromosome territories (80, 81), there seems to be little requirement for genes to locate to the surface of a chromosome to facilitate transcription.

A recent study has investigated the organization of the MHC II (major histocompatability complex II) locus on human chromosome 6 (HSA6p) (31). Several megabases of DNA were observed looping out from the surface of the chromosome. The frequency of a particular region localizing externally was cell-type dependent, and related to the number of genes in that region. There is evidence that other gene-rich regions of the genome exist in similar configurations (Mahy and Bickmore, unpublished). Thus, studying individual genes is insufficient to discern patterns of organization, and sequence context is very important.

4.5 Positioning of specific genes relative to nuclear bodies

Several genes appear to be specifically co-localized with different types of nuclear domains, including interchromatin granules, splicing speckles, coiled bodies, cleavage bodies, and PML (promyelocytic leukaemia) nuclear bodies. Active genes may be associated preferentially with clusters of interchromatin granules (87), which are enriched in factors required for the formation and maturation of mRNA (88). Many coiled bodies are also associated with specific gene loci, including the histone gene clusters, U1 and U2 small nuclear RNA genes, and the U3 small nucleolar RNA genes (89, 90), but the functional significance of this association is unclear.

Cleavage bodies containing the RNA 3'-processing factors CstF and CPSF can be associated with a coiled body. This association changes during the cell cycle and can be correlated with both the spatial juxtaposition and transcription activity of the cell-cycle-regulated histone gene cluster (91). The PML nuclear body is also frequently closely associated with the coiled body/cleavage body doublet (92) and with DNA that is replicated in a particular stage late in S phase (93).

4.6 Sites of transcriptional repression within the nucleus

A number of studies suggest that some transcriptionally repressed domains occupy distinct positions with respect to the nuclear envelope or to other transcriptionally silent sequences within the nucleus (48, 57, 94–96). The LBR and HP1-type chromodomain proteins provide a molecular link between the nuclear periphery and hetero-chromatin (66). In *Saccharomyces cerevisiae*, heritable inactivation of genes occurs at the silent mating-type loci and at telomeres, which are found clustered at the nuclear periphery (as discussed above). Experimental sequestration of gene loci to the nuclear periphery can facilitate establishment of transcriptional silencing in the absence of an effective silencing protein complex (97). The functional significance of positioning telomeres at the nuclear periphery is not known, but this compartmentalization may provide an effective local concentration of silencing factors at the nuclear periphery, also serving to protect the rest of the genome from transcriptional silencing.

In mouse lymphoid cells, the protein Ikaros may mediate specific gene silencing. Ikaros localizes to centromeric heterochromatin foci in the nucleus and is dynamic-ally redistributed during the cell cycle. Genes differentially expressed during differ-ent stages of B-cell maturation show a correlation of expression status with Ikaros association: Ikaros association is evident only in cells in which those genes are not transcribed (98). Recent evidence indicates that transcriptional enhancers may main-tain gene expression by preventing (*trans*)genes from localizing close to centromeric heterochromatin in the nucleus, and/or by recruiting (*trans*)genes to places in the nucleus that are permissive for stable transcription (99). Therefore, it is possible that genetic domains may traffic between repressive (heterochromatic) and transcrip-tionally liberal (euchromatic) compartments. These observations raise the question of which genes are sequestered in these transcriptionally repressed domains, as topo-graphical constraints on chromatin structure would prevent all silenced genes from localizing to pericentric heterochromatin.

However, the precise spatial positioning of genes may merely be the consequence of regulatory events that lead to the repressed state, and it remains to be proven that the subnuclear localization of a gene contributes actively to the establishment and or inheritance of a particular chromatin state (Fig. 3). It is also not clear what proportion or type of genes are amenable to regulation at this level.

5. Dynamics of large-scale organization

5.1 Cell-cycle changes

Large-scale chromatin structure is subject to dynamic change during the cell cycle, with compaction ratios falling from $20\,000:1$ in the metaphase chromosome to as low as $40:1$ within individual 30 nm chromatin fibres. Changes of large-scale chromatin structure during the cell cycle have been observed by light microscopy for more than a century, where a progressive decondensation, accompanied by increases in nuclear volume, through G_1 into early S phase is followed by a progressive condensation during late S and G_2 (100). A similar trend is observed by TEM in mouse cells (101) and a more detailed TEM study of chromosome decondensation reveals the uncoil-ing and straightening of a 100–130 nm chromonema fibre during early G_1, followed by a transition to a 60–80 nm fibre in late G_1 and early S phase (17).

The structure at a specific locus may follow a precise choreography of cell-cycle-

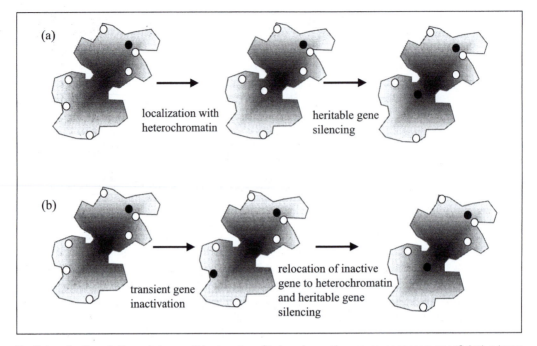

Fig. 3 Localization of silenced genes within domains of heterochromatin—cause or consequence? Active (open circles) and many inactive (filled circles) genes are distributed within the euchromatic portion of chromosome territories. However, some genes change their position within the nucleus/chromosome, dependent on their activity, and become co-localized with heterochromatin when inactive. (a) Recruitment of an inactive gene to a site of heterochromatin brings about its silencing, perhaps by exposure to a concentrated pool of silencing factors or to a particular replication timing domain, or by removal from a pool of available transcription factors. (b) A transient gene inactivation allows a gene to be drawn into heterochromatin, where it is then subject to heritable gene silencing.

associated changes, suggesting an ordered and reproducible organization of DNA within large-scale chromatin domains. The use of a lac repressor–lac operator recognition system has identified cell-cycle-specific conformations of a large heterochromatic HSR (homogeneously staining region) (102); an abrupt decondensation in mid to late S phase appears to slightly precede the onset of DNA replication of this amplified chromosome arm. Interestingly, this decondensation was accompanied by a repositioning from the nuclear periphery to the nuclear interior.

Changes in intranuclear localization of native chromosome regions also occur at distinct stages of the cell cycle. Early replicating, gene-rich R-bands localize preferentially in the nuclear interior, while late replicating, gene-poor G/Q-bands are found preferentially at the nuclear periphery (5, 63). This pattern is established during the first few hours of G_1, which seems to be a particularly dynamic time (65, 77) (see above). The intriguing observation that targeting a transcriptional activator to a specific chromosome site changes its normal cell-cycle programme of intranuclear positioning suggests a possible link among nuclear positioning of large-scale chromatin domains, transcriptional activation, and the cell cycle (Tumbar and Belmont, unpublished).

The mechanism underlying these chromosomal movements remains unclear. In general, studies on chromosome dynamics have shown that DNA is relatively immobile in interphase yeast and mammalian cells over a period of several hours (28, 103, Belmont and Li, unpublished). *In vivo* observation of two adjacent chromosome sites in yeast suggested a 'constrained diffusional motion' of the chromatin fibre on the order of $0.5\,\mu m/min$ (32); in *Drosophila* embryonic nuclei, the pairing of the histone loci between homologous chromosomes has been modelled by a random-walk, diffusion process (104). However, infrequent motion of centromeres during G_1 has been observed (1). Determining whether these intranuclear chromosome movements occur through a passive or active mechanism is of keen interest.

5.2 Quiescence and senescence

While most studies of large-scale chromatin and nuclear architecture have focused on proliferating, transformed cells in culture, it is likely that the nuclear organization in these cells differs significantly from terminally differentiated, quiescent cells, which comprise the majority of tissue mass in multicellular organisms. Quiescence, also known as G_0, is a non-dividing growth state that occurs when cells cease DNA synthesis and withdraw from the cell cycle, either temporarily or permanently in terminal differentiation. Major changes in chromatin organization occur on transition to G_0. In quiescent lymphocytes, a sheet of compact chromatin covers the nuclear envelope, and large masses of chromatin associate with the nucleolus (105).

Re-entry of cells into the cell cycle is also associated with changes in large-scale chromatin structure and changes in specific chromosome positioning. Human chromosome 18, relatively gene-poor and located at the nuclear periphery in proliferating diploid fibroblasts, is positioned in the nuclear interior during G_0. Interestingly, a return to a peripheral localization does not occur during the G_0–G_1 transition, but instead occurs in early G_1 of the next cell division (65). A change in location relative to the nuclear periphery has also been seen for the centromere of human chromosome 11 during the G_0–G_1 transition in peripheral lymphocytes (106). Large-scale changes in positioning of individual genes have also been reported. Several inactive genes, which are found away from centromeres in nuclei of quiescent lymphocytes, are repositioned to heterochromatic centromeres on transition from G_0 to G_1 (107). The mechanism of this relocation is not yet known.

Changes in nuclear organization have also been reported in senescent cells. An age-related reactivation of genes from the inactive X chromosome is observed in mice (108), suggesting a general decondensation of chromatin in senescence. A connection between metabolism, loss of gene silencing, and increased rates of ageing has also been implied for yeast and perhaps higher eukaryotes (109). Other studies, conversely, suggest an increase in DNA condensation with age (110).

Gene-poor human chromosome 18 is preferentially localized towards the nuclear periphery in young, proliferating fibroblasts but is found to relocate to the nuclear interior in senescence (65). This change in position, however, is not accompanied by any detectable change in chromosome volume, so chromatin compaction status in senescence remains open to debate. Using genetic tools, it should be possible to investigate the direct impact of heterochromatin maintenance on life span, providing further insight into this exciting research area.

5.3 Differentiation

Global changes in chromatin compaction and distribution within the nucleus are hallmarks of differentiation. In some cell types, these features are so reproducible that identification is possible from nuclear morphology alone: during mouse oogenesis, four stages of chromatin organization are distinguishable, which correlate with specific stages of the ovarian follicle development (111). A link between increased chromatin compaction and terminal stages of differentiation in vertebrate haematopoietic cells has been identified in the form of MENT (myeloid and erythroid nuclear termination stage-specific protein). MENT is a heterochromatin-associated, serpin-family protein, which associates with, and perhaps facilitates compaction of, peripheral heterochromatin during haematopoiesis (112). In *Drosophila*, instability of heterochromatin-mediated position effect variegation (PEV) has also been observed in post-mitotic cells (113), further evidence that changes in chromatin compaction occur during differentiation.

In vitro differentiation of myoblasts into myotubes results in the redistribution of centromeres from a random location within the nucleus to the nuclear periphery. This correlates with the increase in heterochromatin at the nuclear periphery observed by TEM (114). Chromocentres, formed by centromeres from different combinations of human chromosomes, are also characteristic of terminally differentiated myeloid or lymphoid cells (115).

To date, very little work has been carried out at a molecular level to examine how individual genes or chromosome loci are packaged within the larger chromatin masses of differentiated cells, and whether changes in chromatin condensation, or intranuclear chromosome position, directly influence gene regulation during development.

6. Final comments

Chromatin organization up to the level of the 30 nm fibre has become a central player in the regulation of gene expression over recent years. There seems no reason to believe that the impact of chromatin structure on transcriptional regulation will stop there. The influences of large-scale chromatin organization and spatial nuclear organization, on genome functions such as gene regulation, imprinting, LCR function, and boundary elements, is an area of intense research. In discussion with authors of other chapters it was pointed out that the genome sequencing projects would have a potential impact on this area of research. The availability of finely mapped probes to all regions of genomes and access to large, contiguous stretches of sequence will facilitate the correlation of linear sequence to higher-order chromatin structure and three-dimensional spatial distribution within the nucleus. The regulation of large-scale chromatin structure is believed to be mediated by *cis*-acting sequences that have been defined largely through biochemical and functional assays. In the future it may be possible to identify regulatory sequences that mediate their effect through higher-order chromatin structure by examination of sequence databases.

So far, work relating chromatin compaction and genome function has been largely correlative. Experiments that directly assay and modulate large-scale chromatin structure and gene expression at specific loci are needed. There is also likely to be an increasing impact from mutations in complex eukaryotes, including humans, which disrupt normal chromosome or nuclear organization.

Most of the studies cited in this chapter were performed on fixed material and provide only snapshots of the genome during the cell cycle and during development. Recent advances in live cell analysis will enable questions of chromatin dynamics in real time to be investigated. There will probably be active movement of domains against a background of passive diffusion; these processes, and the mechanisms that govern them, must be distinguished.

Acknowledgements

NLM is supported by a Medical Research Council (UK) PhD studentship and WAB is a James S. McDonnell Foundation Centennial Fellow. TT and ASB are supported by grants to ASB from the National Institutes of Health (NIH).

References

1. Shelby, R. D., Hahn, K. M., and Sullivan, K. F. (1996) Dynamic elastic behaviour of alpha-satellite DNA domains visualised *in situ* in living human cells. *J. Cell Biol.*, **135**, 545.
2. Robinett, C. C., Straight, A., Li, G., Willhelm, C., Sudlow, G., Murray, A., and Belmont, A. S. (1996) *In vivo* localisation of DNA sequences and visualisation of large-scale chromatin organisation using lac operator/repressor recognition. *J. Cell Biol.*, **135**, 1685.
3. Li, G., Sudlow, G., and Belmont, A. S. (1998) Interphase cell cycle dynamics of a late-replicating, heterochromatic homogeneously staining region: precise choreography of condensation/decondensation and nuclear positioning. *J. Cell Biol.*, **140**, 975.
4. Tumbar, T., Sudlow, G., and Belmont, A. S. (1999) Large-scale chromatin unfolding and remodelling induced by VP16 acidic activation domain. *J. Cell Biol.*, **145**, 1341.
5. Croft, J. A., Bridger, J. M., Boyle, S., Perry, P., Teague, P., and Bickmore, W. A. (1999) Differences in the localisation and morphology of chromosomes in the human nucleus. *J. Cell Biol.*, **145**, 1119.
6. Woodcock, C. L. and Horowitz, R. A. (1995) Chromatin organisation re-viewed. *Trends Cell Biol.*, **5**, 272.
7. Ericsson, C., Mehlin, H., Bjorkroth, B., Lamb, M. M., and Daneholt, B. (1989) The ultrastructure of upstream and downstream regions of an active Balbiani ring gene. *Cell*, **56**, 631.
8. Callan, H. G. (1982) Lampbrush chromosomes. *Proc. R. Soc. Lond B Biol. Sci.*, **214**, 417.
9. Jackson, D. A., Iborra, F. J., Manders, E. M., and Cook, P. R. (1998) Numbers and organisation of RNA polymerases, nascent transcripts, and transcription units in HeLa nuclei. *Mol. Biol. Cell*, **9**, 1523.
10. Rattner, J. B. and Lin, C. C. (1985) Radial loops and helical coils coexist in metaphase chromosomes. *Cell*, **42**, 291.

11. Razin, S. V. (1996) Functional architecture of chromosomal DNA domains. *Crit. Rev. Eukaryot. Gene Expr.*, **6**, 247.

12. Bickmore, W. A. and Oghene, K. (1996) Visualising the spatial relationships between defined DNA sequences and the axial region of extracted metaphase chromosomes. *Cell*, **84**, 95.

13. Yokota, H., Singer, M. J., van den Engh, G. J., and Trask, B. J. (1997) Regional differences in the compaction of chromatin in human G0/G1 interphase nuclei. *Chromosome Res.*, **5**, 157.

14. Manuelidis, L. and Chen, T. L. (1990) A unified model of eukaryotic chromosomes. *Cytometry*, **11**, 8.

15. Manuelidis, L. (1990) A view of interphase chromosomes. *Science*, **250**, 1533.

16. Sedat, J. and Manuelidis, L. (1978) A direct approach to the structure of eukaryotic chromosomes. *Cold Spring Harbor Symp. Quant. Biol.*, **42**, 331.

17. Belmont, A. S. and Bruce, K. (1994) Visualisation of G1 chromosomes: a folded, twisted, supercoiled chromonema model of interphase chromatid structure. *J. Cell Biol.*, **127**, 287.

18. Hirano, T. (2000) Chromosome cohesion, condensation and separation. *Annu. Rev. Biochem.*, **69**, 115.

19. Cremer, T., Kurz, A., Zirbel, R., Dietzel, S., Rinke, B., Schrock, E., Speicher, M. R., Mathieu, U., Jauch, A., and Emmerich, P. (1993) Role of chromosome territories in the functional compartmentalisation of the cell nucleus. *Cold Spring Harbor Symp. Quant. Biol.*, **58**, 777.

20. Manuelidis, L. (1985) Individual interphase chromosome domains revealed by *in situ* hybridisation. *Hum. Genet.*, **71**, 288.

21. Schardin, M., Cremer, T., Hager, H. D., and Lang, M. (1985) Specific staining of human chromosomes in Chinese hamster _ man hybrid cell lines demonstrates interphase chromosome territories. *Hum. Genet.*, **71**, 281.

22. Heslop-Harrison, J. S. and Bennett, M. D. (1990) Nuclear architecture in plants. *Trends Genet.*, **6**, 401.

23. Bridger, J. M. and Bickmore, W. A. (1998) Putting the genome on the map. *Trends Genet.*, **14**, 403.

24. Dietzel, S., Jauch, A., Kienle, D., Qu, G., Holtgreve-Grez, H., Eils, R., Munkel, C., Bittner, M., Meltzer, P. S., Trent, J. M., *et al.* (1998) Separate and variably shaped chromosome arm domains are disclosed by chromosome arm painting in human cell nuclei. *Chromosome Res.*, **6**, 25.

25. Scherthan, H., Eils, R., Trelles-Sticken, E., Dietzel, S., Cremer, T., Walt, H., and Jauch, A. (1998) Aspects of three-dimensional chromosome reorganisation during the onset of human male meiotic prophase. *J. Cell Sci.*, **111**, 2337.

26. Visser, A. E., Eils, R., Jauch, A., Little, G., Bakker, P. J., Cremer, T., and Aten, J. A. (1998) Spatial distributions of early and late replicating chromatin in interphase chromosome territories. *Exp. Cell Res.*, **243**, 398.

27. Zink, D., Bornfleth, H., Visser, A., Cremer, C., and Cremer, T. (1999) Organisation of early and late replicating DNA in human chromosome territories. *Exp. Cell Res.*, **247**, 176.

28. Zink, D., Cremer, T., Saffrich, R., Fischer, R., Trendelenburg, M. F., Ansorge, W., and Stelzer, E. H. (1998) Structure and dynamics of human interphase chromosome territories *in vivo*. *Hum. Genet.*, **102**, 241.

29. Visser, A. E. and Aten, J. A. (1999) Chromosomes as well as chromosomal subdomains constitute distinct units in interphase nuclei. *J. Cell Sci.*, **112**, 3353.

30. Belmont, A. S., Dietzel, S., Nye, A. C., Strukov, Y. G., and Tumbar, T. (1999) Large-scale chromatin structure and function. *Curr. Opin. Cell Biol.*, **11**, 307.

31. Volpi, E. V., Chevret, E., Jones, T., Vatcheva, R., Williamson, J., Beck, S., Campbell, R. D.,

Goldsworthy, M., Powis, S. H., Ragoussis, J., *et al.* (2000) Large-scale chromatin organisation of the major histocompatibility complex and other regions of human chromosome 6 and its response to interferon in interphase nuclei. *J. Cell Sci.*, **113**, 1565.

32. Marshall, W. F., Straight, A., Marko, J. F., Swedlow, J., Dernburg, A., Belmont, A., Murray, A. W., Agard, D. A., and Sedat, J. W. (1997) Interphase chromosomes undergo constrained diffusional motion in living cells. *Curr. Biol.*, **7**, 930.

33. Henderson, A. S., Warburton, D., and Atwood, K. C. (1972) Location of ribosomal DNA in the human chromosome complement. *Proc. Natl Acad. Sci., USA*, **69**, 3394.

34. Bobrow, M. and Heritage, J. (1980) Non-random segregation of nucleolar organising chromosomes at mitosis? *Nature*, **288**, 79.

35. Roeder, G. S. (1995) Sex and the single cell: meiosis in yeast. *Proc. Natl Acad. Sci., USA*, **92**, 10450.

36. Metz, C. W. (1916) Chromosome studies on the Diptera II. The paired association of chromosomes in the Diptera and its significance. *J. Exp. Zool.*, **21**, 213.

37. Tartof, K. D. and Henikoff, S. (1991) Trans-sensing effects from *Drosophila* to humans. *Cell*, **65**, 201.

38. Gemkow, M. J., Verveer, P. J., and Arndt-Jovin, D. J. (1998) Homologous association of the Bithorax-Complex during embryogenesis: consequences for transvection in *Drosophila melanogaster*. *Development*, **125**, 4541.

39. Henikoff, S. (1997) Nuclear organisation and gene expression: homologous pairing and long-range interactions. *Curr. Opin. Cell Biol.*, **9**, 388.

40. Wu, C. T. and Morris, J. R. (1999) Transvection and other homology effects. *Curr. Opin. Genet. Dev.*, **9**, 237.

41. Manuelidis, L. and Borden, J. (1988) Reproducible compartmentalisation of individual chromosome domains in human CNS cells revealed by *in situ* hybridisation and three-dimensional reconstruction. *Chromosoma*, **96**, 397.

42. Leitch, A. R., Brown, J. K., Mosgoller, W., Schwarzacher, T., and Heslop-Harrison, J. S. (1994) The spatial localisation of homologous chromosomes in human fibroblasts at mitosis. *Hum. Genet.*, **93**, 275.

43. LaSalle, J. M. and Lalande, M. (1996) Homologous association of oppositely imprinted chromosomal domains. *Science*, **272**, 725.

44. Riesselmann, L. and Haaf, T. (1999) Preferential S-phase pairing of the imprinted region on distal mouse chromosome 7. *Cytogenet. Cell Genet.*, **86**, 39.

45. Stout, K. van der, Maarel S., Frants, R. R., Padberg, G. W., Ropers, H. H., and Haaf, T. (1999) Somatic pairing between subtelomeric chromosome regions: implications for human genetic disease? *Chromosome. Res.*, **7**, 323.

46. Rabl, C. (1885) Uber Zelltheilung. *Morphologisches Jahrbuch*, **10**, 214.

47. Hochstrasser, M., Mathog, D., Gruenbaum, Y., Saumweber, H., and Sedat, J. W. (1986) Spatial organisation of chromosomes in the salivary gland nuclei of *Drosophila melanogaster*. *J. Cell Biol.*, **102**, 112.

48. Marshall, W. F., Dernburg, A. F., Harmon, B., Agard, D. A., and Sedat, J. W. (1996) Specific interactions of chromatin with the nuclear envelope: positional determination within the nucleus in *Drosophila melanogaster*. *Mol. Biol. Cell*, **7**, 825.

49. Csink, A. K. and Henikoff, S. (1998) Large-scale chromosomal movements during interphase progression in *Drosophila*. *J. Cell Biol.*, **143**, 13.

50. Chung, H. M., Shea, C., Fields, S., Taub, R. N., Van der Ploeg, L. H., and Tse, D. B. (1990) Architectural organisation in the interphase nucleus of the protozoan *Trypanosoma brucei*: location of telomeres and mini-chromosomes. *EMBO J.*, **9**, 2611.

51. Funabiki, H., Hagan, I., Uzawa, S., and Yanagida, M. (1993) Cell cycle-dependent specific positioning and clustering of centromeres and telomeres in fission yeast. *J. Cell Biol.*, **121**, 961.

52. Heslop-Harrison, J. S., Leitch, A. R., and Schwarzacher, T. (1993) The physical organisation of interphase nuclei. In *The chromosomes*, (ed. J. S. Heslop Harrison and R. B. Flavell), p. 221. Bios Scientific Publishers, Oxford.

53. Dong, F. and Jiang, J. (1998) Non-Rabl patterns of centromere and telomere distribution in the interphase nuclei of plant cells. *Chromosome Res.*, **6**, 551.

54. Bourns, B. D., Alexander, M. K., Smith, A. M., and Zakian, V. A. (1998) Sir proteins, Rif proteins, and Cdc13p bind *Saccharomyces* telomeres *in vivo*. *Mol. Cell Biol.*, **18**, 5600.

55. Gravel, S., Larrivee, M., Labrecque, P., and Wellinger, R. J. (1998) Yeast Ku as a regulator of chromosomal DNA end structure. *Science*, **280**, 741.

56. Galy, V., Olivo-Marin, J. C., Scherthan, H., Doye, V., Rascalou, N., and Nehrbass, U. (2000) Nuclear pore complexes in the organisation of silent telomeric chromatin. *Nature*, **403**, 108.

57. Maillet, L., Boscheron, C., Gotta, M., Marcand, S., Gilson, E., and Gasser, S. M. (1996) Evidence for silencing compartments within the yeast nucleus: a role for telomere proximity and Sir protein concentration in silencer-mediated repression. *Genes Develop.*, **10**, 1796.

58. Billia, F. and De Boni, U. (1991) Localisation of centromeric satellite and telomeric DNA sequences in dorsal root ganglion neurones, *in vitro*. *J. Cell Sci.*, **100**, 219.

59. He, D. and Brinkley, B. R. (1996) Structure and dynamic organisation of centromeres/prekinetochores in the nucleus of mammalian cells. *J. Cell Sci.*, **109**, 2693.

60. Pierron, G. and Puvion-Dutilleul, F. (1999) An anchorage nuclear structure for telomeric DNA repeats in HeLa cells. *Chromosome Res.*, **7**, 581.

61. Griffith, J. D., Comeau, L., Rosenfield, S., Stansel, R. M., Bianchi, A., Moss, H., and de Lange, T. (1999) Mammalian telomeres end in a large duplex loop. *Cell*, **97**, 503.

62. Haaf, T. and Schmid, M. (1991) Chromosome topology in mammalian interphase nuclei. *Exp. Cell Res.*, **192**, 325.

63. Sadoni, N., Langer, S., Fauth, C., Bernardi, G., Cremer, T., Turner, B. M., and Zink, D. (1999) Nuclear organisation of mammalian genomes. Polar chromosome territories build up functionally distinct higher order compartments. *J. Cell Biol.*, **146**, 1211.

64. Spector, D. L. (1996) Nuclear organisation and gene expression. *Exp. Cell Res.*, **229**, 189.

65. Bridger, J. M., Boyle, S., Kill, I. R., and Bickmore, W. A. (2000) Re-modelling of nuclear architecture in quiescent and senescent human fibroblasts. *Curr. Biol.*, **10**, 149.

66. Ye, Q., Callebaut, I., Pezhman, A., Courvalin, J. C., and Worman, H. J. (1997) Domain-specific interactions of human HP1-type chromodomain proteins and inner nuclear membrane protein LBR. *J. Biol. Chem.*, **272**, 14983.

67. Eils, R., Dietzel, S., Bertin, E., Schrock, E., Speicher, M. R., Ried, T., Robert-Nicoud, M., Cremer, C., and Cremer, T. (1996) Three-dimensional reconstruction of painted human interphase chromosomes: active and inactive X chromosome territories have similar volumes but differ in shape and surface structure. *J. Cell Biol.*, **135**, 1427.

68. Willard, H. F. and Salz, H. K. (1997) Remodelling chromatin with RNA. *Nature*, **386**, 228.

69. Gorman, M., Kuroda, M. I., and Baker, B. S. (1993) Regulation of the sex-specific binding of the maleless dosage compensation protein to the male X chromosome in *Drosophila*. *Cell*, **72**, 39.

70. Kelley, R. L., Solovyeva, I., Lyman, L. M., Richman, R., Solovyev, V., and Kuroda, M. I. (1995) Expression of msl-2 causes assembly of dosage compensation regulators on the X chromosomes and female lethality in *Drosophila*. *Cell*, **81**, 867.

71. Lieb, J. D., Capowski, E. E., Meneely, P., and Meyer, B. J. (1996) DPY-26, a link between dosage compensation and meiotic chromosome segregation in the nematode. *Science*, **274**, 1732.

72. Chuang, P. T., Lieb, J. D., and Meyer, B. J. (1996) Sex-specific assembly of a dosage compensation complex on the nematode X chromosome. *Science*, **274**, 1736.

73. Hirano, T. (1998) SMC protein complexes and higher-order chromosome dynamics. *Curr. Opin. Cell Biol.*, **10**, 317.

74. Donze, D., Adams, C. R., Rine, J., and Kamakaka, R. T. (1999) The boundaries of the silenced HMR domain in *Saccharomyces cerevisiae*. *Genes Develop.*, **13**, 698.

75. Ferreira, J., Paolella, G., Ramos, C., and Lamond, A. I. (1997) Spatial organisation of large-scale chromatin domains in the nucleus: a magnified view of single chromosome territories. *J. Cell Biol.*, **139**, 1597.

76. Berezney, R. and Wei, X. (1998) The new paradigm: integrating genomic function and nuclear architecture. *J. Cell Biochem. Suppl*, **30–31**, 238.

77. Dimitrova, D. S. and Gilbert, D. M. (1999) The spatial position and replication timing of chromosomal domains are both established in early G1 phase. *Mol. Cell*, **4**, 983.

78. Zirbel, R. M., Mathieu, U. R., Kurz, A., Cremer, T., and Lichter, P. (1993) Evidence for a nuclear compartment of transcription and splicing located at chromosome domain boundaries. *Chromosome Res.*, **1**, 93.

79. Clemson, C. M., McNeil, J. A., Willard, H. F., and Lawrence, J. B. (1996) XIST RNA paints the inactive X chromosome at interphase: evidence for a novel RNA involved in nuclear/chromosome structure. *J. Cell Biol.*, **132**, 259.

80. Abranches, R., Beven, A. F., Aragon-Alcaide, L., and Shaw, P. J. (1998) Transcription sites are not correlated with chromosome territories in wheat nuclei. *J. Cell Biol.*, **143**, 5.

81. Verschure, P. J., van, Der Kraan, I., Manders, E. M., and van Driel, R. (1999) Spatial relationship between transcription sites and chromosome territories. *J. Cell Biol.*, **147**, 13.

82. Fakan, S. and Nobis, P. (1978) Ultrastructural localisation of transcription sites and of RNA distribution during the cell cycle of synchronised CHO cells. *Exp. Cell Res.*, **113**, 327.

83. Wansink, D. G., Sibon, O. C., Cremers, F. F., van Driel, R., and de Jong, L. (1996) Ultra-structural localisation of active genes in nuclei of A431 cells. *J. Cell Biochem.*, **62**, 10.

84. Kurz, A., Lampel, S., Nickolenko, J. E., Bradl, J., Benner, A., Zirbel, R. M., Cremer, T., and Lichter, P. (1996) Active and inactive genes localise preferentially in the periphery of chromosome territories. *J. Cell Biol.*, **135**, 1195.

85. Dietzel, S., Schiebel, K., Little, G., Edelmann, P., Rappold, G. A., Eils, R., Cremer, C., and Cremer, T. (1999) The 3D positioning of ANT2 and ANT3 genes within female X chromosome territories correlates with gene activity. *Exp. Cell Res.*, **252**, 363.

86. Tajbakhsh, J., Luz, H., Bornfleth, H., Lampel, S., Cremer, C., and Lichter, P. (2000) Spatial distribution of GC- and AT-rich DNA sequences within human chromosome territories. *Exp. Cell Res.*, **255**, 229.

87. Xing, Y., Johnson, C. V., Moen, P. T. Jr, McNeil, J. A., and Lawrence, J. (1995) Non-random gene organisation: structural arrangements of specific pre-mRNA transcription and splicing with SC-35 domains. *J. Cell Biol.*, **131**, 1635.

88. Huang, S. and Spector, D. L. (1991) Nascent pre-mRNA transcripts are associated with nuclear regions enriched in splicing factors. *Genes Develop.*, **5**, 2288.

89. Smith, K. P., Carter, K. C., Johnson, C. V., and Lawrence, J. B. (1995) U2 and U1 snRNA gene loci associate with coiled bodies. *J. Cell Biochem.*, **59**, 473.

90. Frey, M. R., Bailey, A. D., Weiner, A. M., and Matera, A. G. (1999) Association of snRNA genes with coiled bodies is mediated by nascent snRNA transcripts. *Curr. Biol.*, **9**, 126.

91. Schul, W., van, Der Kraan, I., Matera, A. G., van Driel, R., and de Jong, L. (1999) Nuclear domains enriched in RNA 3'-processing factors associate with coiled bodies and histone genes in a cell cycle-dependent manner. *Mol. Biol. Cell*, **10**, 3815.

92. Schul, W., Groenhout, B., Koberna, K., Takagaki, Y., Jenny, A., Manders, E. M., Raska, I., van Driel, R., and de Jong, L. (1996) The RNA 3' cleavage factors CstF 64 kDa and CPSF 100 kDa are concentrated in nuclear domains closely associated with coiled bodies and newly synthesised RNA. *EMBO J.*, **15**, 2883.

93. Grande, M. A., van der Kraan, I., van Steensel, B., Schul, W., de The, H., van der Voort, H. T., de Jong, L., and van Driel, R. (1996) PML-containing nuclear bodies: their spatial distribution in relation to other nuclear components. *J. Cell Biochem.*, **63**, 280.

94. Csink, A. K. and Henikoff, S. (1996) Genetic modification of heterochromatic association and nuclear organisation in *Drosophila*. *Nature*, **381**, 529.

95. Dernburg, A. F., Broman, K. W., Fung, J. C., Marshall, W. F., Philips, J., Agard, D. A., and Sedat, J. W. (1996) Perturbation of nuclear architecture by long-distance chromosome interactions. *Cell*, **85**, 745.

96. Cockell, M. and Gasser, S. M. (1999) Nuclear compartments and gene regulation. *Curr. Opin. Genet. Dev.*, **9**, 199.

97. Andrulis, E. D., Neiman, A. M., Zappulla, D. C., and Sternglanz, R. (1998) Perinuclear localisation of chromatin facilitates transcriptional silencing. *Nature*, **394**, 592.

98. Brown, K. E., Guest, S. S., Smale, S. T., Hahm, K., Merkenschlager, M., and Fisher, A. G. (1997) Association of transcriptionally silent genes with Ikaros complexes at centromeric heterochromatin. *Cell*, **91**, 845.

99. Francastel, C., Walters, M. C., Groudine, M., and Martin, D. I. (1999) A functional enhancer suppresses silencing of a transgene and prevents its localisation close to centrometric heterochromatin. *Cell*, **99**, 259.

100. Kendall, F., Swenson, R., Borun, T., Rowinski, J., and Nicolini, C. (1977) Nuclear morphometry during the cell cycle. *Science*, **196**, 1106.

101. Leblond, C. P. and El Alfy, M. (1998) The eleven stages of the cell cycle, with emphasis on the changes in chromosomes and nucleoli during interphase and mitosis. *Anat. Rec.*, **252**, 426.

102. Li, G., Sudlow, G., and Belmont, A. S. (1998) Interphase cell cycle dynamics of a late-replicating, heterochromatic homogeneously staining region: precise choreography of condensation/decondensation and nuclear positioning. *J. Cell Biol.*, **140**, 975.

103. Abney, J. R., Cutler, B., Fillbach, M. L., Axelrod, D., and Scalettar, B. A. (1997) Chromatin dynamics in interphase nuclei and its implications for nuclear structure. *J. Cell Biol.*, **137**, 1459.

104. Fung, J. C., Marshall, W. F., Dernburg, A., Agard, D. A., and Sedat, J. W. (1998) Homologous chromosome pairing in *Drosophila melanogaster* proceeds through multiple independent initiations. *J. Cell Biol.*, **141**, 5.

105. Lopez-Velazquez, G., Marquez, J., Ubaldo, E., Corkidi, G., Echeverria, O., and Vazquez Nin, G. H. (1996) Three-dimensional analysis of the arrangement of compact chromatin in the nucleus of G0 rat lymphocytes. *Histochem. Cell Biol.*, **105**, 153.

106. Tagawa, Y., Nanashima, A., Yasutake, T., Hatano, K., Nishizawa-Takano, J. E., and Ayabe, H. (1997) Differences in spatial localisation and chromatin pattern during different phases of cell cycle between normal and cancer cells. *Cytometry*, **27**, 327.

107. Brown, K. E., Baxter, J., Graf, D., Merkenschlager, M., and Fisher, A. G. (1999) Dynamic repositioning of genes in the nucleus of lymphocytes preparing for cell division. *Mol. Cell*, **3**, 207.

108. Wareham, K. A., Lyon, M. F., Glenister, P. H., and Williams, E. D. (1987) Age related reactivation of an X-linked gene. *Nature*, **327**, 725.
109. Imai, S., Armstrong, C. M., Kaeberlein, M., and Guarente, L. (2000) Transcriptional silencing and longevity protein Sir2 is an NAD-dependent histone deacetylase. *Nature*, **403**, 795.
110. Preumont, A. M., Capone, B., and van Gansen, P. (1983) Replicative activity and actinomycin binding in mouse diploid fibroblasts (*in vitro* ageing). *Mech. Ageing Dev.*, **22**, 167.
111. Mattson, B. A. and Albertini, D. F. (1990) Oogenesis: chromatin and microtubule dynamics during meiotic prophase. *Mol. Reprod. Dev.*, **25**, 374.
112. Grigoryev, S. A., Bednar, J., and Woodcock, C. L. (1999) MENT, a heterochromatin protein that mediates higher order chromatin folding, is a new serpin family member. *J. Biol. Chem.*, **274**, 5626.
113. Lu, B. Y., Bishop, C. P., and Eissenberg, J. C. (1996) Developmental timing and tissue specificity of heterochromatin-mediated silencing. *EMBO J.*, **15**, 1323.
114. Chaly, N. and Munro, S. B. (1996) Centromeres reposition to the nuclear periphery during L6E9 myogenesis *in vitro*. *Exp. Cell Res.*, **223**, 274.
115. Alcobia, I., Dilao, R., and Parreira, L. (2000) Spatial associations of centromeres in the nuclei of hematopoietic cells: evidence for cell-type-specific organisational patterns. *Blood*, **95**, 1608.

Index